PHYSICAL CHEMISTRY

Chemical Kinetics and Reaction Mechanisms

Research Progress in Chemistry

PHYSICAL CHEMISTRY

Chemical Kinetics and Reaction Mechanisms

Harold H. Trimm, PhD, RSO

Chairman, Chemistry Department, Broome Community College;
Adjunct Analytical Professor, Binghamton University,
Binghamton, New York, U.S.A.

Apple Academic Press

Research Progress in Chemistry Series
Physical Chemistry: Chemical Kinetics and Reaction Mechanisms

First Published in the Canada, 2011
Apple Academic Press Inc.
3333 Mistwell Crescent
Oakville, ON L6L 0A2
Tel. : (888) 241-2035
Fax: (866) 222-9549
E-mail: info@appleacademicpress.com
www.appleacademicpress.com

The full-color tables, figures, diagrams, and images in this book may be viewed at www.appleacademicpress.com

ISBN 978-1-926692-61-6

Harold H. Trimm, PhD, RSO

Printed at Thomson Press (I) Ltd.
9 8 7 6 5 4 3 2 1

Cover Design: Psqua

Library and Archives Canada Cataloguing in Publication Data
CIP Data on file with the Library and Archives Canada

CONTENTS

INTRODUCTION

Chemistry is the science that studies atoms and molecules along with their properties. All matter is composed of atoms and molecules, so chemistry is all encompassing and is referred to as the central science because all other scientific fields use its discoveries. Since the science of chemistry is so broad, it is normally broken into fields or branches of specialization. The five main branches of chemistry are analytical, inorganic, organic, physical, and biochemistry. Chemistry is an experimental science that is constantly being advanced by new discoveries. It is the intent of this collection to present the reader with a broad spectrum of articles in the various branches of chemistry that demonstrates key developments in these rapidly changing fields.

Physical chemistry is the branch of chemistry that develops theoretical and mathematical explanations for chemical behavior. Physical chemists use advanced mathematics and computers to model the behavior of atoms and molecules, and their research has allowed chemists to produce new compounds with desired properties. Physical chemists use principles from quantum theory and thermodynamics to study intermolecular forces, surface catalysis, and kinetics. Current research includes developing better models to predict the properties of compounds before they are even made. Advances in physical chemistry are used in modeling all types of combustion reactions such as fires, explosions, coal fired power plants, and internal combustion engines. Physical chemistry has helped in the design and understanding of nuclear power and is presently being applied to predicting the dispersion and breakdown of pollutants in the environment.

Like all areas of chemistry, physical chemistry is constantly changing, with new discoveries being made all the time. Chapters within this book ensure that the reader will stay current with the latest methods and applications in this important field.

— Harold H. Trimm, PhD, RSO

Virtual Instrument for Determining Rate Constant of Second-Order Reaction by pX Based on LabVIEW 8.0

Hu Meng, Jiang-Yuan Li and Yong-Huai Tang

ABSTRACT

The virtual instrument system based on LabVIEW 8.0 for ion analyzer which can measure and analyze ion concentrations in solution is developed and comprises homemade conditioning circuit, data acquiring board, and computer. It can calibrate slope, temperature, and positioning automatically. When applied to determine the reaction rate constant by pX, it achieved live acquiring, real-time displaying, automatical processing of testing data, generating the report of results; and other functions. This method simplifies the experimental operation greatly, avoids complicated procedures of manual processing data and personal error, and improves veracity and repeatability of the experiment results.

Introduction

LabVIEW is the abbreviation of Laboratory Virtual Instrument Engineering Workbench, which is the innovative software product of National Instruments (NIs) in United States. It is the preferred platform when people exploit virtual instrument as the most functional graphic software at present [1–5].

Ion analyzer is one of the most widely used analysis instruments in scientific research and practice. It is widely applied in biomedical, chemical, environmental protection, and other fields. Manual reading and processing data with traditional ion analyzer is not only cumbersome in operation process but also difficult to avoid personal error. The research about ion analyzer connected to computer has been reported in recent years [6]; however, the quantity is small, and the software is mostly developed in text-based programming language (e.g., C, Turbo Pascal, VB, and so on). The disadvantage of the text-based language is long developing periods, difficult maintenance, and expandedness. The VI of ion analyzer based on LabVIEW has not been reported yet. The VI of ion analyzer developed by authors is developed in graphical programming language—LabVIEW 8.0. It not only has friendly interface, easy operation, and expandability, but also can realize automatic slope, temperature, and positioning calibrations. The system has been applied to determine the rate constant of ethyl acetate saponification reaction when the LabVIEW program for determining the reaction rate constant by pX is embedded into the system. The authors have obtained satisfactory results.

System Principles and Configuration

Actually, Ion analyzer, as a high input impedance millivoltmeter, is widely used in measuring the EMF of the battery which is composed by ion selective electrode, reference electrode, and solution. The concentration of the unknown solution is acquired through measuring the EMF margin between standard solution and tested solution under the same electrode system. By the Nernst equation the authors have inferred

$$E_x - E_s = \pm\frac{2.303RT}{zF}(pX - pX_s),$$
$$\text{If } S = \frac{2.303RT}{zF}, \qquad pX = pX_s \pm \frac{(E_x - E_s)}{S}, \tag{1}$$

where E_x is the potential, and pX is the pX value of tested solution.

It is obvious that the relation between pX and ΔE(ΔE=Ex-Es) is linear. S is theoretic slope of the electrode system; however, the actual electrode slope is also influenced by temperature, ion valence, manufacturing process, surface structure, activation state of the electrode, and other factors. Although the electrodes are of the same type, the S values of the electrodes may be different. It is why ion analyzer must be proofreaded by standard solution before being used. Traditional ion analyzer needs transform switches and rheostats to complete positioning, ion valence, polarity, electrode slope, and temperature calibrations for proofreading the instrument. In this condition, operators must read and process the data artificially during measurements. The virtual instrument system based on LabVIEW 8.0 for ion analyzer brings the advantage of virtual instrument that "software is instrument" into full play. It achieves proofreading the instrument and data processing automatically by LabVIEW program, so the authors can easily get real-time pX~t and mV~t curve.

System structure of the virtual instrument is shown in Figure 1. Ion selecting electrode and reference electrode are immersed into the tested solution constituting electrode system. The potential signal enters into computer via the PCI-6014 Data Acquisition (DAQ) after impedance transformation and filtering by high input impedance conditioning circuit. Precise digital temperature thermoscope connects the computer by RS-232 serial port which makes the automatical temperature compensation easily.

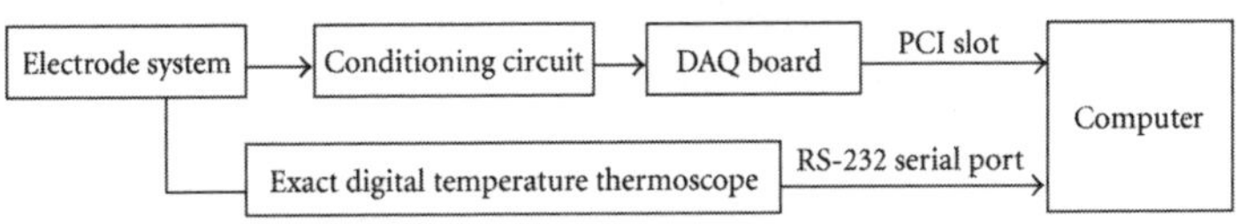

Figure 1: The system structure of VI.

The homemade conditioning circuit is shown in Figure 2. It is made up of one high input impedance voltage follower with CA3140 and second-order active low-pass filter with low drift OP07. Specially the CA3140 is used to accomplish impedance transformation of electrode system, then the low drift OP07 accomplishes the output signal filtering.

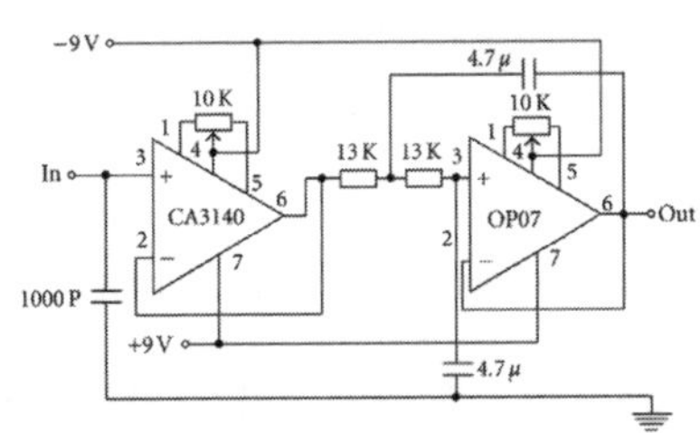

Figure 2: The condition circuit.

Design and Implementation of the Main LabVIEW Program for System

The system software consists of four main parts, including operation panel, data acquiring and real-time displaying, data storage, and data processing and result displaying. Thereinto, the former three parts of the system software are necessary, and the last one can be adjusted according to the purpose of users' measurement.

Program for Operation Panel

Virtual instrument operation panel is shown in Figure 3. To enable users to adjust sampling interval and channels according to their needs, sampling interval knob and channels selecting controller are set on the left of the panel. Users can also decide whether to save the date synchronously by synchronous memory controller or not. In this LabVIEW program, Event Structure which comprises data acquiring, data saving, data processing, result printing, and exit five subevents is used. Each subevent corresponds with a control button on the panel. When clicking the control button, corresponding event will be called. Operation panel adopts Tab Control on the whole to achieve the switch between mV stall and pX stall and connects Tab Control with Selection Terminal of a Case Structure on block panel. Clicking the mV stall and pX stall button on top of the operation panel can complete the switch between mV stall and pX stall easily.

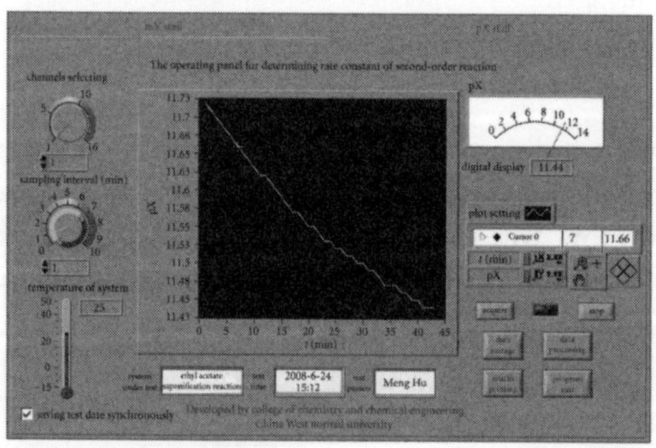

Figure 3: The operating panel.

Program for Data Acquiring and Real-Time Displaying

Program for data acquiring and real-time displaying of pX measurement is the key program of the system which can automatically complete temperature, slope, orientation, and ion valence correction. The program consists of a Stacked Sequence Structure, which includes seven frameworks. While the program is running the structure will gradually execute the framework program according to the order of selection for labels 0,1,2….

Figure 4 shows the block program of the last framework. It executes real-time acquiring anddisplaysthe pX value of the solution under test and data provisional storage, which is mainly constituted by three While Structure. The upmost While Structure accomplishes the driving of data acquisition card and data acquiring continuously. Channels selecting controller select channel, AI Config function sets up the size of buffer, AI Read function sets up sampling rate, and Max & Min function controls the amount lagging to avoid data coverage. To enhance the stability and veracity of the result, 300 acquired data will be switched to pX value after averaged and then real-time displayed by the control Meter and transmitted to the downmost While Structure by native variable to save. At the same time the textboxes on the right of the operation panel display pX value, electrode slope, mV, ion concentration of X, and other data. The While Structure in the middle real time acquires the temperature value of exact digital temperature thermoscope by the serial communication VI and then transmits it to the upmost While Structure by native variable to accomplish temperature compensation automatically. The temperature of solution is real-time displayed by the control Thermometer in the operation panel. The While Structure downmost writes the discrete pX value to file according to the sampling interval set by users and displays it in Express XY Graph. Because LabVIEW program adopts parallel operation mechanism, the three While Structure can run independently.

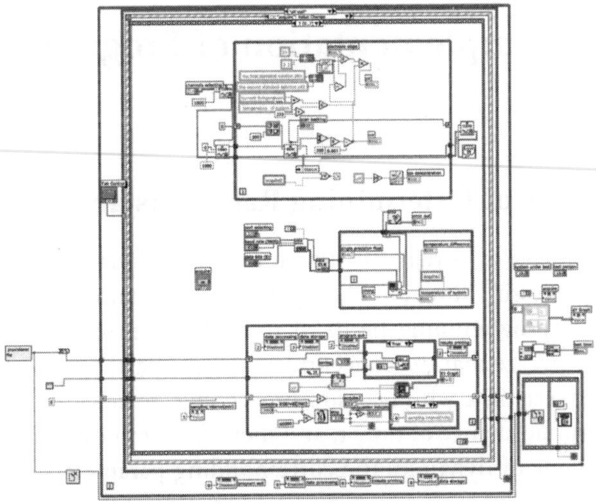

Figure 4: Block program of data acquiring and real-time displaying.

The program of mV stall is similar to this and simpler relatively; the only difference is real-time displays are potential -time curve.

Program for Date Storage

Its block program is shown in Figure 5. First, open a file dialog box with the function File Open to offer the storage path for users. Then copy the data saved in provisional file to appointed file with the function Copy and superadd initial experiment conditions to the file which is saved for the use of data processing at the same time. If any failure during data storage, it will pop up invalid save and give error reasons to users, allowing users to save again until success.

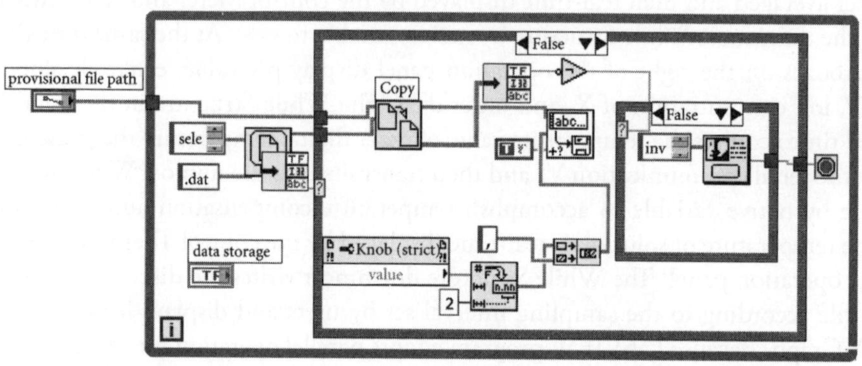

Figure 5: Block program of data storage.

Design and Implementation of Program for Data Progressing

Basic Principles of Determining Rate Constant by pX

The biggest advantage of virtual instrument is easy expandability, fully playing the powerful data processing function of computer to deal with the test data collected according to the programming needs of users. Basic principles of determining rate constant by pX are that the concentration of the participation in the reaction system changes over time with reaction proceeding, so tracking and recording a curve of pX with the change of time t, and then by corresponding data processing under the principle of dynamics the corresponding parameters of dynamics can be determined.

For easy presentation, take a typical second-order reaction the reaction of ethyl acetate saponification as an example:

$$CH3COOC2H5+OH- \rightarrow CH3COO-+C2H5OH. \quad (2)$$

With reaction proceeding, OH- gradually diminishes, and CH3COO- gradually increases in the system, so the PH value of the system gradually reduces. The information of reaction course is obtained by determining the PH value with the time. So the rate constant of this reaction can be determined by the method of pX.

For second-order reaction, data processing formulae are different for the equal and unequal initial concentration of reactants. In accordance with the principle of dynamics, according to [7–9], when the initial concentrations of reactants are equal, setups of the initial concentration are both c0, then the dynamic formula for the reaction

$$\frac{1}{c} = kt + \frac{1}{c_0} \quad (3)$$

where c is the concentration of NaOH at the time t; k is rate constant, for strong base BOH, $[OH^-] \approx c_{BOH}$, pOH = -log $[OH^-]$ when the concentration $c_{BOH} \geq 10^{-6}$ M. Integration yields:

$$\frac{1}{c} = \frac{1}{[OH^-]} = \frac{1}{10^{-pOH}} = 10^{pOH} = 10^{14-pH} \quad (4)$$

So we can deduce the relationship formula between the pH value of the system and time as (5)

$$10^{14-pH} = kt + \frac{1}{c_0} \quad (5)$$

It shows that by plotting $10^{14\text{-pH}}$ against t, the slope of the line is rate constant, k.

When the initial concentrations of reactants are unequal, setups a,b (a>b) are, respectively, the initial concentration of NaOH and CH3COOC2H5, then the concentration of each component of the reaction system at different reaction time can be expressed as

$$CH_3COOC_2H_5 + NaOH \rightarrow CH_3COONa + C_2H_5OH$$

$$\begin{array}{llll} t=0 & b & a & o\ o \\ t=t & b-x & a_t=a-x & x\ x \\ t=\infty & 0 & a_\infty=a-b & b\,b, \end{array} \tag{6}$$

where at, a_∞ are, respectively, the concentration of NaOH at the time of t and t=∞, and x is the concentration of product CH_3COONa and C_2H_5OH at the time of t.

When a≠b the dynamic formula for second-order reaction is

$$\ln\frac{(a-x)}{(b-x)}=(a-b)kt+\ln\frac{a}{b} \tag{7}$$

Because $a-b=a_\infty, a-x=a_t, b-x=a_t-a_\infty$,

$$\ln\frac{a_t}{a_t-a_\infty}=a_\infty kt+\ln\frac{a}{b} \tag{8}$$

Set up $[OH^-]_t$, $[OH^-]_\infty$ are $[OH^-]$ of the system at the time of t=t and t=∞, then obtain

$$-\ln\left(1-\frac{[OH^-]_\infty}{[OH]_t}\right)=[OH^-]_\infty kt+\ln\frac{a}{b} \tag{9}$$

Also because $[OH^-]=10^{pH-14}$, (5) can be expressed with pH value as

$$-\ln\left(1-\frac{10^{pH_\infty-14}}{10^{pH_t-14}}\right)=10^{pH_\infty-14}kt+\ln\frac{a}{b} \tag{10}$$

Set up $A=-\ln\left(1-\left(10^{pH_\infty-14}/10^{pH_t-14}\right)\right)$, then

$$A=10^{pH_\infty-14}kt+\ln\frac{a}{b} \tag{11}$$

Equation (11) is the relationship formula between pH value and time, which shows that by plotting A against t a straight line is obtained, and after determining [OH-]t, [OH-]∞ of the system at the time of t=t and t=∞ rate constant k can be obtained from the slope of the line.

These methods do not require determining the accurate initial concentration of reactants, which simplifies the experimental operation. By Arrhenius equation $\ln(k_1/k_2)=E_a(1/T_2-1/T_1)/R$, the activation energy can be obtained by the rate constant of different temperature.

Program for Data Processing and Results Displaying

The block program is shown in Figure 6. The corresponding program in the Sequence Structure on the far left firstly judges whether the initial concentration of reactant is equal, then chooses the corresponding data file which is waiting for processing. The program performs the corresponding subdiagram of Case Structure according to judging result. Data processing includes linear transforming of data, transforming the label of axis, straight line fitting, and correlation coefficient R calculating, and so forth. A group of data read out from data file enters For Loop structure, by using the Loop Iteration; the automatically matching of data of X axis and data of Y axis is achieved. The autoindex of For Loop makes the results to be an array. After this array enters subprogram for straight line fitting, the unqualified data is deleted and the line slope, intercept, and other data are obtained. A cluster array is produced by the function Build Array after the data "after deleted" and "the data after fitted" are, respectively, bundled by the function Bundle, and is displayed by the control XY Graph. At the same time, by using Xscale.NameLb1.Text and Yscale.NameLb1.Text in Property Node of graph and condition of initial concentration of reactant, the label of axis is automatically transformed with different beginning condition. Correlation coefficient of straight line fitting is obtained after the data enters subprogram for R. The system will remind users whether or not to continue data processing after a set of data is processed. Another data file will be selected if the user continues to obtain rate constant at different temperature. The k value at different temperature can be used to calculate apparent activation energy E_a by the function Formula Node in Case Structure on the far right. The results and necessary parameters are added into array by the function Insert Into Array. Then each element of array is indexed and transformed into specified precise characters by function Number To Fractional String. The characters are concatenated into a string by function of Concatenate Strings, and at last the string is displayed by String Indicator.

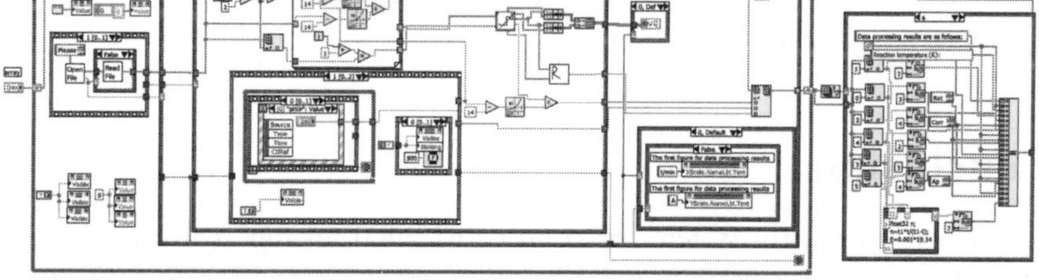

Figure 6: Block program of data processing and results displaying.

Example of Application

This system is used to determine the rate constant of ethyl acetate saponification reaction at the temperature of 298 K and 308 K when the initial concentration of reactants is equal and unequal. In the measuring process the system automatically collects testing data and real-time displays pH~t curve, as shown in Figure 3. It also automatically processes testing data and generates the report of result. Figure 7 is the data processing panel under equal initial reactants concentration, and the two XY Graphs on each side, respectively, display testing data at the temperature of 298 K and 308 K. The underside of panel is a testing conditions and results showing box which can display reaction temperature, rate constant, correlation coefficient, and apparent activation energy together. Relevant data of test is shown in Table 1.

Table 1: The data and results of test. Notes: a is the concentration of NaOH; b is the concentration of $CH_3COOC_2H_5$.

c_0/mol·dm^{-3}		T/K	k/dm^3·mol^{-1}·min^{-1}		δ%	E_a/kJ·mol^{-1}		δ%
a	b		Experimental	Reference [10, 11]		Experi-mental	Reference [12]	
0.0490	0.0490	298	6.864	6.85	0.20	33.48	33.39	0.27
0.0490	0.0490	308	10.643	10.50	1.36			
0.0245	0.0196	298	6.926	6.85	1.11	34.14		2.25
0.0245	0.0196	308	10.834	10.50	3.18			

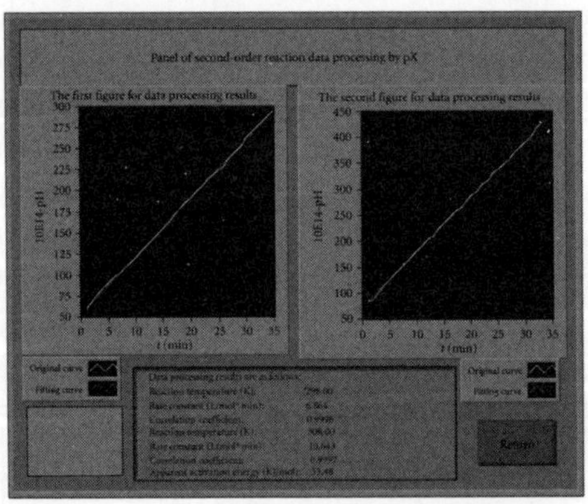

Figure 7: The panel of data processing at a=b.

Conclusion

The virtual instrument system based on Labview8.0 for ion analyzer, which has advantages of easy operation and friendly interface, can determine pX, realize real-time displaying and drawing a curve of pX or mV with the change of time t, process data, and generate result report automatically. The relative error of result is less than 3.5%. For any second-order reaction system, the virtual instrument system can track the concentration change of the corresponding ion and then determine its rate constant if appropriate ion-selective electrode is used. For other series of reactions, the system can also determine their rate constants by amending data-processing program under the principle of dynamics. With its easy expandability, the system can be used in autodetermination studies such as potential titration, potential analysis, and potential equilibrium constant measurement through embedding different data-processing programs of LabVIEW. What is more, the system provides users with great convenience for tracking and measuring pX or potential application research; the authors believe that it can be widely used in the production and environmental monitoring.

Acknowledgement

The authors are grateful to the important research item of Sichuan provincial Department of education [(2005) 198] for financial support.

References

1. A. P. Beltrá, J. Iniesta, L. Gras, et al., "Development of a fully automatic microwave assisted chemical oxygen demand (COD) measurement device," Instrumentation Science and Technology, vol. 31, no. 3, pp. 249–259, 2003.
2. J. Ballesteros, J. I. Fernández Palop, M. A. Hernández, et al., "LabView virtual instrument for automatic plasma diagnostic," Review of Scientific Instruments, vol. 75, no. 1, pp. 90–93, 2004.
3. National Instruments Corporation, LabVIEW User Manual, 2007.
4. J.-Y. Li, Y.-W. Li, and D.-P. Guo, "The design and application of conductance rate virtual instrument based on LabVIEW 8.0 express," Match, vol. 60, no. 2, pp. 325–331, 2008.
5. W.-B. Wang, J.-Y. Li, and Q.-J. Wu, "The design of a chemical virtual instrument based on LabVIEW for determining temperatures and pressures," Journal of Automated Methods and Management in Chemistry, vol. 2007, Article ID 68143, 2007.
6. K. Li and Y. Z. Ye, "The design of intelligent ion analyzer based on microcomputer and its application," Journal of Fuzhou University, vol. 33, no. 1, pp. 54–57, 2005.

7. S. Y. Shao, X. R. Liu, and L. X. Pang, "Measurement of rate constant for saponification reaction of ethyl acetate by acidometer," Journal of Xi'an University of Science and Technology, vol. 24, no. 2, pp. 196–199, 2004.
8. A. C. Feng, "Simple measurement of reaction rate constant of ethyl acetate," Chemistry Online, vol. 9, pp. 36–37, 1983.
9. S. X. Li, "pH measurement of the rate constant of ethyl acetate saponification," Journal of Southwest Nationalities College, vol. 21, no. 2, pp. 232–234, 1995.
10. Fudan University, Physical Chemistry Experiments, Higher Education Press, Beijing, China, 3rd edition, 2004.
11. L. Kuang and L. Wu, "Studying kinetics of saponification of ethyl acetate with computer," University Chemistry, vol. 19, no. 4, pp. 39–42, 2004.
12. J. X. Lu, W. J. Liu, and L. W. Chen, "The determining to kinetic data of saponification of ethyl acetate by ultraviolet spectrophotometry," Chemistry Online, vol. 11, pp. 55–56, 1988.

Electrochemical Behavior of Titanium and Its Alloys as Dental Implants in Normal Saline

Rahul Bhola, Shaily M. Bhola, Brajendra Mishra and David L. Olson

ABSTRACT

The electrochemical behavior of pure titanium and titanium alloys in a simulated body fluid (normal saline solution) has been tested, and the results have been reported. The significance of the results for dental use has been discussed. The tests also serve as a screening test for the best alloy system for more comprehensive long-term investigations.

Introduction

Commercially pure titanium and titanium alloys are known for their use in dental practice owing to their good corrosion resistance, biocompatibility, and biofunctionality in the human body [1].

Titanium is a thermodynamically reactive metal as suggested by its relatively negative reversible potential in the electrochemical series [2]. It gets readily oxidized during exposure to air and electrolytes to form oxides, hydrated complexes, and aqueous cationic species. The oxides and hydrated complexes act as barrier layers between the titanium surface and the surrounding environment and suppress the subsequent oxidation of titanium across the metal/barrier layer/solution interface. Even if the barrier layer gets disrupted, it can get reformed very easily, leading to spontaneous repassivation. A review of literature reveals that extensive work has been done on the use of titanium alloys as prostheses when the alloy comes in contact with saliva [3–5]. Whereas when titanium alloys are used as implants, the alloy remains in contact with saliva only during the transition period of healing of the bony socket and gingiva (gum) but when it is capped with the crown, it is in contact with only bone cells for the rest of its life. The aim of this study is to keep in focus the use of titanium alloy as an implant inside the bone tissue and hence normal saline solution was used to simulate the extracellular interstitial environment. The decrease in pH can occur inside the bone tissue owing to surgical trauma, infection [6, 7] as well as the healing process of the bony socket. Ti3+ ions, if produced, can hydrolyze to further acidify the area [8]. Acidification can also be observed if crevice corrosion occurs in the clefts between the implant and the crown [5]. According to Black [9], the potential of a metallic biomaterial can range from –1 to 1.2 V versus SCE in the human body. It is, thus, significant to study the electrochemical behavior of titanium in this potential range.

Electrochemical corrosion behavior of commercially pure titanium (gradesTi1 & Ti2) and its alloys, Ti64, TiOs, Ti15Mo, and TMZF in normal saline solution was analyzed at the body temperature of 37°C.

Experimental

Materials Preparation

Titanium alloy grades Ti1, Ti2, Ti64, TiOs, Ti15Mo, and TMZF of compositions Ti1 (0.1%C, 0.2%Fe, 0.015%H, 0.03%N, 0.18%O, and 99.47% Ti), Ti2 (0.1%C, 0.3%Fe, 0.015%H, 0.03%N, 0.25%O, and 99.30%Ti), Ti64 (0.1%C, 0.2%Fe, 0.015%H, 0.03%N, 0.2%O, 6%Al, 4%V, and 89.45%Ti), TiOs (0.05%C, 0.05%Fe, 0.015%H, 0.02%N, 0.15%O, 35.5% Nb, 5.7%Ta,

7.3%Zr, and 51.21%Ti), Ti15Mo (0.05%C, 0.1%Fe, 0.015%H, 0.01%N, 0.15%O, 15%Mo, and 84.67%Ti), and TMZF (0.02%C, 2%Fe, 0.02%H, 0.01%N, 0.18%O, 6%Zr, 12%Mo, and 79.77%Ti) were used for the present investigation. Available cuboidal and cylindrical rods were cut to expose cross-section areas of 0.855 cm^2 for Ti1, 1.0 cm^2 for Ti2, 1.5525 cm^2 for Ti64, 1.3270 cm^2 for TiOs, 0.4869 cm^2 for Ti15Mo, and 2.7606 cm^2 for TMZF as working electrodes. Normal saline solution (5.26 gL^{-1} sodium chloride, 2.22 gL^{-1} anhy. sodium acetate, 5.02 gL^{-1} sodium gluconate, 0.37 gL^{-1} potassium chloride, and 0.3 gL^{-1} magnesium chloride hexahydrate) having pH 6.6 was used to carry out the electrochemical testing of alloys.

Measurements

A Princeton Applied Research Potentiostat 273A was used for the electrochemical measurements. A three-electrode cell assembly consisting of titanium alloy as the working electrode, platinum wire as the counter electrode, and a saturated calomel electrode as the reference electrode was used. 450 mL of normal saline solution was used for the immersion and electrochemical testing of the alloys.

Open Circuit Potential (OCP)

Open circuit potential values of the alloys were measured just after immersion, followed by a duration of 2, 6, and 24 hours up to seven days.

Tafel

After the end of the seven days, Tafel curves of various alloys were obtained by polarizing each electrode from –250 mV versus OCP to 2 V with an ASTM scan rate of 1 mV/s [10].

Cyclic Polarization

Cyclic polarization measurements were extended after Tafel measurements by choosing a vertex potential of 2 V, at which the scan was reversed, and the final potential was –250 mV versus OCP.

Results and Discussion

It can be seen from Figure 1 and Table 1 that the OCP values for all alloys have shown an overall increase for the time duration of seven days, despite showing a small decrease in a few intervals.

Table 1: Open circuit potential values of alloys with change in time at 37°C.

Material	Open circuit potential (mV)									
	On immersion	2 hrs.	6 hrs.	1 day	2 days	3 days	4 days	5 days	6 days	7 days
Ti1	−107	−144	−156	−152	−148	−131	−121	−125	−117	−120
Ti2	−168	−143	−140	−134	−131	−113	−111	−111	−110	−105
Ti64	−147	−135	−118	−118	−89	−65	−56	−46	−44	−41
TiOs	−202	−248	−252	−223	−151	−175	−164	−149	−142	−140
Ti15Mo	−148	−179	−181	−179	−142	−68	−82	−57	−47	−27
TMZF	−184	−218	−227	−235	−208	−175	−160	−133	−120	−109

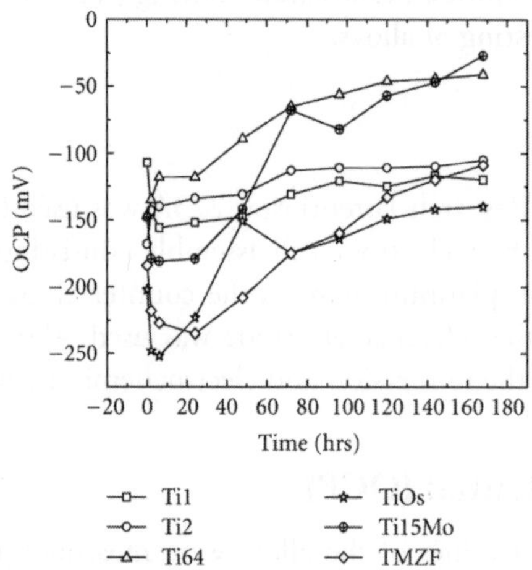

Figure 1: Open circuit potential values of alloys as a function of time at 37°C.

The OCP shift in the noble direction for the alloys suggests the formation of a passive film that acts as a barrier for metal dissolution and reduces the corrosion rate. The shift can also suggest that the composition of corrosive medium might be changing in the implant cavity but since large volume of solution is used in testing, that can be neglected. The potential increase shows that the alloys become thermodynamically more stable with time. According to Blackwood et al. [11], the shift in OCP to positive values lowers the corrosion rate by reducing the driving force of the cathodic reaction and increasing the thickness of the passive oxide film. The initial decrease observed in the OCP values for Ti1, TiOs, and TMZF suggests the initial dissolution of the air-formed oxide film till the formation of a new oxide in solution [12].

The pH values of all alloys in normal saline solution of pH 6.6 were less than 8 till seven days of immersion. Figure 2 shows the pH-potential Pourbaix diagram for titanium-water system at 37°C [13], with potential also shown with respect to SCE on another y-axis. The corresponding region of pH (6.6–8) and potential to which the present system is studied is similar to that of TiO2 in the Pourbaix diagram indicating the fact that under steady-state conditions, the alloys form a stable oxide layer of TiO2.

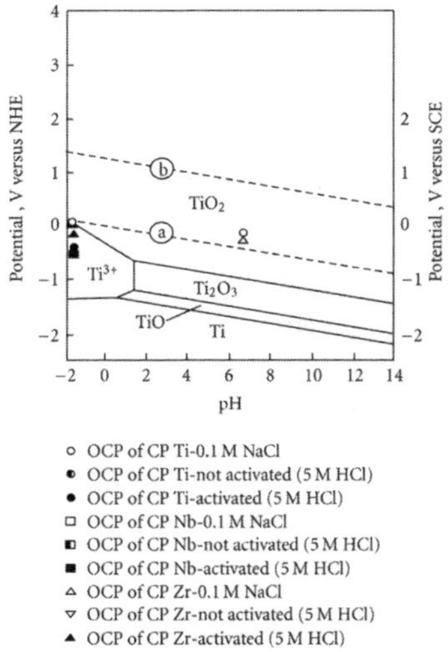

Figure 2: Pourbaix diagram for titanium at 37°C.

Figure 3 shows the potentiodynamic curves for the alloys at 37°C. The corresponding corrosion parameters are given in Table 2, and passivation parameters are given in Table 3.

Table 2: Corrosion parameters for the forward and reverse scans for the alloys at 37°C after seven days.

Material	Forward scan			Reverse scan	
	E_{corr} (mV)	I_{corr} (nA/cm^2)	Corr. rate $*$ 10^{13} (mm/y)	E_{corr} (mV)	I_{corr} (nA/cm^2)
Ti1	−298.2	0.934	3.43	369.6	2.27
Ti2	−69	1.6	5.88	338	1.8
Ti64	−220	0.708	2.65	427.3	0.597
TiOs	−193.3	0.222	0.64	322.1	0.00531
Ti15Mo	−139.9	4.68	15.67	133.7	3.19
TMZF	−256.7	1.26	4.19	225.9	0.658

Table 3: Passivation parameters for the alloys at 37°C after seven days.

Material	E_{pass} (mV)	I_{pass} (μA/cm^2)	E_{corr} (mV)	E_{pass} − E_{corr} (mV)
Ti1	146	0.013	−298.2	442.2
Ti2	359	0.006	−69	428
Ti64	−104	0.00173	−220	116
TiOs	−72	0.00143	−193.3	121.3
Ti15Mo	569	10.02	−139.9	708.9
TMZF	644	2.37	−256.7	900.7

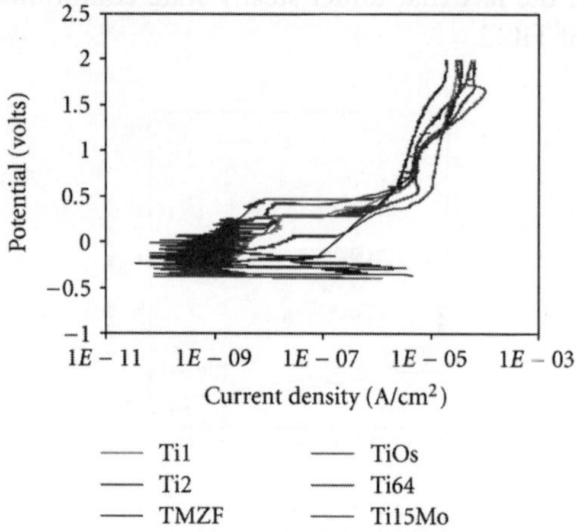

Figure 3: Potentiodynamic curves for the alloys at 37°C.

The corrosion rate was calculated using the expression [14]

$$\text{Corrosion rate}\left(\frac{mm}{year}\right)=\frac{0.00327\times EW.\times I_{corr}}{\rho} \tag{1}$$

where E.W. = equivalent weight, I_{corr}= current density in μA/cm^2, and ρ = density in g/cm^3.

The corrosion rates observed are in the order Ti15Mo > Ti2 > TMZF > Ti1 > Ti64 > TiOs. The corrosion rates observed are very low and belong to the outstanding corrosion resistance category of less than 0.02 mm/year as classified by Fontana [15]. Studies have further revealed that to prevent tissue damage and irritation, the corrosion rate of a metallic implant should be less than 2.5×10−4 mm/year [16]. The difference between the passivation potential and Ecorr can provide an indication of the ease of passivation of the alloys. The lower the difference is, the easier the passivation will be. The difference follows the order TMZF > Ti15Mo > Ti1 > Ti2 > TiOs > Ti64.

Cyclic polarization method is a highly useful method for determining the susceptibility of a metal or alloy to pitting [17]. From Figures 4(a)–4(f), it can be seen that at the vertex potential of 2 V when the scan reverses its direction, the reverse scan starts left of the forward scan curve, that is, towards the low current density region. This type of the cyclic polarization curve is known to resist localized corrosion [18]. It is also observed that the reverse scan curves meet the forward scan curve along the passive range. The reverse scan curves show lower current densities as can be seen in Table 2 except for Ti1 and Ti2. The potentials for the reverse scan curves are more positive than those for the forward scan. These results show that a stable oxide film is formed during the forward scan.

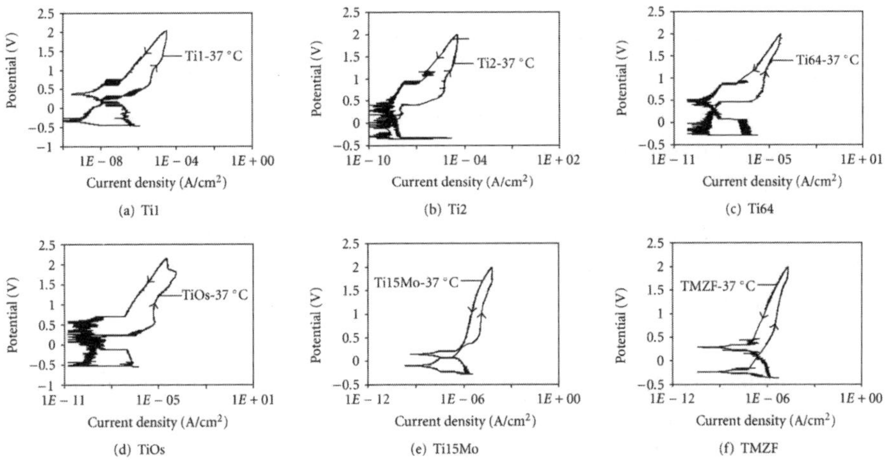

Figure 4: Cyclic polarization curves for alloys (a)–(f) at 37°C.

Conclusions

(1) In normal saline solution, all alloys exhibit a high corrosion resistance, and the corrosion rates observed fall in the acceptable range for biocompatibility of metallic implants.

(2) All alloys show considerable ennoblement and form a stable oxide film of TiO2 as indicated by their corresponding pH and potential position in the Pourbaix diagram.

(3) From the cyclic polarization curves obtained, it can be concluded that the alloys resist localized corrosion.

References

1. J. Lemons, R. Venugopalan, and L. Lucas, "Corrosion and biodegradation," in Handbook of Biomaterials Evaluation, A. F. von Recum, Ed., pp. 155–167, Taylor & Francis, New York, NY, USA, 1999.

2. S. Yu, Corrosion Resistance of Titanium Alloys, Corrosion: Fundamentals, Testing and Protection, vol. 13A of ASM Handbook, ASM International, Materials Park, Ohio, USA, 2003.

3. A. M. Al-Mayouf, A. A. Al-Swayih, N. A. Al-Mobarak, and A. S. Al-Jabab, "Corrosion behavior of new titanium alloy for dental implant applications," The Saudi Dental Journal, vol. 14, no. 3, pp. 118–125, 2002.
4. M. Sharma, A. V. Ramesh Kumar, N. Singh, N. Adya, and B. Saluja, "Electrochemical corrosion behavior of dental/implant alloys in artificial saliva," Journal of Materials Engineering and Performance, vol. 17, no. 5, pp. 695–701, 2008.
5. K. Elagli, M. Traisnel, and H. F. Hildebrand, "Electrochemical behaviour of titanium and dental alloys in artificial saliva," Electrochimica Acta, vol. 38, no. 13, pp. 1769–1774, 1993.
6. M. L. Escudero, M. F. López, J. Ruiz, M. C. García-Alonso, and H. Canahua, "Comparative study of the corrosion behavior of MA-956 and conventional metallic biomaterials," Journal of Biomedical Materials Research Part A, vol. 31, no. 3, pp. 313–317, 1996.
7. T. Eliades, "Passive film growth on titanium alloys: physicochemical and biologic considerations," International Journal of Oral and Maxillofacial Implants, vol. 12, no. 5, pp. 621–627, 1997.
8. D. G. Kolman and J. R. Scully, "On the repassivation behavior of high-purity titanium and selected α, β, and β + α titanium alloys in aqueous chloride solutions," Journal of the Electrochemical Society, vol. 143, no. 6, pp. 1847–1860, 1996.
9. J. Black, in Biological Performance of Materials: Fundamentals of Biocompatibility, pp. 38–60, Marcel Decker, New York, NY, USA, 1992.
10. ASTM Standard F2129-06, ASTM International, USA.
11. D. J. Blackwood, A. W. C. Chua, K. H. W. Seah, R. Thampuran, and S. H. Teoh, "Corrosion behaviour of porous titanium-graphite composites designed for surgical implants," Corrosion Science, vol. 42, no. 3, pp. 481–503, 2000.
12. U. R. Evans, The Corrosion of Metals, Edward Arnold, London, UK, 1960.
13. S. Yu, Corrosion: Fundamentals, Testing, and Protection, vol. 13A of ASM Handbook, ASM International, Materials Park, Ohio, USA, 2003.
14. D. A. Jones, Principles and Prevention of Corrosion, Prentice Hall, Upper Saddle River, NJ, USA, 2nd edition, 1995.
15. M. G. Fontana, Corrosion Engineering, McGraw-Hill, New York, NY, USA, 3rd edition, 1986.
16. W. D. Callister, Jr., Materials Science and Engineering: An Introduction, John Wiley & Sons, New York, NY, USA, 7th edition, 2007.
17. P. R. Roberge, Handbook of Corrosion Engineering, McGraw Hill, New York, NY, USA, 2000.

18. D. C. Silverman, "Tutorial on cyclic potentiodynamic polarization technique," in Proceedings of the CORROSION/98 Research Topical Symposia, NACE International, San Diego, Calif, USA, March 1998, paper no. 299.

Electron-Collision-Induced Dissociative Ionization Cross Sections for Silane

Satyendra Pal, Neeraj Kumar and Anshu

ABSTRACT

Secondary electron energy and angle dependent differential cross sections for the production of cations SiH_n^+ (n=0–3), H_2^+ and H^+ resulting from dissociative ionization of SiH_4 by electron collision have been evaluated at fixed incident electron energies of 100 and 200 eV. The semiempirical formulation of Jain and Khare which requires the oscillator strength data as a major input has been employed. In the absence of experimental data for differential cross sections, the corresponding derived integral partial and total ionization cross sections in the energy range varying from ionization threshold to 1000 eV revealed a satisfactory agreement with the available experimental and theoretical data. We have also evaluated the ionization rate coefficients on the basis of calculated partial ionization cross sections and Maxwell-Boltzmann energy distributions.

Introduction

This work is a part of our project on electron impact ionization on technological molecules in low-energy regimes. Our aim is to determine the differential and integral cross sections corresponding to the production of molecular and atomic cations in electron impact ionization of the SiH_4 molecule, which is widely used in plasma deposition of silicon containing thin films. Electron ionization cross sections of SiH_4 are needed for the modeling of charge carrier balance in plasma and gas phase media [1, 2]. There is dearth to elucidate the atomic and molecular properties and their interaction with photons and electrons. Photoabsorption and photoionization studies of silane and its radicals have already been made by Gallagher et al. [3], Johnson III et al. [4], and Cooper et al. [5, 6]. The experimental determination of partial and total electron ionization cross sections includes those of Perrin et al. [7], Chatham et al. [8], Krishnakumar and Srivastava [9], and Basner et al. [10]. Haaland [11] estimated the partial ionization cross sections for the formation of Si containing radicals by scaling the data of Chatham et al. [8] and the differential data of Morrison and Traeger [12] to his absolute values at 50 eV. From a theoretical standpoint, calculations for SiH_4 are particularly challenging. The rigorous quantum mechanical approach for the calculations for molecules is limited to the application of simple molecules. Contrary to it, there now exist the binary encounter Bethe formalism by Ali et al. [13], the semiempirical formalism by Khare et al. [14], DM-formalism by Deutsch et al. [15], and complex potential model calculations by Joshipura et al. [16, 17].

This paper reports the spectrum of the single differential cross sections with the energy of secondary electron produced in the ionization of silane molecule by electron collision at fixed incident electron energies of 100 and 200 eV, employing a semiempirical formalism based on Jain and Khare approach [18–21]. At these fixed incident electron energies, double differential cross sections as a function of angle and secondary electron energy have also been calculated. This is the only formulation that enables us to evaluate the energy and angle dependent cross sections for molecules corresponding to the formation of cations in electron-molecule collisions. To the best of our knowledge, no experimental and/or theoretical data is available for comparison to the present calculations for differential cross sections. Thus the corresponding derived integral cross sections in terms of the partial ionization cross sections leading to the formation of various cations SiHn+ (n= 0–3), H2+, and H+ through dissociation of SiH4 by electron collision are compared with the available experimental and theoretical data. The present results along with the total ionization cross sections show good agreement with the experimental and theoretical results. In addition, we have calculated the ionization rate coefficients corresponding to the produced cations using the computed ionization cross sections and Maxwell-Boltzmann distribution for the electrons as a function of electron temperature/energy. Instead of cross sections itself, ionization rate coefficients corresponding to the produced cations are more important in various plasma applications, gas discharge, and flowing afterglow studies [22, 23].

Theoretical

The present calculations are carried out using the modified semiempirical formalism developed by Pal et al. (see discussion in [18–21]). In brief, the single differential cross sections in the complete solid angle ($\Omega=4\pi=\int 2\pi \sin\theta d\theta$) as a function of secondary electron energy ε corresponding to the production of ith type of ion in the ionization of a molecule by incident electron of energy E is given by

$$\begin{aligned} &Q_i(E,W) \\ &= \frac{a_o^2 R}{E}\left[\left(1-\frac{\varepsilon}{(E-I_i)}\right)\frac{R}{W}\frac{df_i(W,0)}{dW}\ln\left[1+C_i(E-I_i)\right]\right. \\ &\left.+\frac{R}{E}S_i\frac{(E-I_i)}{\left(\varepsilon_o^3+\varepsilon_i^3\right)}\left(\varepsilon-\frac{\varepsilon^2}{(E-\varepsilon)}+\frac{\varepsilon^3}{(E-\varepsilon)^2}\right)\right] \\ &\times\int 2\pi\sin\theta d\theta, \end{aligned} \tag{10}$$

where W ($=\varepsilon+I_i$) is defined as energy loss suffered by the incident electron. I_i, a_0, ε_0, C_i, S_i, and R are the ionization threshold for the production of ith type of ion, the Bohr radius, energy parameter, collision parameter, number of ionizable electrons, and the Rydberg constant, respectively.

In the present formulation, the dipole oscillator strengths dfi/dW are the key parameters. The oscillator strength is directly proportional to the photoionization cross section [3]. We have used partial photoionization cross section data set in the energy range from 12 to 52 eV provided by Brion et al. [6] using (e,e) spectroscopy. The accuracy of the determined oscillator strength scales was estimated to be better than ±5%. In the photon energy range 52–180 eV, we have used their measured total valence photoabsorption oscillator strength data [5], and for higher photon energy range W > 180 eV the same were extrapolated by Thomas-Reiche-Kuhn (TRK) sum rule (within ±10% error bars) (see, e.g., [5, 6]). The total photoabsorption cross sections have been distributed into ionic fragments considering the constant ionization efficiency to be 1.0 above the dipole breakdown limit of ~25 eV. However, its evaluation is possible quantum mechanically using the suitable wave functions and transition probabilities corresponding to the production of cations. In case of dissociative ionization of polyatomic molecule SiH4, we have no reliable probabilities corresponding to different dissociative ionization processes. The collision parameter Ci (=0.08621/eV) and energy parameter ε0 (=30 eV) are evaluated as for other polyatomic molecules [18–21]. The vertical onsets or the ionization potentials corresponding to the various cations are also given alongwith the photoionization measurements [5, 6]. In the present evaluations of cross sections, the estimated uncertainty is more or less the same as for the measurement of photoionization cross sections.

The double differential cross sections as a function of energy and angle were evaluated by the differentiation of (1) with respect to the solid angle Ω as follows:

$$Q_i(E,W,\theta) = \frac{dQ_i(E,W)}{d\Omega} \tag{2}$$

The double differential cross sections are angular dependant in all the scattering geometries, and hence the oscillator strength must be angle dependent. In this context, we have used the angular oscillator strengths that were derived in the optical limit (Bethe regime) where angular-momentum-transfer K→0 (e.g., see [19] for a detailed discussion):

$$\frac{df_i(W,0,\theta)}{dWd\Omega} = \frac{1}{4\pi}\frac{df_i(W,0)}{dW}\left[1+\frac{\beta}{2}\left(3\cos^2\theta-1\right)\right] \tag{3}$$

where β is an energy dependent asymmetric parameter. Its evaluation is difficult due to the lack of wave functions of molecular ions in ground and excited states. In valance shell ionization of SiH_4, we have computed β as the ratio of the Bethe spectral transitions Si(W) to the dipole matrix squared Mi_2(W) [24, 25]. The oscillator strength appeared in (1) is simply a derived form of (3) in the forward scattering corresponding K→0 and θ→0.

The partial ionization cross section is obtained by the integration of the energy dependent single differential cross sections (1) over the entire energy loss as follows:

$$Q_i(E) = \int_{I_i}^{E} Q_i(E,W)\,dW \tag{4}$$

and the counting or total ionization cross section is obtained by

$$Q^T(E) = \sum_i Q_i(E) \tag{5}$$

In plasma processes, the ionization rate coefficients are important quantities which are determined by using our calculated partial and total ionization cross sections and Maxwell-Boltzmann distribution of temperature/energy [26, 26] as follows:

$$R_i = \int_{-\infty}^{+\infty} 4\pi \left(\frac{1}{2\pi mkT}\right)^{3/2} me^{(-E/kT)} Q_i(E) EdE \quad (6)$$

where k, T, and m are the Boltzmann constant, absolute temperature, and mass of the electron, respectively.

Results and Discussion

The electronic configuration of the ground state of silane is $K^2L^8(3a_1)^2(2t_2)^6\ {}^1A_1$. Ionization of the outermost $2t_2$ electrons results in a degenerate ${}^2T_2$ ion state, which is subject to the Jahn-Teller effect. Experimentally, $2t_2$ electron ionization or the parent ion formation, observed by Berkowitz et al. [27] at adiabatic ionization potential, is the best evidence for the photogeneration of a bound SiH_4^+ species. The amount of molecular ions detected experimentally were approximately two orders of magnitude lower than the intensities of the dissociative ionization products SiH_4^+ (n = 0–3). On the other hand, due to the instability of SiH_4^+, it decomposes into other stable ions. The present calculations for cross sections are based on the experimental data for the oscillator strengths (photoionization cross sections) [5, 6]. The calculations for differential and integral cross sections were carried out for the produced cations SiH_n^+ (n = 0–3), H_2^+, and H^+ through electron dissociation of SiH_4.

The partial single differential cross sections as a function of secondary electron energy in terms of energy loss W at fixed incident electron energies of 100 and 200 eV are presented in Figure 1. More qualitative results are also presented in the Platzman plot of Y(W). The Y(W) parameter is the ratio of the calculated differential cross section and the Rutherford cross section with energy loss W in the dipole energy range. Qualitatively, Y(W) corresponds to the effective number of electrons participating in ionizing collisions. In the present formulation (1), the first Born-Bethe part for slow secondary electron, corresponds to the growing contribution of the dipole-allowed interaction (known as glancing collision) and resembles the photoionization cross section. The second part accounts for the electron exchange effect and corresponds to the nondipole part that defines knock-on collisions. For fast secondary electrons, it is an adaptation of the Rutherford cross sections for the free electrons.

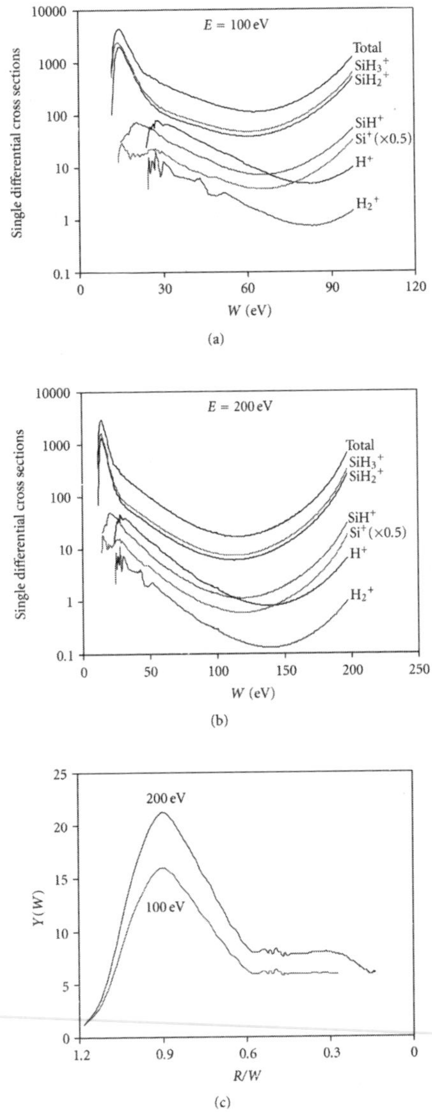

Figure 1: (a) Single differential cross sections (in the units of 10^{-20} cm^2/eV) as a function of energy loss W (=Ii+ε), for the production of cations from electron impact ionization of SiH_4 at constant electron impact energy of 100 eV. (b) Same as (a) but at E=200 eV. (c) The solid lines show the trends of the calculated ratios of cross sections in Platzman plot Y(W) in half energy range at 100 and 200 eV.

In Figure 2, we show the behavior of double differential cross sections with angle varying from 10° to 180° at constant secondary electron energies of 10 and 20 eV and fixed primary electron energies of 100 and 200 eV. The calculations as a function of energy loss W, at constant angles of 30° and 60° at the same incident electron energies, are presented in Figure 3. The 3D profiles of the total double cross sections as a function of secondary electron energy (in the range of 5 eV to W/2) and angle (10° to 180°) at 100 and 200 eV are also shown in Figure 4. To the best of our knowledge, no experimental data is available to compare the present results for differential cross sections. However, the qualitative behavior of the cross section is the same as for other molecules investigated [18–21]. The energy dependent cross sections are symmetric at W/2 where the energies of primary and the secondary electrons are almost equal. The figures clearly show the weight contribution of the molecular and atomic cations. The cross sections for molecular ions are much larger than the atomic ions. In dipole-induced breakdown scheme for the photoionization of silane molecule, the major ions SiH3+ and SiH2+ are produced from 2t2–1 state, and the H+ and H2+ ions are produced from 3a1–1 state, while the production of the SiH+ and Si+ ions comes from the contribution of the both states. In the threshold energy range, the atomic photoionization cross sections include the contribution of the structures and many body states produced near onsets [5, 6] which are reflected in the present calculations for energy dependent differential cross sections.

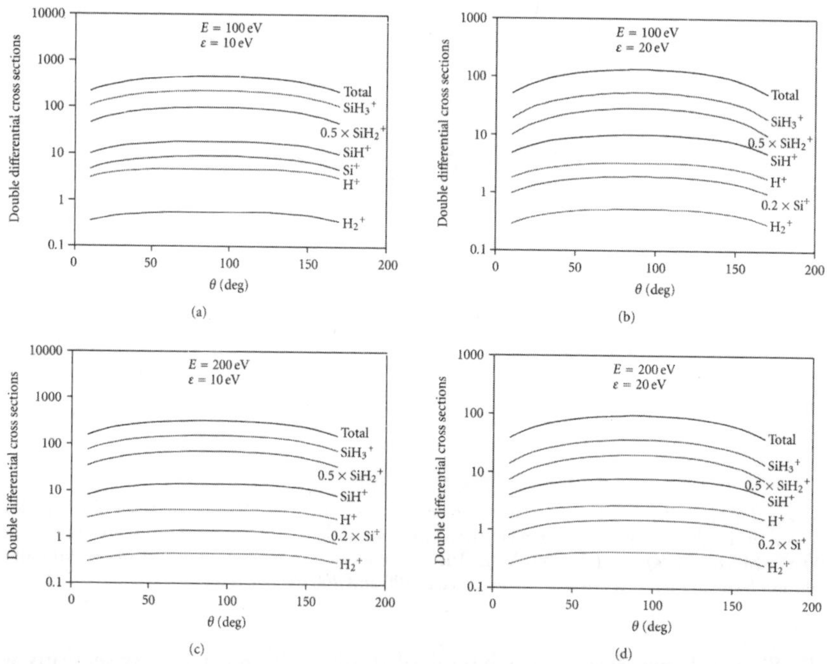

Figure 2: (a) The double differential cross sections (in the units of 10–20 cm^2/eV-sr) as a function of angle θ at (ε=10 eV and E=100 eV) corresponding to the production of various cations from electron impact ionization of SiH4. (b) Same as (a) but at (ε=20 eV and E=100 eV). (c) Same as (a) but at (ε = 10 eV and E=200 eV). (d) Same as (a) but at (ε=20 eV and E=200 eV).

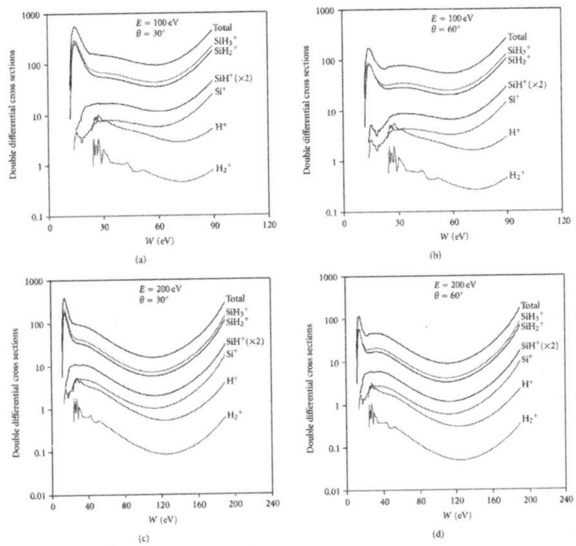

Figure 3: (a) The double differential cross sections (in the units of 10–20 cm^2/eV-sr) as a function of energy loss W at (θ=30° and E=100 eV) for the production of cations from electron impact ionization of SiH_4. (b) Same as (a) but at (θ=60° and E=100 eV). (c) Same as (a) but at (θ=30° and E=200 eV). (d) Same as (a) but at (θ=60° and E=200 eV).

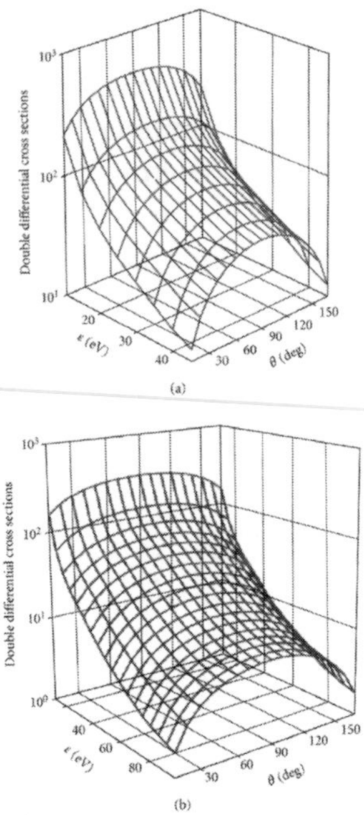

Figure 4: (a) 3D profiles of the double differential cross sections (in the units of 10–20 cm^2/eV-sr) as a function of secondary electron energy ε and angle θ at E=100 eV. (b) Same as (a) but at E=200 eV.

Because of the lack of experimental data for differential cross sections, the corresponding derived partial cross sections and their sum (the total ionization cross sections), from ionization threshold to 1000 eV, become important. The numerical values of our calculated partial and the total ionization cross sections are presented in Table 1. In Figure 5, we have presented the comparison of our partial ionization cross sections with the established experimental data sets of Chatham et al. [8], Krishnakumar and Srivastava [9], and Basner et al. [10]. It is noted that for the SiHn+ (n = 0–3) ions, our results are in good agreement with the experimental data [8, 10] within their composite error bars (The measurements of Chatham et al., Krishnakumar and Srivastava, and Basner et al. account the experimental uncertainties 30%–40%, 15%, and 10%–20%, resp.). On the other hand the experimental data of Haaland [11] which is a scaled value of Chatham et al. [8] and the Morrison and Traeger [12] are lower by a factor of two to three than the present calculations as well as the other experimental data sets [8–10]. Hence for the sake of reasonable shape and size of the figure, the data of Haaland is not shown in the figure. For the H2+ and H+ ions, a considerable disagreement with the experimental data has been noticed. However, our calculations for H2+ agree well with recent data of Basner et al., while the data set of Chatham (not shown) is much higher than the present calculations and the data of Basner et al. The data of Chatham et al. is about 3 and 4 times higher than Basner et al. and our calculations for H+, respectively. No identification of H+ and H2+ ions was made in earlier investigations [7, 11, 12]. In case of all dissociative fragment ions, the experimental data of Krishnakumar and Srivastava [9] show disagreement with our calculations as well as other experimental data [8, 10] in the energy range from 40 to 500 eV. Nevertheless, considering that this is a comparison with the absolute data where error bars of these sets of data are easily in the 10% to 40% regime (in particular taking into account that the calculations are depending on the accuracy of the experimental input parameters), the agreement is acceptable.

Table 1: Table for the partial ionization cross sections for SiH_4 by electron impact.

E (eV)	$Q_i(E)(10^{-16}\ cm^2)$						
	SiH_3^+	SiH_2^+	SiH^+	Si^+	H_2^+	H^+	Total
20	0.71	1.00	0.14	0.17			2.02
30	1.39	1.81	0.43	0.34	0.01	0.04	4.02
40	1.63	2.10	0.56	0.45	0.02	0.11	4.87
50	1.71	2.18	0.60	0.49	0.02	0.15	5.15
60	1.70	2.16	0.60	0.50	0.03	0.16	5.16
70	1.66	2.11	0.59	0.50	0.03	0.17	5.05
80	1.61	2.03	0.57	0.48	0.03	0.17	4.89
90	1.54	1.95	0.54	0.47	0.03	0.17	4.70
100	1.48	1.87	0.52	0.45	0.03	0.17	4.51
125	1.34	1.69	0.46	0.41	0.02	0.16	4.07
150	1.22	1.53	0.41	0.37	0.02	0.14	3.69
200	1.03	1.29	0.34	0.31	0.02	0.12	3.11
250	0.89	1.12	0.29	0.26	0.02	0.11	2.69
300	0.79	0.99	0.26	0.23	0.01	0.09	2.37
400	0.64	0.81	0.21	0.19	0.01	0.08	1.93
500	0.55	0.68	0.17	0.16	0.01	0.06	1.63
600	0.48	0.60	0.15	0.14	0.01	0.06	1.42
700	0.42	0.53	0.13	0.12	0.01	0.05	1.26
800	0.38	0.48	0.12	0.11	0.01	0.04	1.14
900	0.35	0.44	0.11	0.10	0.01	0.04	1.03

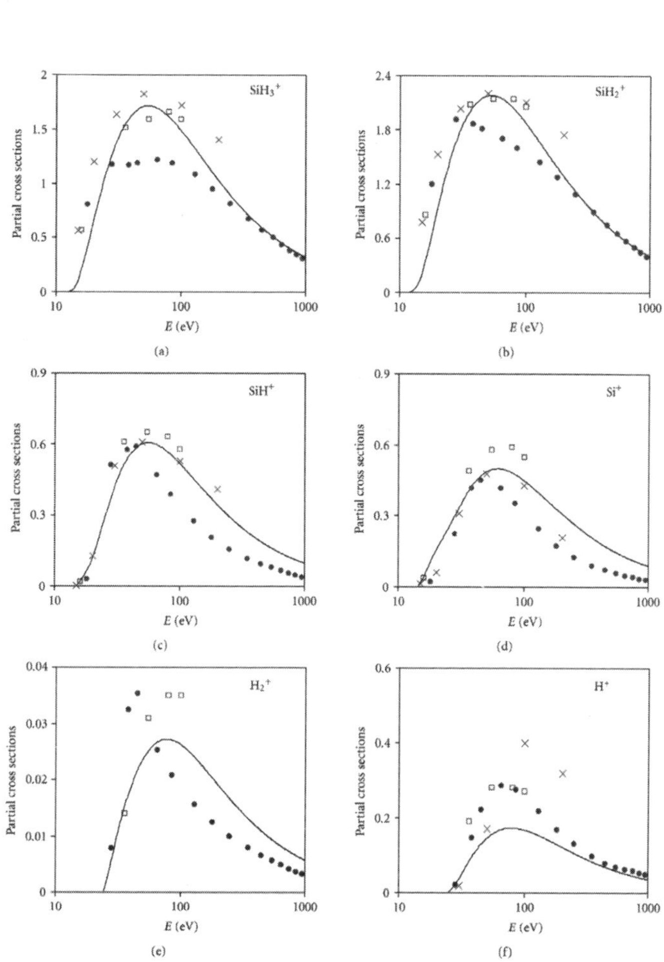

Figure 5: Partial ionization cross sections (in the units of $10^{-16}cm^2$) for electron impact ionization of SiH_4 (designated with solid lines) in comparison with the experimental data designated by ×—Chatham et al. [8], green-circle—Krishnakumar and Srivastava [9], and blue-square—Basner et al. [10].

Figure 6 shows good agreement of the partial cross sections when added up to a total cross section with the experimental data sets [8–10] along with the experimental data of Perrin et al. [7] which is available only at 100 eV and theoretical cross section data sets [13, 16, 17] in the complete energy range covered in the calculations. Recently, Malcolm and Yeager [28] have reviewed the BEB model [13] employing the more accurate ionization potential evaluation from Hartree-Fock theorem using cc-pVDZ and cc-pVTZ basis sets with multiconfigurational self-consistent field. This approach is a viable method for calculating electron impact ionization cross sections for systems where Koopman's theorem is known to be unreliable, or no experimental data is available. However, in case of this open channel molecule SiH4, these calculations for total ionization cross sections do not have a significant effect on the data derived from BEB (see [28] for detailed discussion). The data of Krishnakumar and Srivastava [9] for total cross sections show similar trends as for dissociative ionization cross sections. The total ionization cross sections reported by Haaland [11] are much smaller than all the measured data sets and the calculations. Hence for the sake of brevity, we have not shown the same data [11] in the comparison.

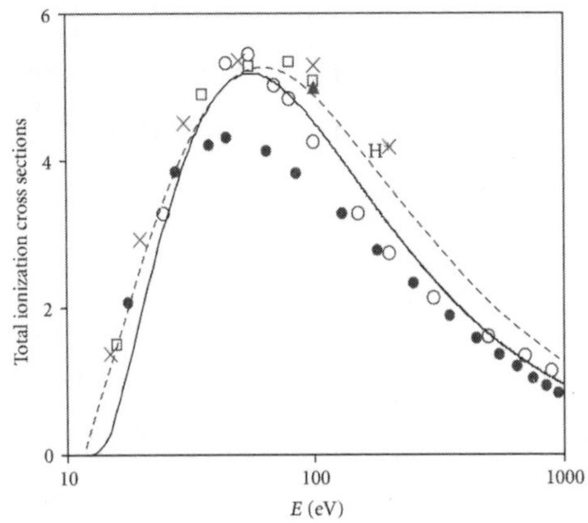

Figure 6: Presently calculated total ionization cross sections (in the units of $10^{-16}cm^2$) for electron impact ionization of SiH_4 (designated by solid line) in comparison with the various experimental data sets designated by red-triangle—Perrin et al. [7], ×—Chatham et al. [8], green-circle—Krishnakumar and Srivastava [9], and blue-square—Basner et al. [10] and theoretical data sets designated by blue-line BEB calculations [13, 28] and purple-circle—Joshipura et al. [16, 17].

The present calculations for partial and total ionization cross sections satisfy the necessary consistency checks to access their consistency and reliability. The consistency checks are derived from the fact that the total electron impact ionization cross section (i) is equal to the charged-weighted sum of the partial ionization cross sections and (ii) may be obtained by integration of the differential cross sections over secondary electron energies and angles. The former condition is used in the summation method for calibration purposes and the later fact in the use of Platzman plots. Both relationships allow one to check the reliability of the absolute magnitude and the energy dependence of ionization cross sections under consideration. In the low-energy limit close to the onset of ionization, the shapes of the partial and total electron impact ionization cross section curves are governed by a threshold law, which is usually expressed in the form of Q(E) ~ (E–Ii)Z, where Z is the charge state of the ion. The precise shape of the cross section in this region is especially important in determination (by extrapolation) of respective ionization thresholds, to compare with those derived by other means [29–31].

In relation to the applications, in particularly to plasma processes, ionization rate coefficients are rather more desirable than ionization cross sections. We have evaluated a set of ionization rate coefficients as a function of electron temperature in the units of energy for the individual cations produced in electron collision with the SiH4 molecule. The calculations are made using the calculated ionization cross sections and Maxwell-Boltzmann energy distribution, and the results are presented in Figure 7 along with Table 2.

Table 2: Table for the ionization rate coefficients (in the units of $10^{-20}cm^3/s$) corresponding to the formation of cations in electron dissociative ionization of SiH_4.

E (eV)	SiH_3^+	SiH_2^+	SiH^+	Si^+	H_2^+	H^+	Total
20	6.22	8.70	0.81	1.15			16.87
30	43.42	56.69	11.36	9.48	0.11	0.50	121.56
40	111.66	143.36	34.35	28.42	0.77	4.45	323.01
50	204.56	260.65	66.52	55.54	2.00	12.21	601.47
55	257.82	327.77	84.98	71.46	2.79	17.27	762.08
60	314.43	399.05	104.55	88.53	3.67	22.97	933.21
70	434.79	550.47	145.91	125.09	5.68	35.81	1297.75
80	560.64	708.68	188.81	163.44	7.88	49.92	1679.38
90	688.18	868.92	231.94	202.30	10.19	64.66	2066.18
100	814.63	1027.73	274.36	240.76	12.52	79.58	2449.58
125	1112.68	1401.83	373.04	330.92	18.15	115.55	3352.17
150	1369.77	1724.34	456.55	407.86	23.10	147.20	4128.82
200	1732.03	2178.41	570.58	514.06	30.22	192.61	5217.90
250	1899.27	2387.60	619.23	560.61	33.63	214.40	5714.74
300	1911.95	2402.83	617.83	561.09	34.09	217.37	5745.16
400	1659.73	2085.17	528.62	481.86	29.72	189.53	4974.64
500	1270.81	1596.30	400.20	365.58	22.74	145.03	3800.66
600	900.73	1131.32	281.06	257.09	16.08	102.55	2688.84
700	605.90	760.96	187.62	171.78	10.79	68.77	1805.81
800	392.47	492.89	120.73	110.62	6.96	44.41	1168.08
900	247.07	310.28	75.57	69.27	4.37	27.87	734.43

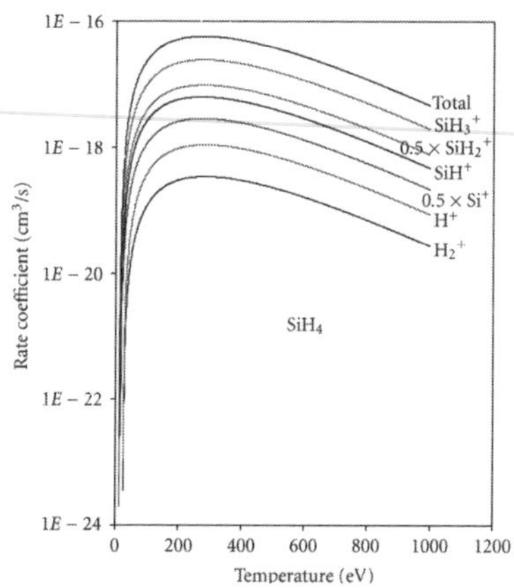

Figure 7: Ionization rate coefficients as a function of the temperature for SiH_4.

Conclusion

The calculations for differential cross sections as a function of secondary electron energy and angle at fixed impinging electron energy, corresponding to the formation of various cations in electron impact dissociative ionization of the SiH4 molecule, have been carried out by employing a semiempirical formalism based on the Jain-Khare approach. The calculations were made for the production of various cations produced via dissociative electron ionization of SiH_4. In absence of any data for differential cross sections, the corresponding derived partial ionization cross sections revealed a good agreement with available experimental data sets. However in case of minor H^+ and H_2^+ ions, some discrepancy was noticed. The ionization rate coefficients, a key parameter in plasma modeling, have been evaluated using a Maxwell energy distribution. The present evaluations for electron ionization cross sections and rate coefficients provide a contribution to the knowledge of various plasma processes.

Acknowledgements

NK is thankful to University Grants Commission, New Delhi for awarding project associateship with Grant no. F.30-10(2004)(SR). Anshu thanks to DAE/BRNS for junior research fellowship award through Grant no. 2007/37/13/BRNS.

References

1. W. L. Morgan, "A critical evaluation of low-energy electron impact cross sections for plasma processing modeling. II: Cl4, SiH4, and CH4," Plasma Chemistry and Plasma Processing, vol. 12, no. 4, pp. 477–493, 1992.
2. P. Kae-Nune, J. Perrin, J. Guillon, and J. Jolly, "Mass spectrometry detection of radicals in SiH4-CH4-H2 glow discharge plasmas," Plasma Sources Science and Technology, vol. 4, no. 2, pp. 250–259, 1995.
3. J. W. Gallagher, C. E. Brion, J. A. R. Samson, and P. W. Langhoff, "Absolute cross sections for molecular photoabsorption, partial photoionization, and ionic photofragmentation processes," Journal of Physical and Chemical Reference Data, vol. 17, no. 1, pp. 9–153, 1988.
4. R. D. Johnson, III, B. P. Tsai, and J. W. Hudgens, "Multiphoton ionization of SiH3 and SiD3 radicals: electronic spectra, vibrational analyses of the ground and Rydberg states, and ionization potentials," The Journal of Chemical Physics, vol. 91, no. 6, pp. 3340–3359, 1989.

5. G. Cooper, T. Ibuki, and C. E. Brion, "Absolute oscillator strengths for photoabsorption, photoionization and ionic photofragmentation of silane. II. The Si 2p and 2s inner shells," Chemical Physics, vol. 140, no. 1, pp. 147–155, 1990.
6. G. Cooper, T. Ibuki, and C. E. Brion, "Absolute oscillator strengths for photoabsorption, photoionization and ionic photofragmentation of silane. I. The valence shell," Chemical Physics, vol. 140, no. 1, pp. 133–145, 1990.
7. J. Perrin, J. P. M. Schmitt, G. de Rosny, B. Drevillon, J. Huc, and A. Lloret, "Dissociation cross sections of silane and disilane by electron impact," Chemical Physics, vol. 73, no. 3, pp. 383–394, 1982.
8. H. Chatham, D. Hils, R. Robertson, and A. Gallagher, "Total and partial electron collisional ionization cross sections for CH4, C2H6, SiH4, and Si2H6," The Journal of Chemical Physics, vol. 81, no. 4, pp. 1770–1777, 1984.
9. E. Krishnakumar and S. K. Srivastava, "Ionization cross sections of silane and disilane by electron impact," Contributions to Plasma Physics, vol. 35, no. 4-5, pp. 395–404, 1995.
10. R. Basner, M. Schmidt, V. Tarnovsky, K. Becker, and H. Deutsch, "Dissociative ionization of silane by electron impact," International Journal of Mass Spectrometry and Ion Processes, vol. 171, no. 1–3, pp. 83–93, 1998.
11. P. Haaland, "Dissociative ionization of silane," Chemical Physics Letters, vol. 170, no. 2-3, pp. 146–152, 1990.
12. J. D. Morrison and J. C. Traeger, "Ionization and dissociation by electron impact III. CH4 and Sih4," International Journal of Mass Spectrometry and Ion Physics, vol. 11, no. 3, pp. 289–300, 1973.
13. M. A. Ali, Y.-K. Kim, W. Hwang, N. M. Weinberger, and M. E. Rudd, "Electron-impact total ionization cross sections of silicon and germanium hydrides," The Journal of Chemical Physics, vol. 106, no. 23, pp. 9602–9608, 1997.
14. S. P. Khare, S. Prakash, and W. J. Meath, "Dissociative ionization of NH3 and H2O molecules by electron impact," International Journal of Mass Spectrometry and Ion Processes, vol. 88, no. 2-3, pp. 299–308, 1989.
15. H. Deutsch, K. Becker, S. Matt, and T. D. Märk, "Theoretical determination of absolute electron-impact ionization cross sections of molecules," International Journal of Mass Spectrometry, vol. 197, no. 1–3, pp. 37–69, 2000.
16. K. N. Joshipura, B. G. Vaishnav, and S. Gangopadhyay, "Electron impact ionization cross-sections of plasma relevant and astrophysical silicon compounds: SiH4, Si2H6, Si(CH3)4, SiO, SiO2, SiN and SiS," International Journal of Mass Spectrometry, vol. 261, no. 2-3, pp. 146–151, 2007.

17. M. Vinodkumar, C. Limbachiya, K. Korot, and K. N. Joshipura, "Theoretical electron impact elastic, ionization and total cross sections for silicon hydrides, SiHx (x=1,2,3,4) and disilane, Si2H6 from threshold to 5 keV," The European Physical Journal D, vol. 48, no. 3, pp. 333–342, 2008.

18. S. Pal, "Determination of single differential and partial cross-sections for the production of cations in electron-methanol collision," Chemical Physics, vol. 302, no. 1–3, pp. 119–124, 2004.

19. S. Pal, J. Kumar, and T. D. Märk, "Differential, partial and total electron impact ionization cross sections for SF6," The Journal of Chemical Physics, vol. 120, no. 10, pp. 4658–4663, 2004.

20. P. Bhatt and S. Pal, "Determination of cross sections and rate coefficients for the electron impact dissociation of NO2," Chemical Physics, vol. 327, no. 2-3, pp. 452–456, 2006.

21. S. Pal, "Differential and partial ionization cross sections for electron impact ionization of plasma processing molecules: CF4 and PF5," Physica Scripta, vol. 77, no. 5, Article ID 055304, 7 pages, 2008.

22. M. Capitelli, R. Celiberto, and M. Cocciatore, "Needs for cross sections in plasma chemistry," Advances in Atomic, Molecular, and Optical Physics, vol. 33, pp. 321–372, 1994.

23. T. Fujimoto, "Semi-empirical cross sections and rate coefficients for excitation and ionization by electron collision and photoionization of Helium," Institute of Plasma Physics Report, IIPJ-AM-8, Nagoya University, Nagoya, Japan, 1978.

24. S. Pal and S. Prakash, "Partial differential cross sections for the ionization of the SO2 molecule by electron impact," Rapid Communications in Mass Spectrometry, vol. 12, no. 6, pp. 297–301, 1998.

25. M. Inokuti, "Inelastic collisions of fast charged particles with atoms and molecules—the bethe theory revisited," Reviews of Modern Physics, vol. 43, no. 3, pp. 297–347, 1971.

26. H. Deutsch, K. Becker, G. Senn, S. Matt, and T. D. Märk, "Calculation of cross sections and rate coefficients for the electron impact multiple ionization of beryllium, boron, carbon, and oxygen atoms," International Journal of Mass Spectrometry, vol. 192, no. 1–3, pp. 1–8, 1999.

27. J. Berkowitz, J. P. Greene, H. Cho, and B. Ruščić, "Photoionization mass spectrometric studies of SiHn (n=1–4)," The Journal of Chemical Physics, vol. 86, no. 3, pp. 1235–1248, 1987.

28. N. O. J. Malcolm and D. L. Yeager, "Purely theoretical electron-impact ionization cross-sections of silicon hydrides and silicon fluorides obtained from explicitly correlated methods," The Journal of Chemical Physics, vol. 113, no. 1, pp. 8–17, 2000.

29. T. D. Märk, "Ionization of molecules by electron impact," in Electron-Molecule Interactions and Their Applications, L. G. Christophorou, Ed., vol. 1, pp. 251–334, Academic Press, Orlando, Fla, USA, 1984.

30. "Secondary electron spectra by charged particles interactions," ICRU, International Commission on Radiation Units and Measurements, Bethesda, Md, USA, 1996.

31. P. G. Burke, "Theory of electron scattering by atoms, ions and molecules," in Proceedings of the 2nd International Conference on Atomic and Molecular Data and Their Applications (ICAMDATA '00), K. A. Berrington and K. L. Bell, Eds., pp. 155–177, AIOP, Oxford, UK, March 2000.

Spatial Heterogeneity and Imperfect Mixing in Chemical Reactions: Visualization of Density-Driven Pattern Formation

Sabrina G. Sobel, Harold M. Hastings and Matthew Testa

ABSTRACT

Imperfect mixing is a concern in industrial processes, everyday processes (mixing paint, bread machines), and in understanding salt water-fresh water mixing in ecosystems. The effects of imperfect mixing become evident in the unstirred ferroin-catalyzed Belousov-Zhabotinsky reaction, the prototype for chemical pattern formation. Over time, waves of oxidation (high ferriin concentration, blue) propagate into a background of low ferriin concentration (red); their structure reflects in part the history of mixing in the reaction vessel. However, it may be difficult to separate mixing effects from reaction *effects. We describe a simpler model system for visualizing density-driven pattern formation in an essentially unmixed chemical system: the reaction of pale yellow Fe^{3+} with colorless SCN^- to form the blood-red $Fe(SCN)^{2+}$ complex ion in aqueous solution. Careful addition of one drop of $Fe(NO_3)_3$ to KSCN yields striped patterns after several minutes. The patterns appear reminiscent of Rayleigh-Taylor instabilities and convection rolls, arguing that pattern formation is caused by density-driven mixing.*

Introduction

Good mixing is a major challenge in many areas of chemistry and chemical engineering, ranging from continuously stirred tank reactors (CSTRs) [1] to industrial level processes [2], where imperfect mixing can adversely affect product quality and yield. Many of us are familiar with mixing problems in making bread, stirring paint, mixing gasoline and air in internal combustion engines, and mixing driven by density differences (salt water-fresh water convection) but even mixing of ordinary liquids can pose challenges. (The density of pure water is 1000 kg/m^3. Ocean water is more dense because of the salt in it. Density of ocean water at the sea surface is about 1027 kg/m^3. There are two main factors that make ocean water more or less dense than about 1027 kg/m^3. From http://www.csgnetwork.com/h2odenscalc.html we derived Table 1 (accessed 1/2009)).

Table 1: Densities of salt water and pure water at various temperatures

Temp (°C)	Ionic strength	Calc. density (g/cm^3)
23	3500	1.000215
	0	0.997568
26	0	0.996814
20	0	0.998234
30	0	0.995678

The effects of imperfect mixing are perhaps most dramatically evident in excitable chemical systems [1, 3] such as the BZ reaction [4–9] and the chlorite-iodide reaction [10–12]. An excitable chemical reaction is a reaction in which suitable small perturbations from steady state generate large excursions (excitations) before the system returns to steady state. Many excitable reactions display auto-oscillatory behavior in which sustained oscillations about an unstable steady state are observed. The effects of imperfect mixing can be seen readily in the unstirred, ferroin-catalyzed Beloushov-Zhabotinsky (BZ) reaction in a Petri disk, a quasi-two-dimensional (2D) system [3–8]. The catalyst in this reaction, ferroin/ferriin, also serves as an indicator; blue/oxidized at high [ferriin] and red/reduced at low [ferriin]. After an induction period of several minutes, one sees the "spontaneous" formation of target patterns of blue/oxidized rings moving outward from oscillatory (red/blue) centers into a red/reduced background.

Menzinger and Dutt [1] first described the effects of imperfect mixing upon reactions involving excitable media. We have explicitly demonstrated nonrandom spatiotemporal order in targetformation in the unstirred BZ reaction, with targets preferentially occurring in excitable regions near existing targets, but too far from pre-existing targets to be generated by wave propagation or chemical diffusion [3]. We explained the observed spatiotemporal order in terms of mixing heterogeneities in preparing the 2D reaction medium likely because of the sensitivity of the dynamics to very small perturbations at the onset of oscillation [13, 14]. In conditions of almost complete mixing, striped patterns reminiscent of convection rolls with spacing approximately equal to the depth of the medium were often observed (Figure 1, panel (a)). Here we describe an even simpler model system which displays analogous pattern formation driven by density differences, c.f. [15–17].

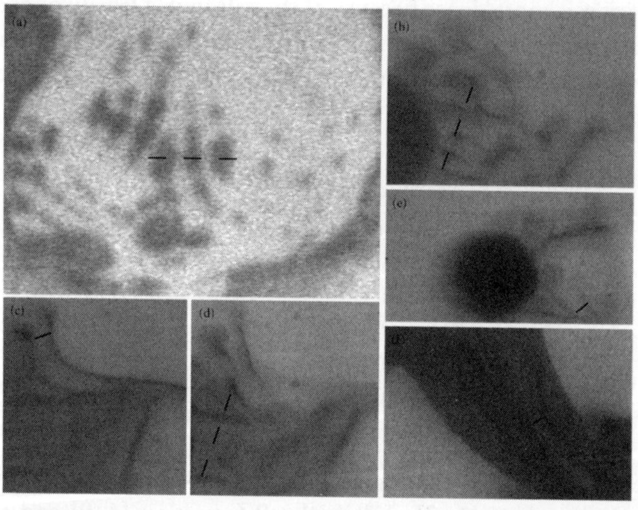

Figure 1: (a) Excitable BZ reaction medium, stirred after initial pattern formation, 2.1 mm depth. From [9, Figure 2c]. Scale bar: alternating 2.1 mm lines and spaces. (b) Chemical model: Fe^{3+} and SCN^-, 2 mm depth. Scale bar: alternating 2 mm lines and spaces. Striped patterns formed outside the visible droplet boundary (lower left of figure). The color corresponds to the concentration of $FeSCN^{2+}$. Resolution ~30 μm in photos (b) through (f). Time: about two minutes after addition of one drop of Fe^{3+} solution. (c), (d) Evolution of panel (b). Same experimental run, 45 and 30 second earlier. Note pattern initiation 2 mm behind visible wavefront (of diffusion wave) and effect of diffusion in blurring visible boundaries as concentration gradients are reduced. (e) Same model, but 3 mm depth, early stage of pattern formation. Note visible droplet boundary in center of figure, and larger scale that corresponds to greater depth as in (b) through (d). (f) Same model, 1 mm depth, early stage of pattern formation. Scale bar: 1 mm. Note fine structure (<<1 mm) transverse to a spatial filament of the reaction, and initial excitation of 1 mm scale.

Pattern Visualization with $Fe(SCN)^{2+}$

The present experiment arose from trying to understand how slight density differences might drive this pattern formation. Since BZ dynamics are relatively complex, we sought a simpler, minimal example to better visualize and understand the effects of mixing heterogeneity. We used a simple chemical reaction, in which two nearly colorless solutions, one containing Fe^{3+} (pale yellow) and the second containing SCN^- (clear), are mixed, producing a deep reddish color where Fe^{3+} and SCN^- combine to form the complex ion $FeSCN^{2+}$ in an essentially irreversible reaction. We made a simple modification—instead of mixing the two solutions, we added a small drop (<1 μL) of 0.1 M Fe^{3+} (as $Fe(NO_3)_3$) to a 1–4 mm deep solution of 0.02 M SCN^- (as KSCN) optionally containing a very small amount of surfactant (which had no apparent effect). The deep reddish color where Fe^{3+} and SCN^- combine to form $FeSCN^{2+}$ clearly labels interfaces between Fe^{3+}-dominated and SCN^--dominated regions. We found patterns reminiscent of convective rolls with spacing equal to the depth of the corresponding reaction medium (Figure 1, panels (b)–(e), below, similar to patterns seen under conditions of almost complete mixing in the BZ reaction, panel (a)). Note that the $Fe(NO_3)_3$ solution is denser than the KSCN solution: 1.010(4) versus 1.001(2) g/mL.

The resulting patterns demonstrate behavior expected in the Rayleigh-Taylor instability [18, 19], which occurs when a heavier fluid (here the Fe(NO3)3 solution) is placed on top of a lighter fluid (here the KSCN solution). Gravitational forces drive the heavier solution downward through the lighter solution in an unstable, turbulent manner, until concentration (and thus density) differences are destroyed by diffusion. In our model system, one sees initial small scale instabilities (panel (f)) which are gradually destroyed by diffusion as longer length scales are excited, with one added twist—selection of a length scale corresponding to the depth of the reaction medium after a time delay a few multiples of the time it takes a drop to fall the depth of the solution. From Stokes law, the net force on a small drop depends upon its volume, velocity, and density difference (ρ1–ρ2). In our model a drop of radius 50 μm rapidly reaches terminal velocity of ~50 μm/s, making the natural time scale (time to fall 2 mm) ~2 mm/ (50 μm/s) = 40 second, consistent with the formation of relatively stable patterns over ~3 minutes (~4 times this natural time scale).

Conclusion

This simple minimal model, in a nonexcitable system, demonstrates excitation of the most unstable mode (the only one that survives over minutes). It visually demonstrates the persistence of spatial heterogeneities in liquids. The problem of mixing is important in real-world situations often encountered by chemical engineers (industrial chemical reactions), process/flow engineers, and even fields such as hydrology and geoscience–temperature and salinity differences can cause density differences in water similar in magnitude to the density differences in this experimental model. This simple model provides a mechanism to experimentally visualize and study the patterns formed by density differences.

Acknowledgements

This material is based upon work supported by the US Department of Energy under Award no. DE-FG02-08ER64623 for the Hofstra University Center for Condensed Matter Research and by the US National Science Foundation Grant CHE-0515691. Neither the United States Government nor any agency thereof, nor any of their employees, makes any warranty, expressed or implied, or assumes any legal liability or responsibility for the accuracy, completeness, or usefulness of any information, apparatus, product, or process disclosed, or represents that its use would not infringe privately owned rights. Reference herein to any specific commercial product, process, or service by trade name, trademark, manufacturer, or otherwise does not necessarily constitute or imply its endorsement, recommendation, or favoring by the United States Government or any agency thereof. The views and opinions of authors expressed herein do not necessarily state or reflect those of the United States Government or any agency thereof.

References

1. M. Menzinger and A. K. Dutt, "The myth of the well-stirred CSTR in chemical instability experiments: the chlorite/iodide reaction," Journal of Physical Chemistry, vol. 94, no. 11, pp. 4510–4514, 1990.
2. E. L. Paul, V. Atiemo-Obeng, and S. M. Kresta, Eds., Handbook of Industrial Mixing: Science and Practice, E. L. Paul, V. Atiemo-Obeng, and S. M. Kresta, Eds., Wiley-Interscience, Hoboken, NJ, USA, 2004.
3. H. M. Hastings, S. G. Sobel, A. Lemus, et al., "Spatiotemporal clustering and temporal order in the excitable BZ reaction," Journal of Chemical Physics, vol. 123, no. 6, Article ID 064502, 6 pages, 2005.
4. B. P. Belousov, "A periodic reaction and its mechanism," in Sbornick Referatov po Radiatsionni Meditsine, pp. 145–160, Medgiz, Moscow, Russia, 1958.

5. A. M. Zhabotinsky, "Periodic processes of the oxidation of malonic acid in solution (study of the kinetics of Belousov's reaction)," Biofizika, vol. 9, p. 306, 1964.

6. R. J. Field, E. Körös, and R. M. Noyes, "Oscillations in chemical systems. II. Thorough analysis of temporal oscillation in the bromate-cerium-malonic acid system," Journal of the American Chemical Society, vol. 94, no. 25, pp. 8649–8664, 1972.

7. R. J. Field and R. M. Noyes, "Oscillations in chemical systems. V. Quantitative explanation of band migration in the Belousov-Zhabotinskii reaction," Journal of the American Chemical Society, vol. 96, no. 7, pp. 2001–2006, 1974.

8. B. Z. Shakashiri, Chemical Demonstrations, Vol. 2, University of Wisconsin Press, Madison, Wis, USA, 1985.

9. S. K. Scott, Oscillations, Waves, and Chaos in Chemical Kinetics, Oxford Chemistry Primers no. 18, Oxford University Press, New York, NY, USA, 1994.

10. C. E. Dateo, M. Orban, P. De Kepper, and I. R. Epstein, "Systematic design of chemical oscillators. 5. Bistability and oscillations in the autocatalytic chlorite-iodide reaction in a stirred-flow reactor," Journal of the American Chemical Society, vol. 104, no. 2, pp. 504–509, 1982.

11. M. Orban, C. Dateo, P. De Kepper, and I. R. Epstein, "Systematic design of chemical oscillators. 11. Chlorite oscillators: new experimental examples, tristability, and preliminary classification," Journal of the American Chemical Society, vol. 104, no. 22, pp. 5911–5918, 1982.

12. O. Citri and I. R. Epstein, "Systematic design of chemical oscillators. 42. Dynamic behavior in the chlorite-iodide reaction: a simplified mechanism," Journal of Physical Chemistry, vol. 91, no. 23, pp. 6034–6040, 1987.

13. H. M. Hastings, R. J. Field, and S. G. Sobel, "Microscopic fluctuations and pattern formation in a supercritical oscillatory chemical system," Journal of Chemical Physics, vol. 119, no. 6, pp. 3291–3296, 2003.

14. H. M. Hastings, S. G. Sobel, R. J. Field, et al., "Bromide control, bifurcation and activation in the Belousov-Zhabotinsky reaction," Journal of Physical Chemistry A, vol. 112, no. 21, pp. 4715–4718, 2008.

15. Y. Wu, D. A. Vasquez, B. F. Edwards, and J. W. Wilder, "Convective chemical-wave propagation in the Belousov-Zhabotinsky reaction," Physical Review E, vol. 51, no. 2, pp. 1119–1127, 1995.

16. D. Zhang, W. R. Peltier, and R. L. Armstrong, "Buoyant convection in the Belousov-Zhabotinsky reaction. I. Thermally driven convection and

distortion of chemical waves," The Journal of Chemical Physics, vol. 103, no. 10, pp. 4069–4077, 1995.

17. D. Zhang, W. R. Peltier, and R. L. Armstrong, "Buoyant convection in the Belousov-Zhabotinsky reaction. II. Chemically driven convection and instability of the wave structure," The Journal of Chemical Physics, vol. 103, no. 10, pp. 4078–4089, 1995.
18. A. W. Cook and D. Youngs, "Rayleigh-Taylor instability and mixing," Scholarpedia, vol. 4, no. 2, p. 6092, 2009.
19. M. J. Andrews and D. B. Spalding, "A simple experiment to investigate two-dimensional mixing by Rayleigh-Taylor instability," Physics of Fluids A, vol. 2, no. 6, pp. 922–927, 1990.

Fischer-Tropsch Synthesis over Iron Manganese Catalysts: Effect of Preparation and Operating Conditions on Catalyst Performance

Ali A. Mirzaei, Samaneh Vahid and Mostafa Feyzi

ABSTRACT

Iron manganese oxides are prepared using a coprecipitation procedure and studied for the conversion of synthesis gas to light olefins and C_5^+ hydrocarbons. In particular, the effect of a range of preparation variables such as [Fe]/[Mn] molar ratios of the precipitation solution, pH of precipitation, temperature of precipitation, and precipitate aging times was investigated in detail. The results are interpreted in terms of the structure of the active catalyst and it has been generally concluded that the calcined catalyst (at 650°C for 6 hours) containing 50%Fe/50%Mn-on molar basis which is the most active catalyst *for the conversion of synthesis gas to light olefins. The effects of different promoters and supports with loading of optimum support on the catalytic performance of catalysts are also studied. It was found that the catalyst containing 50%Fe/50%Mn/5 wt.%Al_2O_3 is an optimum-modified catalyst. The catalytic performance of optimal catalyst has been studied in operation conditions such as a range of reaction temperatures, H_2/CO molar feed ratios and a range of total pressures. Characterization of both precursors and calcined catalysts is carried out by powder X-ray diffraction (XRD), scanning electron microscopy (SEM), BET specific surface area and thermal analysis methods such as TGA and DSC.*

Introduction

Fischer-Tropsch (FT) synthesis is of great industrial importance due to the great variety of products obtained such as paraffins, olefins and alcohols. An approach to improve the selectivity of the classical Fischer-Tropsch (FT) process for conversion of synthesis gas to hydrocarbons involves the use of a bifunctional catalyst system containing a metal catalyst (FT catalyst) combined with a support. There has been renewed interest in recent years in FT synthesis, especially for the selective production of petrochemical feedstocks such as ethylene, propylene, and buthylene (C_2–C_4 olefins) directly from synthesis gas [1–7]. Compared to other metal catalysts for Fischer-Tropsch (FT) synthesis, an iron-based catalyst is distinguished by higher conversion, selectivity to the lower olefins, and flexibility to the process parameters [8, 9]. However, the use of iron catalyst does not solve the problem of insufficient selectivity, which represents a general limitation of FT synthesis. Manganese has been widely used as one of the promoters for FTS on iron catalyst, particularly in producing low olefins [10–13]. Large efforts have also been exerted on the individual effect of manganese promotion on supported or unsupported iron catalysts [13, 14]. Fe–Mn and Co–Mn catalysts favor C_2–C_4 olefins [15, 16]. High C_2–C_4 selectivity for the iron-rich Fe–Mn solid has been correlated with the iron manganese oxides phase and two carbide phases, while the manganese-rich solid has been correlated with two spinel phases and different carbide phases [11, 17]. The Fe–Mn catalyst, as one of the most important catalyst systems, has received extensive attention in recent years because of the higher olefin and middle distillation cut selectivities which allow their products to be used as a feedstock for the chemical industry. Therefore, the Fe–Mn catalyst has a promising industrial application [18–22], and it is well known that higher selectivity of alkenes can be obtained on Fe/Mn catalysts than on other iron-based catalysts [23, 24]. The aim of this research work was to investigate the effect of a range of preparation variables including the precipitate aging time, pH, temperature of precipitation, and the [Fe]/[Mn] molar ratio of the precipitation solution of mixed iron manganese oxide catalysts. We also report further results concerning the effects of different promoters and supports along with loadings of Al_2O_3 as an optimum support on catalytic performance of this catalyst for Fischer-Tropsch synthesis. In addition, the catalyst structural and morphological was investigated by XRD, SEM, BET, and thermal analysis methods such as TGA and DSC. Also, the effects of operation conditions such as H_2/CO molar feed ratios, a range of reaction temperatures and total pressures for conversion of synthesis gas to light olefins have been studied.

Experimental

Catalyst Preparation

All the catalysts were prepared using the coprecipitation procedure. Aqueous solutions of $Fe(NO_3)_3 \cdot 9H_2O$ (0.5 M) (99%, Merck, Germany) and $Mn(NO_3)_2 \cdot 4H_2O$ (0.5 M) (99.5%, Merck, Germany) with different molar ratios were premixed and the resulting solution heated to 70°C in a round-bottomed flask fitted with a condenser. Aqueous Na_2CO_3 (0.5 M) (99.8%, May & Baker, France) was added dropwise to the mixed nitrate solution, which was continuously stirred whilst the temperature was maintained isothermally in the range of 40–80°C. The final pH achieved was varied between 6.3 and 10.3. This procedure took approximately 10 minutes to complete. The resulting precipitate was then left in this medium at the required pH and temperature used for the precipitation for times ranging from 0 to 5 hours. The precipitate was first filtered and then washed several times with warm distilled water until no further Na^+ was observed in the washings tested by flam atomic absorption. The precipitate was then dried at 110°C for 16 hours to give a material denoted as the catalyst precursor which was subsequently calcined in static air in the furnace (650°C, 6 hours) to give the final catalyst. For preparation of the supported catalysts, the same amount (20 wt%) of each support such as TiO_2 (98%, May & Baker), SiO_2 (98%, Merck), γ-Al_2O_3 (98%, Merck), MgO (98%, Merck), and ZSM-5 zeolite (99%, Aldrich, UK) has been added separately to the mixed solution of iron and manganese nitrates with nominal ratio of Fe/Mn=1/1. After the test of all these supported catalysts, it was found that Al_2O_3 is the best support than the others, so the Al_2O_3 loading of 5, 10, 15, 20, and 25 wt% based on the total catalyst weight, were used to obtain the best loading of Al_2O_3 support. The supported catalyst was then promoted with different promoters (Li, K, Rb, and Mg) by adding a small amount (1.5 wt%) of $LiNO_3$ (99.5%, Prolabo, France), KNO3 (99%, Merck), RbCl (99%, Prolabo), and $Mg(NO_3)_2 \cdot 6H_2O$ (99%, May & Baker) as separately to the suspension containing 50%Fe/50%Mn/5 wt.%Al_2O_3.

Catalyst Characterization

X-Ray Diffraction (XRD)

Powder X-ray diffraction (XRD) measurements were performed using a Bruker Axs Company, D8 Advance diffractometer (Germany). Scans were taken with a 2θ stepsize of 0.02 and a counting time of 1 second using CuK_{α} radiation source generated at 40 kV and 30 mA. Specimens for XRD were prepared by compaction into a glass-backed aluminum sample holder. Data was collected over a 2θ range from 4° to 70° and phases were identified by matching experimental patterns to entries in the Diffrac^{plus} Version 6.0 indexing software.

BET Measurements

Brunauer-Emmett-Teller surface area BET measurements were conducted using a micrometrics adsorption equipment (Quantachrome instrument, model Nova 2000, USA) determining nitrogen (99.99% purity) as the analysis gas and the catalyst samples were slowly heated to 300 for 3 hours under nitrogen atmospheric. Prior to analysis each precursors and catalyst and after reaction catalysts measurements specific surface area was evacuated at –196°C for 66 minutes.

Thermal Gravimetric Analysis (TGA) and Differential Scanning Calorimetriy (DSC)

The TGA and DSC were carried out using simultaneous thermal analyzer apparatus of Rheometric Scientific Company (STA 1500+ Model, England) under a flow of dry air. The temperature was raised from 25°C to 800°C using a linear programmer at a heating rate of 10°C/min. The sample weight was between 10 and 20 mg.

Scanning Electron Microscopy (SEM)

The morphology of catalysts and their precursors was observed by means of an S-360 Oxford Eng scanning electron microscopy (made in USA). All of the SEM images in this study are taken at the same magnification of 20 μm.

Catalyst Testing

The catalyst tests were carried out in a fixed bed stainless steel microreactor at different operation conditions (Figure 1). All gas lines to the reactor bed were made from 1/4″ stainless steel tubing. Three mass flow controllers (Brooks, Model 5850E) equipped with a four-channel control panel (Brooks 0154) were used to adjust automatically the flow rate of the inlet gases (CO, H_2, and N_2 with purity of 99.999%). The mixed gases passed into the reactor tube, which was placed inside a tubular furnace (Atbin, Model ATU 150-15) capable of producing temperature up to 1300°C and controlled by a digital programmable controller (DPC). The reactor tube was constructed from 1/2″ stainless steel tubing; internal diameter of 9 mm, with the catalyst bed situated in the middle of the reactor. The reaction temperature was controlled by a thermocouple inserted into catalyst bed and visually monitored by a computer equipped with software. The meshed catalyst (1.0 g) was held in the middle of the reactor with 110 cm length using quartz wool. It consists of an electronic back pressure regulator which can control the total pressure of the desired process using a remote control via the TESCOM software package integration that improve or modify its efficiency that capable of working on pressure ranging from atmospheric pressure to 100 bar. The catalyst was pre-reduced in situ atmospheric pressure in a flowing H_2–N_2 stream (N_2/H_2=1, flow rate of each gas=30 mL/min) at 400°C for 6 hours before synthesis gas exposure. The FT reactions was carried out at 260–420°C (P=1–15 bar, H_2/CO=1/1–3/1, GHSV=2700–5400 h-1). Reactant and product streams were analyzed online using a gas chromatograph (Varian, Model 3400 Series) equipped with a 10-port sampling valve (Supelco company, USA, Visi Model), a sample loop, and thermal conductivity detector (TCD). The contents of sample loop were injected automatically into a packed column (Hayesep DB, Altech Company, USA, 1/8″ OD, 10 meters long, and particle mesh 100/120). Helium was employed as a carrier gas for optimum sensitivity (flow rate=30 mL/min). The calibration was carried out using various calibration mixtures and pure compounds obtained from American Matheson Gas Company (USA). GC controlling and collection of all chromatograms was done via an IF-2000 single channel data interface (TG Co, Tehran, Iran) at windows environment. The results in terms of CO conversion, selectivity, and yield of products are given at each space velocity. The CO conversion (%) is calculated according to the normalization method [25]:

$$\text{CO conversion (\%)} = \frac{(MolesCO_{in}) - (MolesCo_{out})}{MolesCo_{in}} \times 100 \cdot \qquad (1)$$

The selectivities (%) toward the individual components on carbon basis are calculated according to the same principle [26]:

$$\text{Selectivity of j product (\%)} = \frac{(Moles \text{ of j product})}{(MolesCO_{in}) - (MolesCO_{out})} \times 100. \qquad (2)$$

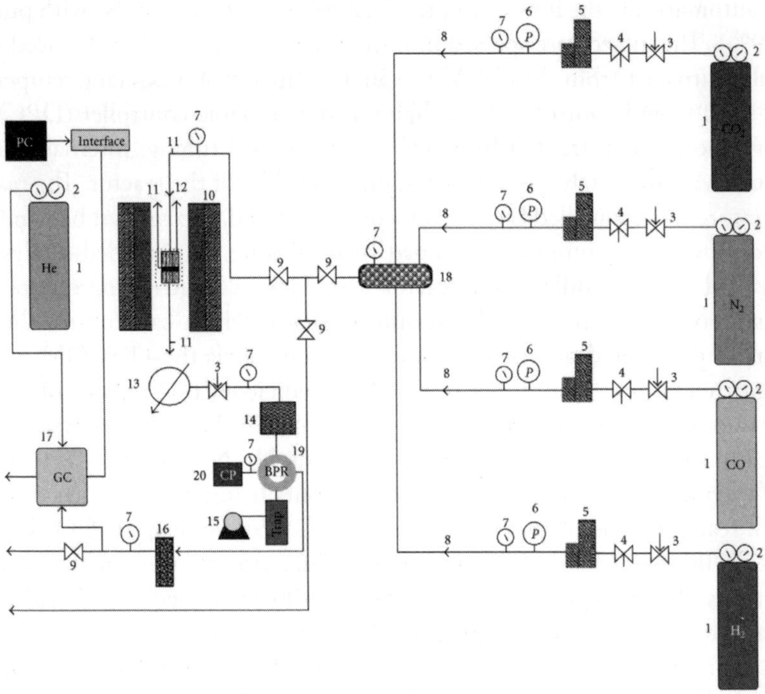

Figure 1: Schematic representation of the reactor in a flow diagram used. (1) Gas cylinders, (2) pressure regulators, (3) needle valves, (4) valves, (5) mass flow controllers (MFCs), (6) digital pressure controllers, (7) pressure Gauges, (8) nonreturn valves, (9) ball valves, (10) tubular furnace, (11) temperature indicators, (12) tubular reactor and catalyst bed, (13) condenser, (14) trap, (15) air pump, (16) silica gel column, (17) gas chromatograph (GC), (18) mixing chamber, (19) (BPR) back pressure regulator (electronically type), (20) control panel (CP).

Results and Discussion

Effect of Preparation Conditions

In this part of study, we have investigated the effect of a range of iron manganese oxide catalysts preparation variables at the precursor stage upon the structure of these materials, and the subsequent influence these structural effects have on the activity of the final calcined catalysts. The optimum preparation conditions were identified with respect to the catalytic activity for the conversion of synthesis gas to light olefins.

Effect of Aging Time

Aging time is one of the most important factors on the catalytic performance of the catalyst and it was defined as the time between the formation of precipitate and the removal of solvent. In our previous study, we demonstrated the importance of aging time with respect to catalyst activity for oxidation of CO by mixed copper manganese oxide and mixed copper zinc oxide catalysts [27–35] and for hydrogenation of CO by mixed of cobalt-iron oxide, cobalt-manganese oxide, and cobalt-cerium oxide catalysts for Fischer-Tropsch synthesis [36–40]. In all of these investigations, our results have shown that the aging of the precipitates obtained by coprecipitation leads to phase changes toward the forms, which are more stable thermodynamically. In this study, to examine the effect of aging on the performance of iron manganese oxide catalysts for the hydrogenation of CO, a series of mixed iron manganese oxide catalysts were prepared by coprecipitation method ([Fe]/[Mn]=1/1, 70°C, pH=8.3) with a range of aging times between 0 minute (unaged) and 300 minutes for the precipitate. The catalysts were prepared by calcination at 650°C for 6 hours and then were tested for hydrogenation of CO. The effect of aging time on catalytic performance is shown in Table 1. These results show that there is a considerable variation in the catalyst performance with respect to aging time and the sample aged for 3 hours gave the optimal catalytic performance for CO conversion in FTS. Thus aging time is a parameter of crucial importance in the preparation of active mixed iron manganese oxide catalysts for the hydrogenation of CO. Characterization studies were carried out using various techniques for both the precursors and calcined catalysts. Characterization of precursors was studied to establish the importance of the structure of the catalyst precursor in controlling the structure of the final catalyst and consequently activity. The catalyst precursors which were prepared using coprecipitation method in different aging times were characterized by thermal gravimetric analysis (TGA) and the TGA curves of these precursors are displayed in Figure 2. For these precursors, the thermogravimetric curves seem to indicate three-stage decomposition. The first stage is considered to be due to the removal of adsorbed water (40–110°C) and the second stage is due to the decomposition of hydroxyl bimetallic or nitrate precursor (330–420°C), respectively. The peak around 550–630°C is due to the decomposition of $MnCO_3$ or $Fe_2CO_3(OH)_2$ to oxides. The TGA curves are involved with total overall weight loss of ca. 18–23 wt.%. The catalyst precursors were also characterized by XRD and similar phases were identified for these catalyst precursors with preparation ratios (MnCO3 (rhombohedral). The X-ray diffraction patterns for their calcined samples were similar together, although the relative diffracted intensities from the phases were slightly different and their patterns are illustrated in Figure 3. The actual phases identified in these calcined catalysts under the specified preparation conditions were Mn_2O_3 (cubic) and Fe_2O_3 (rhombohedral). Note that during the calcination of the precursors, the carbonate phases disappeared and oxide phases were formed. Some of these calcined catalysts were characterized by DSC method and their DSC curves are presented in Figure 4. These curves show that the catalysts have high-heat conductivity; this important case for the FTS catalysts is necessary to conduct the produced heat of reaction. Also, the absence of peaks on the DSC curves showed that the calcined catalysts have high-heat stability and it might be a reason why 650°C has been chosen for calcination of catalysts. In order to identify the changes in the tested catalysts during the reaction and to detect the phases formed; these catalysts were characterized by XRD after the test. The actual phases identified in these catalysts are presented in Table 2. As it shown, all of these tested catalysts have the MnO and FeO and iron carbide phases, which all of these phases are active in FTS. MnO is the active phase for production of olefins and iron carbide is the active phase for hydrogenation of CO. Characterization of both precursors and calcined catalysts for the series of differently aged samples (before and after reaction) was carried out using BET surface area and the results are listed in Table 3. These results illustrated that increasing the aging time produced higher surface area materials. The 50%Fe/50%Mn calcined catalyst aged for 180 minutes was the most active for conversion of synthesis gas to light olefins. This catalyst has almost a high-specific surface area; it is, therefore, considered that the higher activity of 50%Fe/50%Mn catalyst may be attributed to its high BET surface area.

Table 1: Effect of aging times on the catalytic performance.

Aging time (minute)	CO conversion (%)	Selectivity (%)				
		CH_4	C_2H_4	C_3H_6	C_4H_8	C_5^+
60	78.2	32.0	24.3	9.0	3.4	7.0
120	87.3	22.2	30.1	10.2	3.7	7.3
180	80.2	21.4	26.2	10.0	3.9	6.2
240	72.1	27.4	23.2	8.5	2.4	6.9
300	69.1	22.9	18.1	9.7	1.5	6.1

Table 2: Different identified phases in the calcined catalysts with different aging times.

Aging time (minute)	Phases
0	MnO (cubic), FeO (cubic), Fe_2C (orthorhombic), $CFe_{2.5}$ (monoclinic)
30	MnO (cubic), FeO (cubic), Fe_2C (orthorhombic), $CFe_{2.5}$ (monoclinic)
60	MnO (cubic), FeO (cubic), Fe_2C (orthorhombic), $CFe_{2.5}$ (monoclinic)
120	$CFe_{2.5}$ (monoclinic), MnO (cubic), FeO (cubic)
180	MnO (cubic), FeO (cubic), $CFe_{2.5}$ (monoclinic), Fe (cubic)
300	$CFe_{2.5}$ (monoclinic), MnO (cubic), FeO (cubic)

Table 3: BET results for catalysts containing 50%Fe/50%Mn with different aging times.

Aging time (minute)	Specific surface area (m^2/g)		
	Precursor	Calcined catalyst (before reaction)	Calcined catalyst (after reaction)
0	85.0	82.3	79.1
30	89.3	84.2	80.5
60	93.2	86.6	82.7
120	107.9	98.6	95.6
180	115.8	113.1	109.5
300	119.0	115.6	112.9

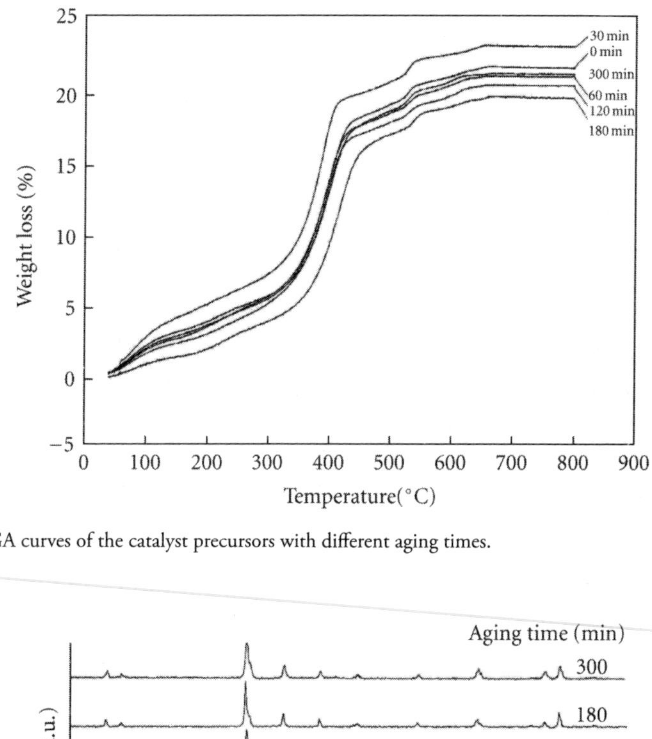

Figure 2: TGA curves of the catalyst precursors with different aging times.

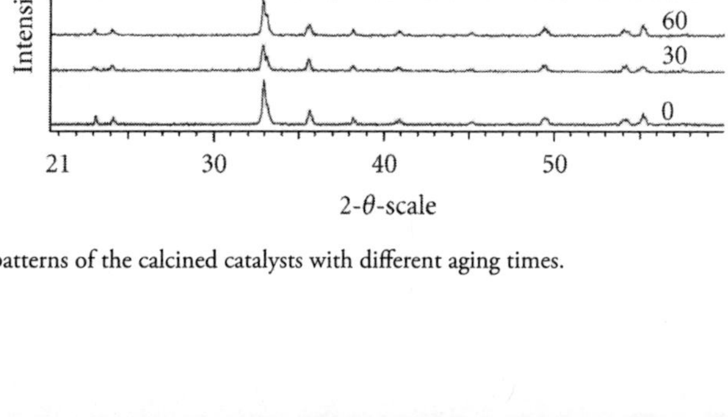

Figure 3: XRD patterns of the calcined catalysts with different aging times.

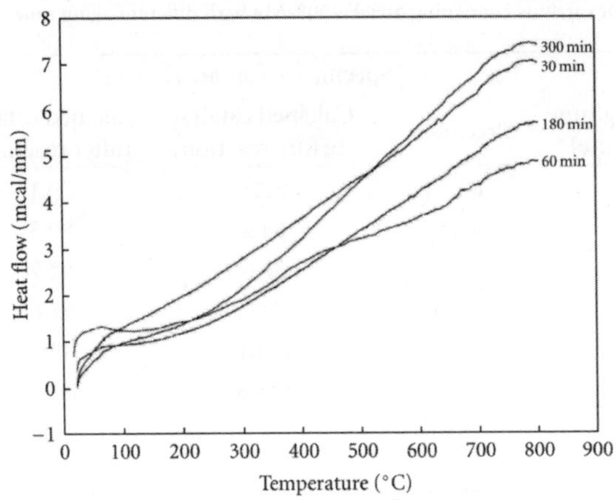

Figure 4: DSC curves of some calcined catalysts with different aging times.

Effect of Precipitation pH

A series of iron manganese oxide catalysts were prepared by coprecipitation method ([Fe]/[Mn]=1/1, 70°C, 3 hours aging time) with a range of precipitation pH from 6.3 to 10.3. The catalysts precursors, prepared using coprecipitation procedure in different pH, were characterized by XRD and showed the $MnCO_3$ phase as rhombohedral structure. The catalytic activity for the Fischer-Tropsch synthesis was investigated for the materials following calcination (650°C, 6 hours) and the effect of precipitation pH on the catalytic performance is shown in Table 4. It is apparent that during precursor calcination, carbonate phase leads to the oxide phases and the sample prepared at pH=8.3 gives the highest activity. These calcined catalysts were characterized by XRD and their patterns are shown in Figure 5. The actual phases identified in these catalysts were Mn_2O_3 (cubic) and Fe_2O_3 (rhombohedral). The calcined tested catalysts prepared at pH=8.3 also were characterized by XRD and different phases including MnO (cubic), FeO (cubic), and $CFe_{2.5}$ (monoclinic) were identified in this catalyst. Note that in the tested catalyst, the oxidic and carbide phases are formed which both of them are active in FTS [41, 42].

Table 4: Effect of precipitation pH on the catalytic performance.

pH	CO conversion (%)	Selectivity (%)				
		CH_4	C_2H_4	C_3H_6	C_4H_8	C_5^+
6.3	84.0	43.0	19.6	8.4	1.5	5.3
7.3	79.6	27.9	19.0	9.0	2.7	7.2
8.3	86.3	22.5	32.1	10.8	4.1	6.8
9.3	81.2	20.0	24.1	10.0	4.0	5.9
10.3	80.0	19.0	22.9	10.9	3.2	6.1

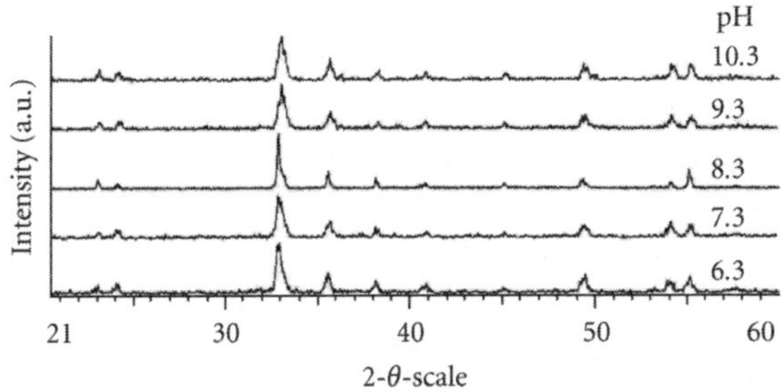

Figure 5: XRD patterns of calcined catalysts with different precipitation pH.

Effect of Solution [Fe]/[Mn] Ratio

Iron manganese oxide catalysts were prepared by coprecipitation method (70°C, pH=8.3, 3 hours aging time) with a range of [Fe]/[Mn] solution ratios varying from 100%Fe to 100%Mn and the catalytic performance for the Fischer-Tropsch synthesis was investigated for the materials following calcination (650°C, 6 hours). The CO conversion and hydrocarbons selectivity percent present on steady-state catalytic performance under comparable reaction conditions for the iron-manganese oxide catalysts with different [Fe]/[Mn] molar ratios are shown in Table 5. The catalyst precursors prepared using coprecipitation method in different [Fe]/[Mn] ratios and their calcined catalysts were characterized by XRD. The catalyst precursor containing 100%Fe-0%Mn was found to be amorphous and the other precursors showed the $MnCO_3$ phase as rhombohedral structure. The XRD patterns of the different calcined catalysts with different [Fe]/[Mn] molar ratios are presented in Figure 6. In the calcined catalyst containing 100%Fe-0-%Mn, the Fe_2O_3 (rhombohedral) phase was observed and in the calcined catalyst containing 0%Fe-100%Mn, the Mn_2O_3 (cubic) phase was identified. For the other calcined catalysts, different phases including Mn_2O_3 (cubic) and Fe_2O_3 (rhombohedral) were identified. In order to identify the changes in the catalyst containing 50%Fe/50%Mn during the reaction and to detect the phases formed, this catalyst was characterized by XRD after the test. Its phases were found to be MnO (cubic), FeO (cubic), and $CFe_{2.5}$ (monoclinic). Characterization of these calcined catalysts was carried out using the BET surface area measurements and obtained results are presented in Table 6. The BET results show that the calcined catalysts prepared with a range of [Fe]/[Mn] solution ratios varying from 100%Fe to 100%Mn have different specific surface areas. The catalyst with 1/1 [Fe]/[Mn] ratio has a higher specific surface area than the other catalysts (Table 6), which is one reason for the enhanced performance of this catalyst [43]. According to the obtained results (Table 5), since the catalyst with preparation ratio of 1/1 [Fe]/[Mn] showed the highest selectivity toward both ethylene and propylene, so the catalyst prepared with this molar ratio was chosen as the best catalyst for the conversion of synthesis gas to ethylene and propylene under reaction conditions (P=1 atm, T=400°C, $H_2/CO=2/1$). The specific surface areas (BET) results of the precursors and calcined catalysts (before and after reaction) for different Fe/Mn molar ratio are given in Table 6. Monometallic 100%Fe is used as the basis for comparing the physical characteristics and CO hydrogenation performance and selectivity for the bimetallic Fe–Mn catalysts. As shown in Table 6, the BET surface areas for the catalysts prepared from different iron/manganese molar ratios are dependent on the Fe/Mn solution ratios. However, the specific surface area of catalyst precursor and calcined catalysts before reaction for each molar ratio were found to be nearly similar, the BET specific surface areas of the catalysts before and after reaction are different and the specific surface areas of all the calcined catalysts after reaction were decreased. The BET data for the catalysts containing 50%Fe/50%Mn showed the high-specific surface area (98.6 m^2/g for calcined catalyst) this might be a reason why the 50%Fe/50%Mn catalyst shows a better catalytic performance than the other catalysts.

Table 5: Effect of Fe/Mn molar ratio on the catalytic performance.

Fe/Mn	CO conversion (%)	Selectivity (%)				
		CH_4	C_2H_4	C_3H_6	C_4H_8	C_5^+
1/0	44.3	33.0	18.6	7.4	2.9	4
4/1	59.1	32.2	20.1	8.6	3.7	4.9
2/1	66.1	25.5	31.2	8.8	3.9	4.5
1/1	81.2	20.0	24.1	10.0	4.0	6.3
1/2	80.7	20.4	22.6	9.6	3.3	5.7
1/4	79.2	21.3	21.8	9.2	3.7	7.1
0/1	72.0	23.4	22.7	9.6	4.0	8.9

Table 6: BET results for different molar ratios of Fe/Mn.

Fe/Mn	Specific surface area (m^2/g)		
	Precursor	Calcined catalyst (before reaction)	Calcined catalyst (after reaction)
1/0	65.1	62.7	59.3
4/1	71.2	68.2	66.4
2/1	98.1	96.7	84.3
1/1	107.9	98.6	95.6
1/2	109.8	104.7	102.1
1/4	111.3	108.9	106.1
0/1	110.8	107.9	105.6

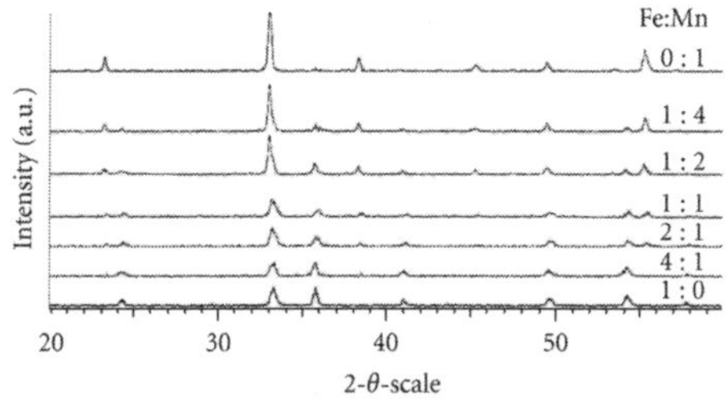

Figure 6: XRD patterns of calcined catalysts with different Fe/Mn molar ratios.

Effect of Precipitation Temperature

Iron manganese oxide catalysts were prepared by coprecipitation ([Fe]/[Mn]=1/1, pH=8.3, 3 hours aging time) with a range of solution temperature from 40 to 80°C. The catalytic performance of these series catalysts for the conversion of synthesis gas to light olefins was investigated for the materials following calcination (650°C, 6 hours). The CO conversion and hydrocarbons selectivity percent present on steady-state catalytic performance for the iron manganese oxide catalysts with different solution temperature in reaction conditions (P=1 atm, T=400°C, H_2/CO=2/1) are shown in Table 7. According to the obtained results, the percent of CO conversion was partially changed with the change of precipitation temperature and 70°C is considered to be practical maximum precipitation temperature and in this study, it was chosen as the optimum temperature for the catalysts preparation. The XRD patterns of catalyst precursors prepared by varying the temperature of the aging solution all showed similar diffraction patterns, the materials were poorly crystalline and comprised the $MnCO_3$ phase with rhombohedral structure. The calcined catalysts were characterized by XRD and their patterns are presented in Figure 7. The XRD patterns of the calcined samples were similar to each other, although the relative diffracted intensities from the phases were slightly different. The actual phases identified in these catalysts under the specified preparation conditions were Mn_2O_3 (cubic), Fe_2O_3 (rhombohedral), and MnO_2 (tetragonal).

Table 7: Effect of preparation temperature on the catalytic performance.

T (°C)	CO conversion (%)	Selectivity (%)				
		CH_4	C_2H_4	C_3H_6	C_4H_8	C_5^+
40	52.3	29.7	18.5	8.5	2.8	5.4
50	61.5	29.5	22.8	10.1	3.5	6.1
60	71.0	32.5	27.5	10.0	4.1	5.9
70	81.2	20.0	24.1	10.0	4.0	6.3
80	78.1	25.9	30.3	11.1	4.0	7.2

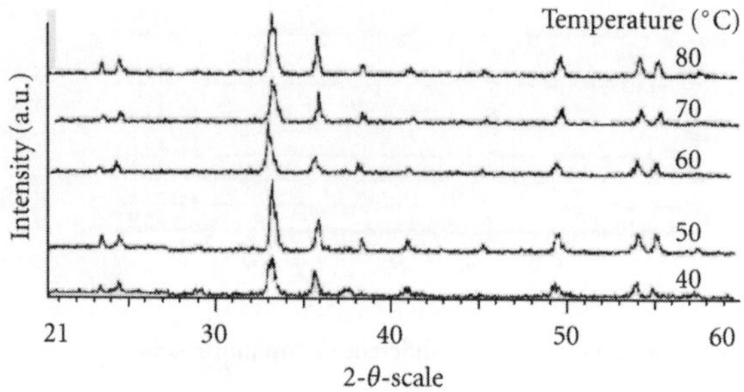

Figure 7: XRD patterns of calcined catalysts with different preparation temperatures.

Effect of Different Supports and Support Loadings

Hydrogenation of carbon monoxide is susceptible to metal support effects, and both specific activity and selectivity can be markedly influenced by both the metal and the support [44, 45]. In order to study the effect of some supports such as TiO_2, SiO_2, Al_2O_3, MgO, and zeolite into the iron manganese oxide catalysts ([Fe]/[Mn]=1/1), the same amount (10 wt.%) from each support has been added separately to a solution containing iron and manganese with above molar ratio. All of the different supported catalysts were tested for the selectivity of hydrogenation of CO. The CO conversion and hydrocarbons selectivity of the catalysts containing different supports are shown in Table 8. As it can be observed, the catalyst supported by Al_2O_3 is more active than the other supported catalysts and shows a high selectivity toward hydrocarbons. To understand the influence of loading of Al_2O_3 on the catalytic activity of mixed iron manganese oxide catalysts, a series of different 50%Fe/50%Mn/wt.%Al_2O_3 catalysts were prepared by coprecipitation procedure outlined above. The Al_2O_3 loadings were 5, 10, 20, 25, and 30 wt% based on the total catalyst weight and the catalytic performance results are shown in Table 9. According to these results, the catalysts loaded with 5 wt.%Al_2O_3 showed the optimal catalytic performance for conversion of synthesis gas to light olefins. This catalyst was characterized by XRD and its patterns on different stages are shown in Figure 8. The actual phases identified in this catalyst were Mn_5O_8 (monoclinic), Fe_2O_3 (rhombohedral), and Mn_2O_3 (cubic) and its precursors comprised the $MnCO_3$ phase with rhombohedral structure. In the tested catalyst, different phases including FeO (cubic), MnO (cubic), Fe_3O_4, and Fe_2C (orthorombic) were identified. Characterization of both precursors and calcined-supported catalysts was carried out using BET surface area measurement and obtained results are presented in Table 14. In general, the BET results show that the catalyst precursors containing different supports and the calcined catalysts derived from these precursors have different specific surface area. However, as shown in Table 14, the catalyst precursors have higher specific surface areas than their calcined catalysts. Furthermore, the Al_2O_3-supported catalyst has a higher specific surface area than the other supported catalysts (Table 14), which is one reason for the better catalytic performance of this catalyst [38]. A detailed SEM study of the precursor calcined and tested catalysts for the sample containing optimum amount of 5 wt%Al_2O_3 was also carried out. The SEM images of these catalysts are presented in Figure 9, and have shown the major differences in their morphology. By addition of Al2O3 as a support to the catalyst, the particle size of some grains has slightly increased and the supported catalyst revealed structural differences, which are probably due to the presence of Al_2O_3. This catalyst is comprised of large grains which are embedded in a mixture consisting of small grains (Figure 9(b)). However, the size of these grains grew larger by agglomeration in the tested catalyst (Figure 9(c)), which may be due to sintering after reactions. This result is in agreement with Galarraga [46], who indicate that high temperature could cause agglomeration of these small grains, which leads to catalyst deactivation under high temperature. The agglomeration is also caused by the imhomogenity distribution of metal precursor (Figure 9(a)).

Table 8: Effect of different supports on the catalytic performance.

Support	CO conversion (%)	Selectivity (%)				
		CH_4	C_2H_4	C_3H_6	C_4H_8	C_5^+
TiO_2	61.3	21.5	23.6	8.2	2.4	5.8
Zeolite	70.5	27.6	22.3	9.3	3.1	6.3
MgO	67.8	36.5	20.9	8.4	2.3	4.7
Al_2O_3	80.9	30.5	34.7	12.5	5.4	7.1
Silica	73.2	29.9	30.1	10.4	4.5	2.1

Table 9: Effect of loading of Al2O3 on the catalytic performance.

		Selectivity (%)				
Al_2O_3 (Wt%)	CO conversion (%)	CH_4	C_2H_4	C_3H_6	C_4H_8	C_5^+
5	80.9	30.5	34.7	12.5	5.4	2.5
10	75.9	51.0	18.4	7.8	3.9	5.0
15	80.0	57.3	15.2	7.5	4.3	4.5
20	82.2	40.6	27.2	10.2	4.2	5.8
25	67.5	50.0	17.5	8.0	3.7	7.8
30	60.9	49.5	16.9	8.6	3.1	9.5

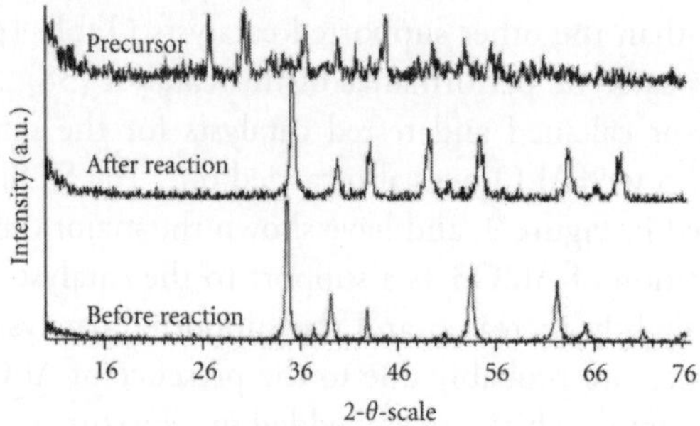

Figure 8: XRD patterns of the catalyst containing Al_2O_3 support in different states (precursor, calcined catalyst before and after the test).

Figure 9: The SEM images of catalyst containing 50%Fe/50%Mn/5 wt%Al_2O_3; (a) precursor, (b) calcined catalyst before reaction, and (c) calcined catalyst after reaction.

Effect of Promoters

Alkali metals have been used widely as promoters to improve the activity and selectivity of the catalysts in the CO hydrogenation [47, 48]. To determine the utility of promoters on the catalytic performance of mixed iron manganese oxide catalysts, as alkali metals promoters, a small amount (1.5 wt.%) of $LiNO_3$, KNO_3, RbCl, and $Mg(NO_3)_2.6H_2O$ were separately introduced to the resulting suspension containing 50%Fe/50%Mn/5 wt%Al_2O_3. Then all of these different promoted catalysts were tested at the same reaction conditions (400°C and feed molar ratio of H_2/CO=1/2 at atmospheric pressure) for the conversion of synthesis gas to light olefins and the results are shown in Table 10. These results indicate that the addition of the K, Mg, Li, and Rb promoters into the catalyst texture leads to a decrease on the selectivity of the catalyst toward light olefins. Thus taking these results into consideration, the catalyst containing 50%Fe/50%Mn/5 wt.%Al_2O_3 without any promoter and aged for 3 hours appears to be the optimum-modified catalyst for the conversion of synthesis gas to light olefins.

Table 10: Effect of different promoters on the catalytic performance of 50%Fe/50%Mn/5 wt%Al_2O_3 catalyst.

Promoter	CO conversion (%)	Selectivity (%)				
		CH_4	C_2H_4	C_3H_6	C_4H_8	C_5^+
Rb	69.0	48.1	19.8	8.5	2.8	6.5
Mg	76.4	50.7	17.0	7.2	2.1	4.1
Li	73.2	45.9	18.5	7.4	1.7	11.8
K	75.5	46.3	15.8	6.0	1.5	15.2

Effect of FTS Reaction Conditions

The other category of factors which have a marked effect on the catalytic performance of a catalyst is the operating conditions. For optimizing of the reaction conditions in this study, the effects of operating conditions such as H_2/CO feed molar ratios, reaction temperatures, and reactor total pressures were examined to investigate the catalyst stability and its performance under different Fischer-Tropsch operating conditions.

Effect of H_2/CO Molar Feed Ratio

The influence of the H_2/CO molar feed ratio on the steady-state catalytic performance of the iron manganese oxide catalyst containing 50%Fe/50%Mn/5 wt.%Al_2O_3 for the Fischer-Tropsch reaction at 400°C under atmospheric pressure was investigated and the CO conversion and light olefins products selectivity percent, present on steady-state catalytic performance, are presented in Table 11. The results showed that with variation in H_2/CO feed ratio from 1/1 to 3/1, different selectivities with respect to light olefins were obtained. However, in the case of the H_2/CO=1/1, the total selectivity of light olefins products was higher and the CH_4 selectivity was lower than the other H_2/CO feed ratios under the same temperature and pressure condition. It is also apparent that, for all of the H_2/CO feed molar ratios, the optimum catalyst shows a high selectivity toward ethylene. Therefore, the H_2/CO=1/1 ratio was chosen as the optimum ratio for conversion of synthesis gas to C_2–C_4 olefins over iron manganese catalysts.

Table 11: Effect of different H_2/CO feed molar ratios on the catalytic performance of 50%Fe/50%Mn/5 wt%Al_2O_3 catalyst.

H_2/CO	CO conversion (%)	Selectivity (%)				
		CH_4	C_2H_4	C_3H_6	C_4H_8	C_5^+
1/1	84.4	27.4	36.8	12.8	5.4	10.8
2/1	80.9	30.5	34.7	12.5	5.4	7.1
3/1	97.2	68.3	14.1	7.5	2.2	5.9

Effect of Reaction Temperature

The effect of reaction temperature on the catalytic performance of the 50%Fe/50%Mn/5 wt.%Al_2O_3 catalyst was studied at a range of temperature between 260–420°C and the results are presented in Table 12, (P=1 atm, H_2/CO=1/1). The results show that as the operating temperature is increased, the CO conversion is increased. In addition, for the reaction temperature at 360°C, the total selectivity of light olefins products was higher than the other reaction temperatures under the same reaction conditions. In general, an increase in the reaction temperature leads to an increase in the catalytic performance; furthermore, it has shown that the reaction temperature should not be too low [14]. At low reaction temperatures, the conversion percentage of CO is too low and so it causes a low-catalytic performance. On the other hand, increasing the reaction temperature leads to the formation of large amounts of coke as an unwanted product, as we found in this work. Therefore, in this study, 360°C is considered to be the optimum operating temperature because of high-CO conversion, total selectivity of light olefins products, low CH_4, and not formation of coke.

Table 12: Effect of different reaction temperatures on the catalytic performance of 50%Fe/50%Mn/5 wt%Al_2O_3 catalyst.

T (°C)	CO conversion (%)	Selectivity (%)				
		CH_4	C_2H_4	C_3H_6	C_4H_8	C_5^+
280	39.2	20.6	16.5	3.1	0.9	6.3
300	49.8	23.5	17.3	3.5	1.8	5.7
320	67.4	29.7	22.1	5.7	2.2	6.9
340	84.4	27.4	36.8	12.8	5.4	5.4
360	87.4	30.5	38.0	13.1	5.5	6.9
380	87.1	34.9	32.2	9.1	2.9	5.3
400	89.4	41.3	27.0	7.9	3.2	4.2
430	80.3	40.5	24.2	8.2	3.7	5.7
450	78.0	40.2	25.1	8.5	2.0	5.1

Effects of Total Pressure

An increase in total pressure would generally result in condensation of hydrocarbons, which are normally in the gaseous state at atmospheric pressure. Higher pressures and higher carbon monoxide conversions would probably lead to saturation of catalyst pores by liquid reaction products [49]. A different composition of the liquid phase in catalyst pores at high syngas pressures could affect the rate of elementary steps and carbon monoxide and hydrogen concentrations. A series of experiments were carried out for the 50%Fe/50%Mn/5 wt.%Al_2O_3 catalyst to investigate on the performance of this catalyst in variation of total pressure in the range of 1–15 bar, at the optimal conditions of H_2/CO=1/1 and 360°C (Table 13). The results indicate that, at the total pressure of 1 bar, the optimal catalyst showed a total selectivity of 42.5% with respect to C_2–C_4 light olefins and 13.8% produce the C_5^+ products. It is also apparent that increasing in total pressure in the ranges of 2–15 bar significantly increases the C_5^+ selectivity and leads to an increase to 43.2% at the pressure of 15 bar. In the other hand, as it can be seen on Table 13 at the ranges of 1–6 bar total pressures, no significant decreasing on CO conversion was observed, however, the light olefins selectivities were increased and the results indicate that at the total pressure of 6 bar, the optimal catalyst containing 50%Fe/50%Mn/5 wt.%Al_2O_3 showed the highest total selectivity of 59.2% with respect to C_2–C_4 light olefins and also led to 20.5% total of selectivity toward the C_5^+ products. The results also indicate that the CO conversion and the total selectivity with respect to C_2–C_4 light olefins were decreased as the total pressures are increased from 6 bar to 15 bar. Hence because of high CO conversion, low CH_4 selectivity, and also higher total selectivity with respect to C_2–C_4 olefins at the total pressure of 6 bar, this pressure was chosen as the optimum pressure.

Table 13: Effect of different total pressures on the catalytic performance of 50%Fe/50%Mn/5 wt%Al_2O_3 catalyst.

		Selectivity (%)				
Pressure (bar)	CO conversion (%)	CH_4	C_2H_4	C_3H_6	C_4H_8	C_5^+
1	84.4	29.8	29.2	11.2	2.1	13.8
2	84.0	26.5	30.7	11.9	2.6	15.6
3	84.0	25.6	34.5	12.5	3.0	16.0
4	84.1	25.1	37.2	12.8	3.2	17.5
5	84.8	23.8	38.9	13.0	3.5	18.7
6	84.0	19.8	40.1	13.2	5.9	20.5
7	83.8	19.7	31.8	12.9	5.7	27.2
8	82.2	19.5	31.2	12.1	5.3	29.6
9	82.8	19.3	30.1	11.0	5.4	30.0
10	78.2	19.0	28.0	10.3	3.5	34.4
11	78.0	18.8	27.7	10.5	2.9	35.6
12	76.6	18.0	27.1	9.2	2.7	38.0
13	71.6	17.2	24.5	7.5	2.1	39.6
14	68.2	16.7	22.2	7.2	2.0	41.7
15	64.2	16.5	21.0	6.7	1.5	43.2

Table 14: BET results for different supports on Fe/Mn catalysts.

Support	Precursor	Specific surface area (m^2/g) Calcined catalyst (before reaction)	Calcined catalyst (after reaction)
TiO_2	112.0	109.8	105.3
Zeolite	145.3	139.7	133.1
MgO	123.6	118.2	115.4
Al_2O_3	157.5	152.3	150.7
SiO_2	165.6	148.2	142.6

Conclusions

Many variables in the preparation of the catalyst during the coprecipitation procedure and the subsequent calcination step are important in controlling the catalytic performance of iron-manganese mixed oxide catalysts for conversion of synthesis gas to light olefins. Preparation conditions for optimum catalytic performance are 1/1:[Fe]/[Mn] ratio at pH 8.3 and 70°C for 3 hours aging time, followed by calcination at 650°C for 6 hours. The optimal-supported catalyst was found to be 50%Fe/50%Mn/5 wt.%Al_2O_3. The optimal reaction conditions were found to be 360°C with molar feed ratio of H_2/CO=1/1, (GHSV=2700 h^{-1}) under the total pressure of 6 bar. The characterization of both precursors and calcined catalysts by powder XRD, SEM, BET specific surface area, and thermal analysis (TGA/DSC) methods showed that the catalyst precursors are sensitive to the preparation conditions. Relationships between bulk phases and catalytic performance were complex, although the catalysts showed X-ray diffraction features which correspond to amorphous mixed iron manganese oxide phases. The SEM results show that longer time enhances agglomerate size growth during reaction conditions which may be due to the formation of iron carbides as active phases for Fischer-Tropsch synthesis. From the results presented in this study, it is clear that the precipitation conditions used in the preparation procedure and also the operation conditions are of crucial importance and control of these parameters should be incorporated into the design of experimental programmers involving precipitation as the method of catalyst preparation and design of catalytic reactor as the method of operation conditions.

References

1. D. J. Duvenhage and N. J. Coville, "Fe:Co/TiO_2 bimetallic catalysts for the Fischer-Tropsch reaction I. Characterization and reactor studies," Applied Catalysis A, vol. 153, no. 1-2, pp. 43–67, 1997.

2. C. Cabet, A. C. Roger, A. Kiennemann, S. Läkamp, and G. Pourroy, "Synthesis of new Fe-Co based metal/oxide composite materials: application to the Fischer-Tropsch synthesis," Journal of Catalysis, vol. 173, no. 1, pp. 64–73, 1998.

3. S. L. González-Cortés, S. M. A. Rodulfo-Baechler, A. Oliveros, et al., "Synthesis of light alkenes on manganese promoted iron and iron-cobalt Fischer-Tropsch catalysts," Reaction Kinetics and Catalysis Letters, vol. 75, no. 1, pp. 3–12, 2002.

4. F. Tihay, G. Pourroy, M. Richard-Plouet, A. C. Roger, and A. Kiennemann, "Effect of Fischer-Tropsch synthesis on the microstructure of Fe-Co-based metal/spinel composite materials," Applied Catalysis A, vol. 206, no. 1, pp. 29–42, 2001.

5. F. Tihay, A. C. Roger, A. Kiennemann, and G. Pourroy, "Fe-Co based metal/spinel to produce light olefins from syngas," Catalysis Today, vol. 58, no. 4, pp. 263–269, 2000.

6. T. V. Reshetenko, L. B. Avdeeva, A. A. Khassin, et al., "Coprecipitated iron-containing catalysts (Fe-Al2O3, Fe-Co-Al2O3, Fe-Ni-Al2O3) for methane decomposition at moderate temperatures I. Genesis of calcined and reduced catalysts," Applied Catalysis A, vol. 268, no. 1-2, pp. 127–138, 2004.

7. V. A. de la Peña O'Shea, N. N. Menéndez, J. D. Tornero, and J. L. G. Fierro, "Unusually high selectivity to C2+ alcohols on bimetallic CoFe catalysts during CO hydrogenation," Catalysis Letters, vol. 88, no. 3-4, pp. 123–128, 2003.

8. M. E. Dry, "The Fischer-Tropsch synthesis," in Catalysis Science and Technology, J. R. Anderson and M. E. Boudart, Eds., vol. 1, pp. 159–255, Springer, New York, NY, USA, 1981.

9. D. L. King, J. A. Cusumano, and R. L. Garten, "A technological perspective for catalytic processes based on synthesis gas," Catalysis Reviews, vol. 23, no. 1-2, pp. 233–263, 1981.

10. G. C. Maiti, R. Malessa, and M. Baerns, "Iron/manganese oxide catalysts for Fischer-Tropsch synthesis—part I: structural and textural changes by calcination, reduction and synthesis," Applied Catalysis, vol. 5, no. 2, pp. 151–170, 1983.

11. R. Malessa and M. Baerns, "Iron/manganese oxide catalysts for Fischer-Tropsch synthesis. 4. Activity and selectivity," Industrial and Engineering Chemistry Research, vol. 27, no. 2, pp. 279–283, 1988.

12. H. Kölbel and K. D. Tillmetz, "Process for the production of hydrocarbons and oxygen-containing compounds and catalysts therefor," US patent no. 4177203, 1979.

13. C. K. Das, N. S. Das, D. P. Choudhury, G. Ravichandran, and D. K. Chakrabarty, "Hydrogenation of carbon monoxide on unsupported Fe-Mn-K catalysts for the synthesis of lower alkenes: promoter effect of manganese," Applied Catalysis A, vol. 111, no. 2, pp. 119–132, 1994.
14. J. Barrault, C. Forquy, and V. Perrichon, "Effects of manganese oxide and sulphate on olefin selectivity of iron supported catalysts in the Fischer-Tropsch reaction," Applied Catalysis, vol. 5, no. 1, pp. 119–125, 1983.
15. G. P. Van der Laan and A. A. C. M. Beenackers, "Kinetics and selectivity of the Fischer-Tropsch synthesis: a literature review," Catalysis Reviews, vol. 41, no. 3-4, pp. 255–318, 1999.
16. S. L. González-Cortés, S. M. A. Rodulfo-Baechler, A. Oliveros, et al., "Synthesis of light alkenes on manganese promoted iron and iron-cobalt Fischer-Tropsch catalysts," Reaction Kinetics and Catalysis Letters, vol. 75, no. 1, pp. 3–12, 2002.
17. J. B. Butt, "Carbide phases on iron-based Fischer-Tropsch synthesis catalysts—part II: some reaction studies," Catalysis Letters, vol. 7, no. 1–4, pp. 83–105, 1990.
18. Y. Liu, B.-T. Teng, X.-H. Guo, et al., "Effect of reaction conditions on the catalytic performance of Fe-Mn catalyst for Fischer-Tropsch synthesis," Journal of Molecular Catalysis A, vol. 272, no. 1-2, pp. 182–190, 2007.
19. T. Li, Y. Yang, C. Zhang, et al., "Effect of manganese on an iron-based Fischer-Tropsch synthesis catalyst prepared from ferrous sulfate," Fuel, vol. 86, no. 7-8, pp. 921–928, 2007.
20. Y. Yang, H.-W. Xiang, L. Tian, et al., "Structure and Fischer-Tropsch performance of iron-manganese catalyst incorporated with SiO2," Applied Catalysis A, vol. 284, no. 1-2, pp. 105–122, 2005.
21. T. Herranz, S. Rojas, F. J. Pérez-Alonso, M. Ojeda, P. Terreros, and J. L. G. Fierro, "Hydrogenation of carbon oxides over promoted Fe-Mn catalysts prepared by the microemulsion methodology," Applied Catalysis A, vol. 311, no. 1-2, pp. 66–75, 2006.
22. C.-H. Zhang, Y. Yang, B.-T. Teng, et al., "Study of an iron-manganese Fischer-Tropsch synthesis catalyst promoted with copper," Journal of Catalysis, vol. 237, no. 2, pp. 405–415, 2006.
23. H. Schultz and H. Gökcebay, "Fischer-Tropsch CO-hydrogenation as a means for linear olefins production," in Catalysis of Organic Reactions, J. R. Kosak, Ed., pp. 153–169, Marcel Dekker, New York, NY, USA, 1984.

24. H. Kölbel and K. D. Tillmetz, "Hydrocarbons and oxygen containing compounds," Belgian Patent, no. 237 628, 1976.

25. C. H. Bartholomew, "History of cobalt catalyst design for FTS," in Proceedings of the National Spring Meeting of the American Institute of Chemical Engineers (AIChE '03), New Orleans, La, USA, March-April 2003.

26. N.-O. Ikenaga, H. Taniguchi, A. Watanabe, and T. Suzuki, "Sulfiding behavior of iron based coal liquefaction catalyst," Fuel, vol. 79, no. 3-4, pp. 273–283, 2000.

27. G. J. Hutchings, A. A. Mirzaei, R. W. Joyner, M. R. H. Siddiqui, and S. H. Taylor, "Ambient temperature CO oxidation using copper manganese oxide catalysts prepared by coprecipitation: effect of ageing on catalyst performance," Catalysis Letters, vol. 42, no. 1-2, pp. 21–24, 1996.

28. G. J. Hutchings, A. A. Mirzaei, R. W. Joyner, M. R. H. Siddiqui, and S. H. Taylor, "Effect of preparation conditions on the catalytic performance of copper manganese oxide catalysts for CO oxidation," Applied Catalysis A, vol. 166, no. 1, pp. 143–152, 1998.

29. S. H. Taylor, G. J. Hutchings, and A. A. Mirzaei, "Copper zinc oxide catalysts for ambient temperature carbon monoxide oxidation," Chemical Communications, no. 15, pp. 1373–1374, 1999.

30. D. M. Whittle, A. A. Mirzaei, J. S. J. Hargreaves, et al., "Co-precipitated copper zinc oxide catalysts for ambient temperature carbon monoxide oxidation: effect of precipitate ageing on catalyst activity," Physical Chemistry Chemical Physics, vol. 4, no. 23, pp. 5915–5920, 2002.

31. A. A. Mirzaei, H. R. Shaterian, R. W. Joyner, M. Stockenhuber, S. H. Taylor, and G. J. Hutchings, "Ambient temperature carbon monoxide oxidation using copper manganese oxide catalysts: effect of residual Na+ acting as catalyst poison," Catalysis Communications, vol. 4, no. 1, pp. 17–20, 2003.

32. A. A. Mirzaei, H. R. Shaterian, M. Habibi, G. J. Hutchings, and S. H. Taylor, "Characterisation of copper-manganese oxide catalysts: effect of precipitate ageing upon the structure and morphology of precursors and catalysts," Applied Catalysis A, vol. 253, no. 2, pp. 499–508, 2003.

33. A. A. Mirzaei, H. R. Shaterian, S. H. Taylor, and G. J. Hutchings, "Co-precipitated copper zinc oxide catalysts for ambient temperature carbon monoxide oxidation: effect of precipitate aging atmosphere on catalyst activity," Catalysis Letters, vol. 87, no. 3-4, pp. 103–108, 2003.

34. S. H. Taylor, G. J. Hutchings, and A. A. Mirzaei, "The preparation and activity of copper zinc oxide catalysts for ambient temperature carbon monoxide oxidation," Catalysis Today, vol. 84, no. 3-4, pp. 113–119, 2003.

35. A. A. Mirzaei, H. R. Shaterian, and M. Kaykhaii, "The X-ray photoelectron spectroscopy of surface composition of aged mixed copper manganese oxide catalysts," Applied Surface Science, vol. 239, no. 2, pp. 246–254, 2005.
36. A. A. Mirzaei, R. Habibpour, and E. Kashi, "Preparation and optimization of mixed iron cobalt oxide catalysts for conversion of synthesis gas to light olefins," Applied Catalysis A, vol. 296, no. 2, pp. 222–231, 2005.
37. A. A. Mirzaei, M. Faizi, and R. Habibpour, "Effect of preparation conditions on the catalytic performance of cobalt manganese oxide catalysts for conversion of synthesis gas to light olefins," Applied Catalysis A, vol. 306, pp. 98–107, 2006.
38. A. A. Mirzaei, R. Habibpour, M. Faizi, and E. Kashi, "Characterization of iron-cobalt oxide catalysts: effect of different supports and promoters upon the structure and morphology of precursors and catalysts," Applied Catalysis A, vol. 301, no. 2, pp. 272–283, 2006.
39. A. A. Mirzaei, M. Galavy, A. Beigbabaei, and V. Eslamimanesh, "Preparation and operating conditions for cobalt cerium oxide catalysts used in the conversion of synthesis gas into light olefins," Journal of the Iranian Chemical Society, vol. 4, no. 3, pp. 347–363, 2007.
40. A. A. Mirzaei, M. Galavy, and V. Eslamimanesh, "SEM and BET methods for investigating the structure and morphology of coCe catalysts for production of light olefins," Australian Journal of Chemistry, vol. 61, no. 2, pp. 144–152, 2008.
41. H.-B. Zhang and G. L. Schrader, "Characterization of a fused iron catalyst for Fischer-Tropsch synthesis by in situ laser Raman spectroscopy," Journal of Catalysis, vol. 95, no. 1, pp. 325–332, 1985.
42. M. D. Shroff, D. S. Kalakkad, K. E. Coulter, et al., "Activation of precipitated iron Fischer-Tropsch synthesis catalysts," Journal of Catalysis, vol. 156, no. 2, pp. 185–207, 1995.
43. Y. Yang, H.-W. Xiang, Y.-Y. Xu, L. Bai, and Y.-W. Li, "Effect of potassium promoter on precipitated iron-manganese catalyst for Fischer-Tropsch synthesis," Applied Catalysis A, vol. 266, no. 2, pp. 181–194, 2004.
44. D. L. King, "A Fischer-Tropsch study of supported ruthenium catalysts," Journal of Catalysis, vol. 51, no. 3, pp. 386–397, 1978.
45. R. C. Reuel and C. H. Bartholomew, "Effects of support and dispersion on the CO hydrogenation activity/selectivity properties of cobalt," Journal of Catalysis, vol. 85, no. 1, pp. 78–88, 1984.

46. C. E. Galarraga, Heterogeneous catalyst for the synthesis of middle distillate hydrocarbons, M.S. thesis, University of Western Ontario, London, Canada, 1998.

47. H. A. Dirkse, P. W. Lednor, and P. C. Versloot, "Alkali metal-naphthalene adducts as reagents for neutralizing oxide surfaces, and the effect of alkali metal treated surfaces in Rh-catalysed synthesis gas (CO + H2) conversion," Journal of the Chemical Society, Chemical Communications, no. 14, pp. 814–815, 1982.

48. J. Gaube, K. Herzog, L. König, and B. Schliebs, "Kinetische untersuchungen der Fischer-Tropsch-Synthese zur klärung der wirkung des alkali als promotor in eisen-katalysatoren," Chemie Ingenieur Technik, vol. 58, no. 8, pp. 682–683, 1986.

49. A. Griboval-Constant, A. Y. Khodakov, R. Bechara, and V. L. Zholobenko, "Support mesoporosity: a tool for better control of catalytic behavior of cobalt supported Fischer-Tropsch catalysts," Studies in Surface Science and Catalysis, vol. 144, pp. 609–616, 2002.

Use of Viscosity to Probe the Interaction of Anionic Surfactants with a Coagulant Protein from *Moringa oleifera* Seeds

Raymond Maikokera and Habauka M. Kwaambwa

ABSTRACT

The intrinsic viscosity of the coagulant protein was evaluated from the flow times of the protein solutions through a capillary viscometer, and the results suggested the coagulant protein to be globular. The interactions of the coagulant protein with anionic surfactant sodium dodecyl sulphate (SDS) and sodium dodecyl benzene sulfonate (SDBS) were also investigated by capillary viscometry. We conclude that there is strong protein-surfactant interaction at very low surfactant concentrations, and the behavior of the anionic surfactants in solutions containing coagulant protein is very similar. The viscometry results of protein-SDS system are compared with surface tension, fluorescence, and circular dichroism reported earlier. Combining the results of the four studies, the four approaches seem to confirm the same picture of the coagulant protein-SDS interaction. All the physical quantities when studied as function of surfactant concentration for 0.05% (w/v) protein solution either exhibited a maximum or minimum at a critical SDS concentration.

Introduction

Moringa oleifera is a multipurpose tropical tree with most of its parts being useful for medicinal and commercial applications in addition to its nutritional value [1–8]. The seeds from this plant contain active coagulating agents characterized as dimeric cationic proteins having a molecular weight of 13 kDa and an isoelectric pH between 10 and 11 [2]. It is suggested that in the near future, coagulating protein extracted from *Moringa oleifera* seeds could be a potential challenger of synthetic coagulants (e.g., iron chloride and aluminium sulphate) for water purification [3] and for primary treatment of industrial and domestic wastewaters [9]. As a result, it has been recommended by the Food and Agricultural Organization (FAO) as a proper and advisable way for treating water [8]. Recently, protein extract from *Moringa oleifera* seeds has been found to be an anionic surfactant-removal agent in aqueous solutions [4]. Thus, it is important to keep on researching on protein extracts from *Moringa's* properties.

The interaction between surfactants and proteins is an active field of research in colloid science because very often in practice the proteins and surfactants are present in the same systems such as industrial, biological, pharmaceutical, and cosmetic systems. In addition, physicochemical characterisation of seed proteins is important in order to understand the nature of the protein and its interaction with other components.

It is known in general that anionic surfactants interact strongly with proteins and form protein-surfactant complexes, which would induce the unfolding of proteins [10]. The sensitivity of viscosity to molecular structure makes it useful for monitoring processes that result in changes in overall protein conformation [11, 12]. Apart from viscosity measurements, other classical techniques such as surface tension and fluorescence spectroscopy have been employed to examine the occurrence of critical phenomena in solution properties of protein-surfactant mixtures [13, 14]. As has been established using surface tensiometry and intrinsic fluorescence spectroscopy, the anionic surfactant sodium dodecyl sulphate (SDS) interacts strongly with the coagulant protein from Moringa oleifera seeds [15].

Similar interactions were observed between the coagulant protein and the anionic surfactant sodium bis(2-ethyl-1-hexyl)sulfosuccinate (AOT) when monitored by surface tension probe [16].

In this work, the effects of the presence of anionic surfactants SDS and SDBS on the viscosity properties of the coagulant protein from Moringa oleifera are reported. The results of solution properties of the protein in the presence of SDS are verified by surface tension, circular dichroism, and fluorescence measurements reported earlier [15–18].

Materials and Methods

Extraction and Purification

The extraction and purification of protein powder was done using the method of Ndabigengesere and Narasiah [3], and the experimental details are described in our previous work [15–18].

Viscosity and Protein-Surfactant Interactions Measurements

The viscosity measurements were carried out using an Ubbelohde suspended-level capillary viscometer. The viscometer was suspended vertically in a thermostat water bath at 25 ± 0.1°C. The solutions were allowed to equilibrate for 10 minutes before beginning timed runs. The flow time of a constant volume of solution through the capillary was measured with a digital stopwatch.

The intrinsic viscosity [η] of the coagulant protein was measured for protein concentration, c, in the range 0.02–0.15 g/mL in 0.1 M NaCl. Solution environment such as presence of surfactants can affect protein conformation. To study interactions SDS and SDBS with the coagulant protein, capillary viscosities of the surfactant/protein solutions were measured using a capillary viscometer. The protein concentration (% w/v) was kept constant at 0.05% whereas the surfactant concentration was varied up to concentrations higher than the critical micelle concentration (CMC). The protein solution was used as the reference standard for surfactant dissolved in 0.05% protein. The SDS (99% purity) was supplied by Sigma-Aldrich whereas SDBS was supplied by Fluka, and both surfactants were used without further purification. The measurements of surface tension, fluorescence, and circular dichroism spectral correlation coefficients of SDS solutions in the presence of protein were done in similar manner, and the details are described elsewhere [15–18].

Results and Discussion

Intrinsic Viscosity

For background ionic strength capable of swamping contributions to the ionic strength from the protein, the Huggins and Kraemer equations, respectively, are

$$\frac{\eta_{sp}}{c} = [\eta]_H + k_H [\eta]_{H^c}^2,$$
$$\ln\frac{\eta_r}{c} = [\eta]_K + k_K [\eta]_{K^c}^2, \tag{1}$$

where η_r is the relative viscosity, η_{sp} (= η_r-1) is the specific viscosity, c is the protein concentration, and k_H and k_K are the Huggins and Kraemer's constants, respectively, [19–21]. η_{sp} is dependent on concentration and interaction forces. Figure 1 shows the plots obtained by fitting the experimental data to (1). The plots are linear with statistical correlation coefficient, R^2, of 0.999 and 0.995 for Huggins plot and Kraemer plot, respectively. Thus, the [η] was estimated from the intercepts, giving $[\eta]_H$ = 2.4 mL/g and $[\eta]_K$ = 2.6 mL/g. Thus, [η] of the coagulant protein was estimated as the average of the two values is 2.5 mL/g. Very high protein concentrations (up to 0.15 g/mL) were used for intrinsic viscosity determination. If the polymer concentration is very high while the intrinsic viscosity is very low, the molecule must be globular with a compact, nearly spherical shape [12]. The [η] has been defined as the shape factor, and a value between 2.5 and 4 mL/g indicates globular particles [22].

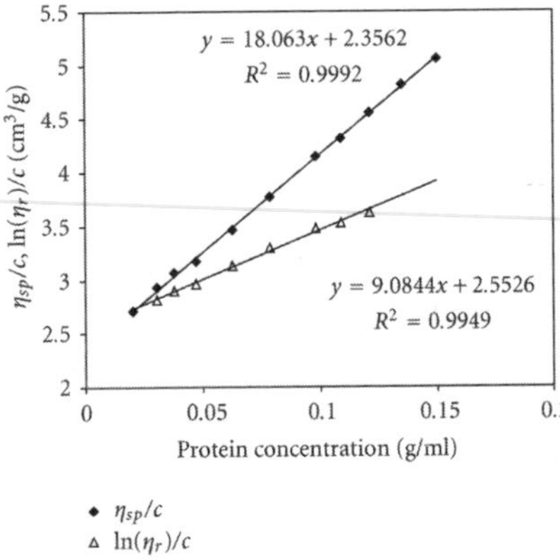

Figure 1: Evaluation of intrinsic viscosity for solutions in 0.1 M NaCl of the coagulant protein at 25°C. The Huggins (filled diamonds) and Kraemer (open triangles) plot give $[\eta]_H$ = 2.4 mL/g and $[\eta]_K$ = 2.6 mL/g, respectively.

Protein-Surfactant Interaction

Viscometry is effective in probing conformational changes due to interaction of protein with ionic surfactants [23]. The measurement of solution viscosity was performed to monitor the interaction of the coagulant protein with anionic surfactants SDS and SDBS in water. Results in Figure 2 show that addition of the anionic surfactants to the protein solutions gives a strong initial increase in viscosity and then a decrease, through a maximum at 0.768 mM surfactant concentration. The maximum suggests a change in the conformation of the protein structure or aggregation [24]. Upon further increase in the surfactant concentration above 2 mM, the viscosity increases gradually (Figure 2), suggesting protein restructuring and a significant expansion of the protein molecule coil in solution [24]. The behavior of the two anionic surfactants in solutions containing coagulant protein is very similar.

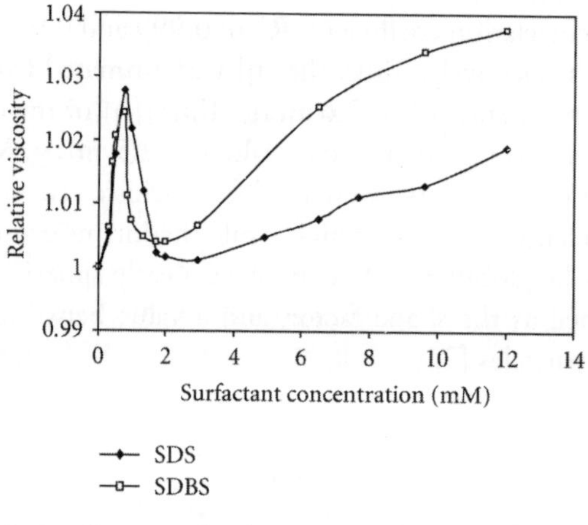

Figure 2: Influence of anionic surfactants on the relative viscosity of solutions containing 0.05% of the coagulant protein in water at 25°C.

It is not possible from viscosity measurements alone to establish the origin of the viscosity changes that occur when surfactants are added to the coagulant protein solutions. A variety of physicochemical phenomena may contribute to the overall viscosity changes, including surfactant binding, protein aggregation, protein conformational changes, and surfactant micellization. The large increase in viscosity at low surfactant concentration may indicate the formation of aggregates. The coagulant protein is highly cationic in water (i.e., in the absence of surfactants) because the protein is below its isoelectric point [2]. Consequently, there is an electrostatic repulsion between the protein molecules that is large enough to prevent them from associating [25]. Thus, at a very low surfactant concentration, there should be strong electrostatic interactions between the negatively charged surfactants and the positively charged groups in the coagulant protein. This reduces the net positive charge on the protein, and eventually, electrical net charge would no longer be large enough to prevent the protein-surfactant complexes formed from aggregating [26]. This is the surfactant concentration where the electrical charge on the protein molecules is close to zero, that is, charge neutralization occurs [26]. Further binding of anionic surfactant to the protein molecules caused them to gain a net negative charge sufficiently large such that electrostatic repulsions between the protein-surfactant molecules oppose aggregation, and so the aggregates would dissolve [25], which could explain why the solution viscosity decreases again (Figure 2).

Combined Analysis of the Coagulant Protein-SDS Interaction

In this section, the viscosity data was directly compared to earlier reported surface tension [15, 16], fluorescence [17], and CD [18] data of the coagulant protein-SDS interaction studies (Figure 3). Similar SDS concentration ranges were used, and it was convenient to divide the profiles into four different regions depending on the SDS concentration.

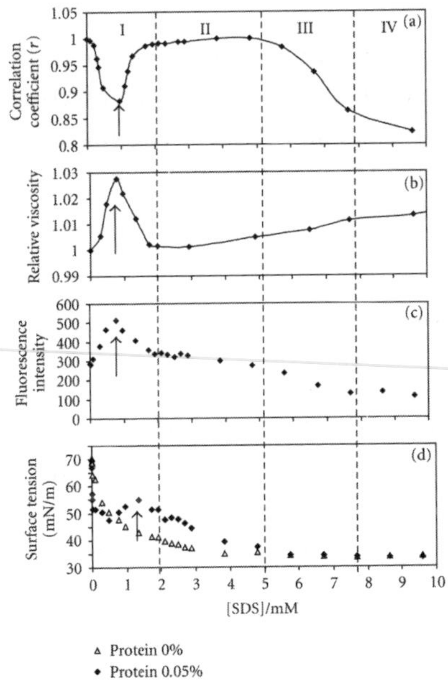

Figure 3: Variation with SDS concentration of (a) spectral correlation coefficient, (b) relative viscosity, (c) protein fluorescence intensity and (d) surface tension in the presence of protein. The protein concentration was 0.05% w/v.

The main point is that in region I, (<2 mM SDS), the bulk solution and interfacial properties exhibit critical behavior (marked by the arrows in Figure 3). It can be seen that the minimum in the spectral correlation coefficients and maxima in the relative viscosity, fluorescence intensity, and surface tension occurs at approximately the same concentration of SDS of about 1 mM. The authors suggest that this marks the stoichiometric balance point, where charge neutralization between the negatively charged SDS and the positively charged fragments in the coagulant protein occur [27–29]. Below 1 mM SDS, it is suggested that the coagulant protein-SDS complexes form aggregates.

Although charge neutralization and aggregation of protein/surfactant mixtures has been reported before [25, 26, 30, 31], the conclusion that the SDS bound to the coagulant protein promotes protein aggregation ([SDS] <1 mM) and opposes protein aggregation (1 mM < [SDS] <2 mM) is presently not well understood. But it does provide the framework for testable models. These issues are being addressed in our on-going research work.

Conclusions

The determined intrinsic viscosity of the coagulant protein is 2.5 mL/g. The intrinsic viscosity value suggests globular nature. The interaction between the coagulant protein and anionic surfactants, SDS and SDBS, in aqueous medium was investigated by viscometry and compared with tensiometry, fluorescence, and CD results previously reported. The occurrence of critical phenomena in solution properties of mixed coagulant protein-anionic surfactants system has been observed. Put together, these four approaches seem to confirm the same picture of the coagulant protein-SDS interaction.

Acknowledgements

The authors wish to acknowledge German Academic Exchange Service (DAAD) and the University of Botswana Office of Research and Development (ORD) for the financial support in form of a fellowship for RM and research grant for HMK, respectively. Dr. C. Nakabale (nakabale@botsnet.bw, Mahalpye, Botswana) is also acknowledged for the supply of *Moringa oleifera* seeds.

References

1. J. P. Sutherland, G. K. Folkard, M. A. Mtawali, and W. D. Grant, "Moringa oleifera as a natural coagulant," in Proceedings of the 20th WEDC Conference

Affordable Water Supply and Sanitation, pp. 297–299, Colombo, Sri Lanka, August 1994.

2. A. Ndabigengesere, K. S. Narasiah, and B. G. Talbot, "Active agents and mechanism of coagulation of turbid waters using Moringa oleifera," Water Research, vol. 29, no. 2, pp. 703–710, 1995.
3. A. Ndabigengesere and K. S. Narasiah, "Quality of water treated by coagulation using Moringa oleifera seeds," Water Research, vol. 32, no. 3, pp. 781–791, 1998.
4. J. Beltrán-Heredia and J. Sánchez-Martín, "Removal of sodium lauryl sulphate by coagulation/flocculation with Moringa oleifera seed extract," Journal of Hazardous Materials, vol. 164, no. 2-3, pp. 713–719, 2009.
5. T. L. Hart, "Natural coagulants: an investigation of Moringa oleifera and Tamarind seeds," MSc Report, University of Texas, Austin, Tex, USA, 2000.
6. G. K. Folkard and J. P. Sutherland, "The use of Moringa oleifera seed as a natural coagulant for water and wastewater treatment," in Proceedings of Simposio Internacional Sobre Tecnologias Deapolo a Gestao De Recursos Hidricos, 2001, May 2007, http://www.isesd.cv.ic.ac.uk/.
7. K. A. Ghebremichael, Moringa oleifera and Pumice as alternative natural materials for drinking water treatment, Ph.D. thesis, Kungl Tekniska Högskolan, Stockholm, Sweden, 2004.
8. S. A. A. Jahn, H. A. Musnad, and H. Burgstaller, "The tree that purifies water: cultivating multipurpose Moringaceae in the Sudan," Unasylva, vol. 38, no. 152, pp. 23–28, 1986.
9. Y. Kalogo, F. Rosillon, F. Hammes, and W. Verstraete, "Effect of a water extract of Moringa oleifera seeds on the hydrolytic microbial species diversity of a UASB reactor treating domestic wastewater," Letters in Applied Microbiology, vol. 31, no. 3, pp. 259–264, 2000.
10. E. D. Goddard and K. P. Ananthapadmanabhan, Interactions of Surfactants with Polymers and Proteins, CRC Press, Boca Raton, Fla, USA, 1993.
11. J. S. Barton, "Protein denaturation and tertiary structure," Journal of Chemical Education, vol. 63, no. 4, pp. 367–368, 1986.
12. J. L. Richards, "Viscosity and the shapes of macromolecules: a physical chemistry experiment using molecular-level models in the interpretation of macroscopic data obtained from simple measurements," Journal of Chemical Education, vol. 70, no. 8, pp. 685–689, 1993.

13. B. Lindman, "Surfactant-polymer systems," in Handbook of Applied Surface and Colloid Chemistry, Vol. 1, R. Holmberg, Ed., pp. 445–463, John Wiley & Sons, New York, NY, USA, 2001.
14. R. Nagarajan, "Polymer-surfactant interactions," in New Horizons: Detergents for the New Millennium Conference Invited Paper, American Oil Chemists Society and Consumer Specialty Products Association, Fort Myers, Fla, USA, 2001.
15. R. Maikokera and H. M. Kwaambwa, "Interfacial properties and fluorescence of a coagulating protein extracted from Moringa oleifera seeds and its interaction with sodium dodecyl sulphate," Colloids and Surfaces B, vol. 55, no. 2, pp. 173–178, 2007.
16. H. M. Kwaambwa and R. Maikokera, "Air-water interface interaction of anionic, cationic, and non-ionic surfactants with a coagulant protein extracted from Moringa oleifera seeds studied using surface tension probe," Water SA, vol. 33, no. 4, pp. 583–588, 2007.
17. H. M. Kwaambwa and R. Maikokera, "A fluorescence spectroscopic study of a coagulating protein extracted from Moringa oleifera seeds," Colloids and Surfaces B, vol. 60, no. 2, pp. 213–220, 2007.
18. H. M. Kwaambwa and R. Maikokera, "Infrared and circular dichroism spectroscopic characterisation of secondary structure components of a water treatment coagulant protein extracted from Moringa oleifera seeds," Colloids and Surfaces B, vol. 64, no. 1, pp. 118–125, 2008.
19. P. A. Lovell, "Dilute solution viscometry," in Comprehensive Polymer Science: Polymer Characterization, Vol. 1, C. Price and C. Booth, Eds., pp. 175–197, Pergamon Press, Oxford, UK, 1989.
20. G. L. Flickinger, I. S. Dairanieh, and C. F. Zukoski, "The rheology of aqueous polyurethane dispersions," Journal of Non-Newtonian Fluid Mechanics, vol. 87, no. 2-3, pp. 283–305, 1999.
21. L. C. Rosenthal, "A polymer viscosity experiment with no right answer," Journal of Chemical Education, vol. 67, no. 1, pp. 78–80, 1990.
22. S. R. Patil, T. Mukaiyama, and A. K. Rakshit, "Interfacial, thermodynamic, and performance properties of α-sulfonato myristic acid methyl ester—hexaoxyethylene monododecyl ether mixed surfactants," Journal of Dispersion Science and Technology, vol. 24, no. 5, pp. 659–671, 2003.
23. R.-C. Lu, A.-N. Cao, L.-H. Lai, B.-Y. Zhu, G.-X. Zhao, and J.-X. Xiao, "Interaction between bovine serum albumin and equimolarly mixed cationic-anionic surfactants decyltriethylammonium bromide-sodium decyl sulfonate," Colloids and Surfaces B, vol. 41, no. 2-3, pp. 139–143, 2005.

24. P. Deo, S. Jockusch, M. F. Ottaviani, A. Moscatelli, N. J. Turro, and P. Somasundaran, "Interactions of hydrophobically modified polyelectrolytes with surfactants of the same charge," Langmuir, vol. 19, no. 26, pp. 10747–10752, 2003.

25. D. Kelley and D. J. McClements, "Interactions of bovine serum albumin with ionic surfactants in aqueous solutions," Food Hydrocolloids, vol. 17, no. 1, pp. 73–85, 2003.

26. S. Magdassi, Y. Vinetsky, and P. Relkin, "Formation and structural heat-stability of β-lactoglobulin/surfactant complexes," Colloids and Surfaces B, vol. 6, no. 6, pp. 353–362, 1996.

27. N. Plucktaveesak, A. J. Konop, and R. H. Colby, "Viscosity of polyelectrolyte solutions with oppositely charged surfactant," The Journal of Physical Chemistry B, vol. 107, no. 32, pp. 8166–8171, 2003.

28. R. J. Green, T. J. Su, H. Joy, and J. R. Lu, "Interaction of lysozyme and sodium dodecyl sulfate at the air-liquid interface," Langmuir, vol. 16, no. 13, pp. 5797–5805, 2000.

29. M. D. Lad, V. M. Ledger, B. Briggs, R. J. Green, and R. A. Frazier, "Analysis of the SDS-lysozyme binding isotherm," Langmuir, vol. 19, no. 12, pp. 5098–5103, 2003.

30. A. D. Nielsen, L. Arleth, and P. Westh, "Analysis of protein-surfactant interactions—a titration calorimetric and fluorescence spectroscopic investigation of interactions between Humicola insolens cutinase and an anionic surfactant," Biochimica et Biophysica Acta, vol. 1752, no. 2, pp. 124–132, 2005.

31. Y. Vinetsky and S. Magdassi, "Properties of complexes and particles of gelatin with ionic surfactants," Biochimica et Biophysica Acta, vol. 276, no. 5, pp. 395–401, 1998.

The Cold Contact Method as a Simple Drug Interaction Detection System

Ilma Nugrahani, Sukmadjaja Asyarie, Sundani Nurono Soewandhi and Slamet Ibrahim

ABSTRACT

The physical interaction between 2 substances frequently occurs along the mixing and manufacturing of solid drug dosage forms. The physical interaction is generally based on coarrangement of crystal lattice of drug combination. The cold contact method has been developed as a simple technique to detect physical interaction between 2 drugs. This method is performed by observing new habits of cocrystal that appear on contact area of crystallization by polarization microscope and characterize this cocrystal behavior by melting point determination. Has been evaluated by DSC, this method is proved suitable to identify eutecticum interaction of pseudoephedrine HCl-acetaminophen, peritecticum interaction of methampyrone-phenylbutazon, and solid solution interaction of amoxicillin-clavulanate, respectively.

Introduction

In the pharmaceutical area, investigation of physical interaction which is very possible to occur along the mixing manufacturing process of dosage forms is an important issue because a lot of physical properties of drugs especially in solid state dosage forms could be influenced. Such changes as stability, drug performance, dissolution profile, pharmacokinetics profile, and, moreover, the pharmacological effect should be much impacted by the interactions [1–3]. Physical interactions in the solid state dosage form frequently occur even in storage and distribution time of dosage forms [4, 5]. Therefore, the simple technique to detect the interaction might be very useful. Along last decades, Kofler's hot stage contact method, which observed the co-recrystallization from 2 compounds from their hot molten state, has been used as a simple method to detect the cocrystal formation as an indicator for physical interaction occurrence [6]. Unfortunately, in pharmaceutical area, there are a lot of thermolabile compounds which cannot be crystallized after melting, because we have arranged a simple method used to detect physical interaction of the thermolabile compounds. This method was conducted by observing the process of cocrystallization and its behavior on crystallization contact area of 2 drugs from their solution under microscope polarization and melting point determination. Briefly, this method is developed from Kofler's contact methods by changing the comelting technique to the cosolvating crystallization. Recently, we have investigated the cold contact suitability to detect and identify the physical interaction of 3 drug combinations which are usually found in the dosage forms. Acetaminophen-pseudoephedrine HCl is found in antiinfluenza dosage forms, methampyrone-phenylbutazon in analgesic dosage forms, while amoxicillin-clavulanate in antibiotic combination dosage forms. All of these combinations have brought some troubles in mixing and compounding which caused high variability on their dosage forms quality. We used the cold contact method to detect the possibility of physical interaction of these drug combinations and evaluated the method by differential scanning calorimeter (DSC) as a primary thermal analysis method.

Method and Results

The cocrystallization from solution in ambient temperature was suggested to be mentioned as the *cold contact method* [7–9]. In this report, differential scanning calorimeter (DSC) was used to evaluate the results by observing the thermal profile and arranging phase diagrams [10–12]. From DSC data of varies molar or weight ratios (0 : 1, 1 : 9, 2 : 8, 3 : 7, 4 : 6, 5 : 5, 6 : 4, 7 : 3, 9 : 1, and 10 : 0), phase diagrams were arranged. In purpose to abbreviate the paper, phase diagrams are not presented.

The first drug combination which was examined is acetaminophen-pseudoephedrine HCl. Under polarization microscope, a black area was observed which melted at 113°C, while pseudoephedrine HCl alone melted at 184°C and acetaminophen alone at 170°C (Figure 1(a)). It could be predicted that the binary system composed an eutectic mixture. By DSC, the prediction was evaluated and proved coherency. To clarify, the thermograms of 3 : 7, 5 : 5, and 7 : 3 weight ratios are described in Figure 1(b) which indicate the eutectic point at 112.3°C, appropriate with cold contact data. Secondly, phenylbutazon-methampyrone combination showed a peritectic interaction which was proved by cold contact method and has been proven to be coherent with DSC analysis data. Last, amoxicillin-clavulanate mixture showed strong interaction with single exothermic transition curve at 202°C, equal to its melting point which was observed by cold contact method.

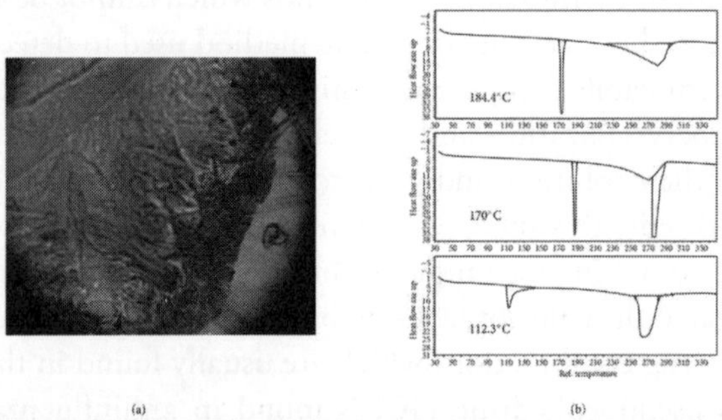

Figure 1: (a) The cold contact area of acetaminophen-pseudoephedrine HCl corecrystallization showed an area which melted at (1) 113°C, compared to the starting materials at (2) 170°C and (3) 184°C. (b) Thermograms of pseudoephedrine HCl showed melting point at 184.4°C (top), acetaminophen at 169.5°C (middle), and the eutectic mixture in weight ratios 5 : 5 at 112.3°C (bottom).

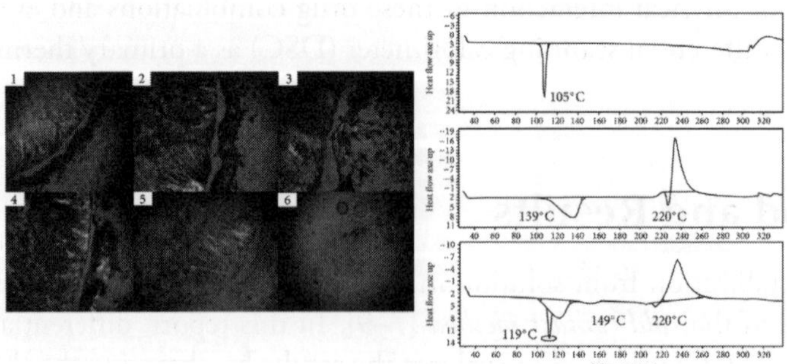

Figure 2: (a) The contact area of methampyrone-phenylbutazon indicated peritecticum mixture: (1) before heating, (2) phenylbutazon melted at 105°C, (3) contact area-1 melted at 119°C, (4) contact area-2 melted at 140°C, (5) contact area-3 melted at 150°C, and (6) methampyron melted followed by oxidation at 220–230°C. (b) DSC data of phenylbutazon (top), methampyrone (middle), and the mixture of methampyrone-phenybutazon 7 : 3 weight ratio.

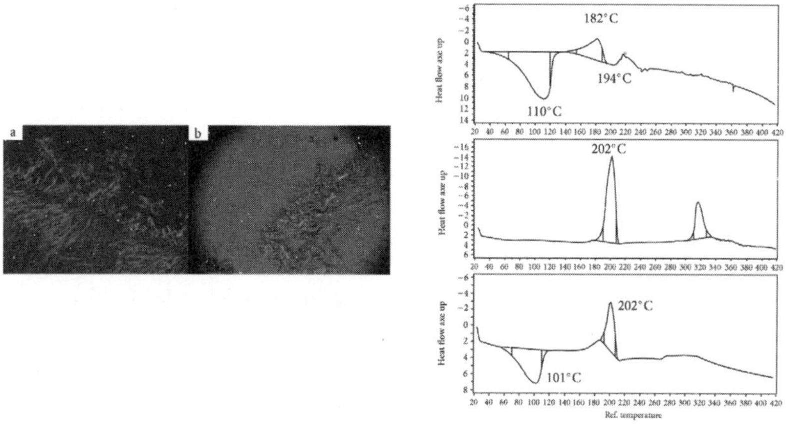

Figure 3: (a) Cold contact observation results: (i) the cocrystal grew from clavulanate crystal to amoxicillin solution in NaOH, (ii) amoxicillin melted at 194°C, clavulanate oxidized at 203°C, while the cocrystal oxidized least. (b) DSC data of amoxicillin trihydrate (top), potassium clavulanate (middle), and the physical mixture 1 : 1 (bottom). The exothermic peak of 1 : 1 mixture shows that amoxicillin and clavulanate overlay and become 1 peak at 202°C which indicates a solid solution interaction.

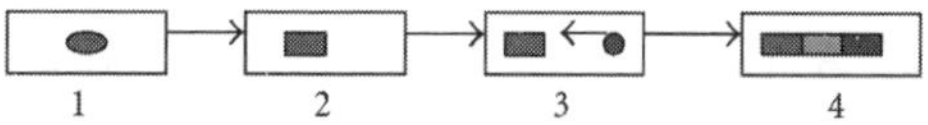

Figure 4: Cold contact preparation: (1) solution A dropped on clean object glass, (2) let it be crystallized, (3) then solution B dropped near the crystal A, (4) if it interacted, the new habit with different melting point will be formed on the contact area.

The results prove strong relation between cold contact method data and DSC. The simple eutectic interaction of pseudoephedrine hcl-acetaminophen, the peritectic interaction between methampyrone-phenylbutazon, and solid solution formation between amoxicillin trihydrate-clavulanate have been early detected by this simple method. Therefore, this method has high possibility to be used as a simple method to evaluate the other drug interactions [10–13].

Briefly, the experiment could be performed as follows.

(i) Each of the components is dissolved in the same solvent. Drop the solution 1 on cleaned object glass, evaporate the solvent and let it crystallize. The second solution is dropped near the formed crystal, let it diffuse slowly toward the crystal then quickly evaporate the solvent. Let second crystallization be performed and observe the contact area.

(ii) Heat the cold contact preparation on hot plate and observe the melting point. Differences of melting points indicate the cocrystallization or physical interaction.

(iii) The observation result could be confirmed with DSC.

Conclusion

The acceptability of the cold contact method as a simple method to identify the physical interaction of drug combination has been proved by coherency between the cold contact method data with the DSC evaluation.

References

1. G. Bettinetti, M. R. Caira, A. Callegari, and M. Merli, "Structure and solid-state chemistry of anhydrous and hydrated crystal forms of the trimethoprim-sulfamethoxypyridazine 1:1 molecular complex," Journal of Pharmaceutical Sciences, vol. 89, no. 4, pp. 478–489, 2000.
2. M. R. Caira, "Sulfa drugs as model co-crystals former," Molecular Pharmaceutics, vol. 4, no. 3, pp. 310–316, 2007.
3. G. P. Stahly, "Diversity in single- and multiple-component crystals. The search for and prevalence of polymorphs and cocrystals," Crystal Growth & Design, vol. 7, no. 6, pp. 1007–1026, 2007.
4. R. E. Davis, K. A. Lorimer, M. A. Wilkowski, and J. H. Rivers, "Studies of phase relationships in cocrystal systems," Transaction of the American Crystallographic Association, vol. 39, pp. 41–61, 2004.
5. N. Rodriguez-Hornedo, "Crystallization and the properties of crystals," in Encyclopedia of Pharmaceutical Technology, J. Swarbrick and J. C. Boylan, Eds., vol. 3, pp. 399–434, Marcel Dekker, New York, NY, USA, 1990.
6. Wikipedia, "Recrystallization," 2007, http://en.wikipedia.org/.
7. I. Nugrahani, S. N. Soewandhi, S. Asyarie, and S. Ibrahim, "The cold contact methods to detect physical interaction of paracetamol-pseudoephedrine HCl," Artocarpus, vol. 6, no. 1, pp. 18–29, 2007.
8. I. Nugrahani, S. N. Soewandhi, S. Asyarie, and S. Ibrahim, "The cold contact method to detect physical interaction of amoxicillin-clavulanate," in Proceeding of International Chemical Conference and Seminar, Yogyakarta, Indonesia, 2007.
9. I. Nugrahani, S. N. Soewandhi, S. Asyarie, and S. Ibrahim, "Study of levodopa-benserazide interaction by cold contact method," Indonesian Journal of Pharmaceutical Science, vol. 18, no. 2, 2007.
10. S. R. Byrn, R. R. Pfeiffer, and J. G. Stowell, Solid State Chemistry of Drugs, SSCI, West Lafayette, Ind, USA, 2nd edition, 2000.

11. J. T. Cartensen, Advanced Pharmaceutical Solids, Taylor & Francis, New York, NY, USA, 2001.

12. F. Giordano and A. Rossi, "Phase diagrams in the binary system," Bollettino Chimico Farmaceutico, vol. 139, no. 4, pp. 345–349, 2000.

13. Drugbank, "Acetaminophen, Pseudoephedrine, Antalgine, Phenylbutazone, Amoxicillin, and Clavulanate," March 2006, http://en.wikipedia.org/wiki/DrugBank/.

The Effect of Tb and Sm Ions on the Photochromic Behavior of Two Spiropyrans of Benzoxazine Series in Solution

Esam Bakeir, G. M. Attia, Maria Lukyanova, Boris Lukyanov and M. S. A. Abdel-Mottaleb

ABSTRACT

The photochromism of [7′-hydroxy-8′-formyl-3-methyl-4-oxospiro[1,3-benzoxazin-2,2′-[2H-1]benzopyran],SP(I),[7′-hydroxy-8′-formyl-3-benzyl-4-oxospiro[1,3-benzoxazin-2,2′-[2H-1]benzopyran] SP(II) and their co-ordination with Tb^{3+} *and* Sm^{3+} *ions have been studied in DMF. UV/vis induced-color development due to heterolytic bond cleavage of SP(I) and SP(II) is greatly influenced by complexation with the lanthanide ions. The irradiation-induced color enhancement due to ring opening and thermal de-coloration of the open forms of SP(I), SP(II) follows first-order kinetics. Physical characteristics of the studied systems such as colorability and relaxation time of thermal bleaching parameters were determined. Moreover, light-energy transfer-induced luminescence of lanthanide ions via coordination with the two spirobenzoxazines was monitored.*

Introduction

Recently, many profitable applications of photochromic dyes, particularly spirooxazines, either as passive or active devices, have been proposed [1–13].

In some cases, it has been reported that thermal equilibrium between the closed (colorless form) and opened form (colored merocyanine quinonoid form) is affected by the change in solvent polarity [14, 15], since polar solvents promote the formation of the colored form at room temperature in the absence of light. The equilibrium between both forms is strongly displaced upon irradiation to the side of open-chain colored photomerocyanine and spontaneously converts to the colorless spiro form to reach thermal equilibrium immediately after removing the light [15]. The metal-ion coordination ability of photochromic spirocyclic compounds adequately substituted is of a great interest and is being the topic of several recent studies [12, 13]. Search for molecules possessing better performances is valuable and it is important to continue to explore this subject. Here, we report on the possibility to stabilize the colored open forms of the recently synthesized spirobenzoxazines SP(I) and SP(II) toward thermal bleaching by coordination with Tb3+ and Sm3+ ions in polar DMF solvent. Moreover, the expected light-energy transfer-induced characteristic luminescence of lanthanide ions via complexation with the merocyanine quinonoid open forms of the two spirobenzoxazines will be explored.

Experimental

Materials

The metal chlorides (Sigma-Aldrich, 99.99%) were used as received. The synthesis of [7′-hydroxy-8′-formyl-3-methyl-4-oxospiro [1,3-benzoxazin-2,2′-[2H-1] benzopyran], SP(I), [7′-hydroxy-8′-formyl-3-benzyl-4-oxospiro [1,3-benzoxazin-2,2′-[2H-1] benzopyran] SP(II) were described previously [16, 17]. Spectroscopic pure grade solvents were used.

Instruments

UV-visible absorption spectra were recorded on a range (200–650 nm) using λ-Helios SP Pye-Unicam spectrophotometer and/or Ocean Optics US4000 fiber optics spectrophotometer. Continuous irradiation experiments were performed using a 150 W xenon arc lamp (PTI-LPS-220 Photon Technology International, USA) operated at 70 W. Photochemical reactions were carried out in the spectrophotometric quartz cell with a homogeneously spread light on the cell window to avoid stirring. Measurements were made on aerated solutions.

Fluorescence spectra were measured in the range (290–750 nm) using Shimadzu RF5301 (PC) spectrofluorophotometer.

Kinetics Measurements

The ring-closure reaction after photocoloration was monitored directly after removal of light at room temperature. First-order rate constants were obtained from the linear ln A versus time descending curves. By extrapolation of the obtained ln A/t plots to zero time, the absorbance A_o of the open form or its complexes at t = 0 was related to their "colorabilities" [18–21] using the expression ($A_o/c_{SP}b$), where c_{SP} is the initial concentration of SP and b is the optical path length.

Results and Discussion

Absorption Spectra

The absorption spectrum of SP(I) and SP(II) displayed two bands at 276 and 370 nm, Table 1, which are solvent independent (methanol, ethanol, and DMF solvents were tried). UV light-induced color development of both compounds was only remarkably observed in DMF solution.

Table 1: The experimental values of absorption maxima of SP and its open form (λ_{SP}, λ_{open}), SP ring opening rate constant (k_1), thermal open form ring closure rate constant (k_{-1}), equilibrium constant of the reversible reaction (K_e), relaxation time of open form ($\tau_{open\text{-}SP}$), and colorability in DMF at 295 ± 1K. Error limits of the kinetics parameters are in the order of about 5%.

Compound	λ_{SP}(nm)	λ_{open}(nm)	$k_{-1} \times 10^3$ (min^{-1})	$k_1 \times 10^3$ (min^{-1})	$\tau_{open-SP}$ (min)	K_e	Colorability $M^{-1}cm^{-1} \times 10^{-2}$
SP(I)	276 370	409	3.16	192	316	60	7.36
SP(II)	276 370	404	7.96	529	125	66	1.30

Upon irradiation of 60 μM SP(I) solutions, the absorption spectra at 276 nm and 370 nm decreased and a new band at 409 nm appeared and its intensity increased by increasing the irradiation time. Addition of lanthanide chloride solution in DMF showed no change in the absorption band of SP(I) in the dark. However, irradiation of the (1 : 1 molar ratio) solution of SP(I) in presence of

Tb3+ and Sm3+, respectively, led to a new band at 412 nm and 414 nm, Figure 1. Similar behavior observed for SP(II) in presence of lanthanide ions (Tb3+, Sm3+) in DMF, Table 1. Thermal bleaching was monitored spectrophotometrically. First-order kinetics rate constants of the reversible close $\rightleftharpoons$ open reactions (see Scheme 2) were determined graphically (Figure 2, as an example) and data are collected in Tables 1 and 2.

Table 2: The experimental values of absorption, M(III)-SP formation rate constant (k_2), thermal ring closure rate constant (k_{-2}), equilibrium (K_e) constant of the reversible reaction, relaxation time $\tau_{M(III)\text{-open-SP}}$, and colorability in DMF at 295 ± 1 K. Error limits of the kinetics parameters are in the order of about 5%.

Complexes	$\lambda_{open-M(III)}$ (nm)	$k_{-2} \times 10^3$ (min^{-1})	$k_2 \times 10^3$ (min^{-1})	$\tau_{M(III)\text{-open-SP}}$ (min)	K_e	Colorability $M^{-1}cm^{-1}10^{-2}$
Tb(III)-SP(I)	412	4.04	661	247	163	7.02
Sm(III)-SP(I)	414	4.55	258	219	56	7.85
Tb(III)-SP(II)	412	8.16	327	122	40	7.73
Sm(III)-SP(II)	414	4.77	225	209	47	7.72

Scheme 1.

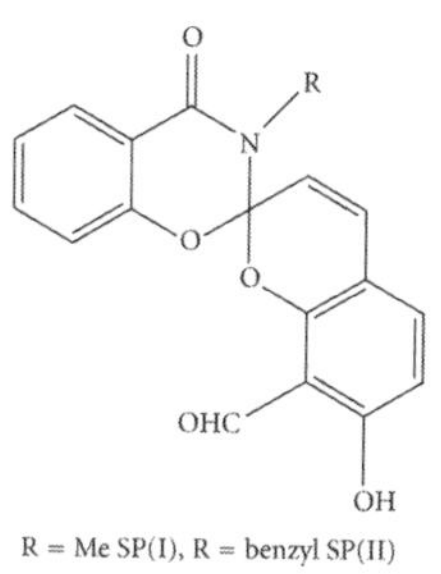

R = Me SP(I), R = benzyl SP(II)

SCHEME 1

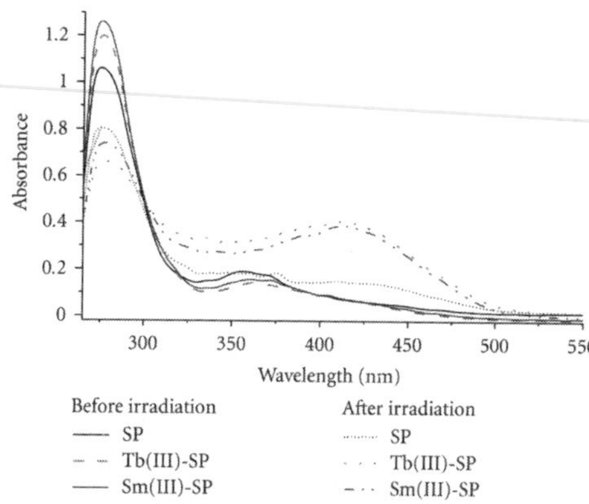

Figure 1: UV/vis absorption spectra of SP(I) and its complexes with Tb^{3+} and Sm^{3+} ions in DMF before and after light irradiation ([SP(I)] = [Ln^{3+}] = 60 μM) (at 295 ± 1K).

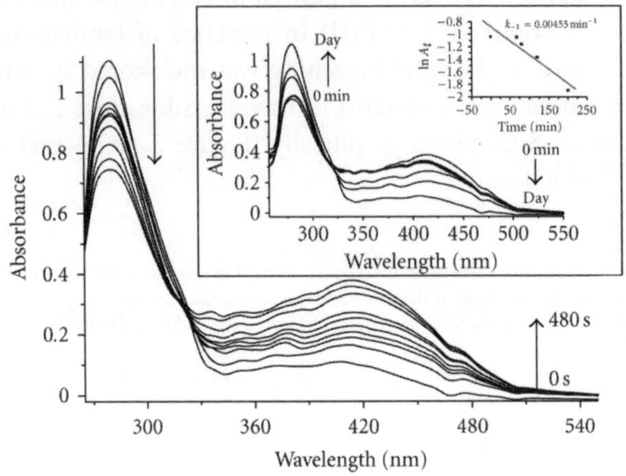

Figure 2: Effect of time of irradiation on the UV/vis absorption spectra of DMF solution of a mixture of 60 μM Sm^{3+} and 60 μM SP(II) (inset back ring closure reaction and the first-order plots of thermal decoloration of SP in dark) after UV irradiation at 295 ±1K.

Scheme 2: Illustration of the reversible structural transformation of SP to the MC form in response to light in the absence and presence of Ln^{3+} metal ion.

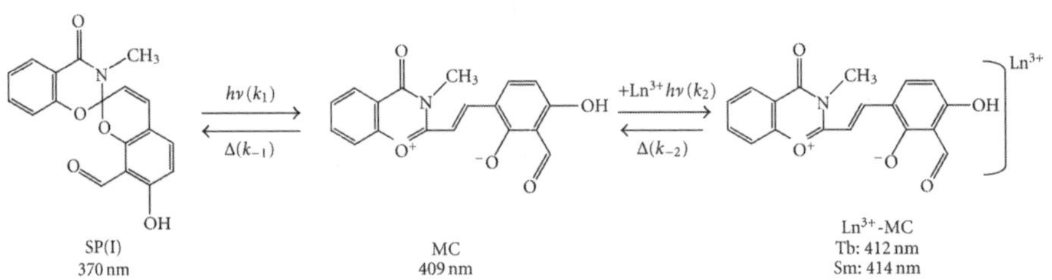

Color development rate constant (k1) of the photomerocyanine form and the thermal bleaching rate constant (k−1) are used to estimate the equilibrium constant (Ke=k1/k−1) of the forward photochemical ring opening reaction and the backward thermal bleaching one. The results of Ke are summarized in Table 1 for the SP(I) and SP(II). Data for Ln3+-SP (I and II) appear in Table 2. The relaxation time of the open form (τopen-SO) given in Table 1 was obtained from the first-order rate constant using the expression τ=1/k−1 [18–21]. The obtained relaxation time of SP and its complexes with lanthanide ions are relatively high [22], reflecting highly stabilized color form. The colored open MC form of the methyl-substituted SP(I) is more stable than that of the bulky benzyl derivative reflecting the influence of the constituent's nature and size. Moreover, the presence of the lanthanide ions generally enhances the colorability of both SP derivatives reflecting the result of coordination with the MC forms.

Benzyl group substituent in SP(II) relative to the smaller methyl group substituent in SP(I) accelerates both rates of photocoloration and thermal bleaching as reflected in the higher values of the rate constants for SP(II) shown in Table 1. A significant decrease in relaxation time and colorability of the open form (MC quinonoid form) of the SP(II) was also induced due to the effect of more bulky benzyl group. Generally speaking, the presence of Ln3+ metal ions induced more bathochromic shift in the absorption band of the MC form, enhance colorability and relaxation time. Exception is the case of Tb3+-SP(I) complexes, where colorability and relaxation time slightly decrease.

Luminescence Spectra of Lanthanide Complexes

The ligand-centered luminescence was not observed in Tb^{3+} and Sm^{3+} complexes, whereas the typical characteristic narrow emission bands of the Tb^{3+} ($^5D_4 \rightarrow {}^7F_6$, $^7F_5$, $^7F_4$, and $^7F_3$) and Sm^{3+} ($^4G_{5/2} \rightarrow {}^6H_{5/2}$, $^6H_{7/2}$, $^6H_{9/2}$, and $^6H_{11/2}$) (at λex = 360 nm) ions can be detected in polar protic solvents upon excitation of the SP absorption band (see Figure 3).This indicates efficient energy transfer from the excited open MC form to the Ln^{3+} ions having lower energy levels [23].

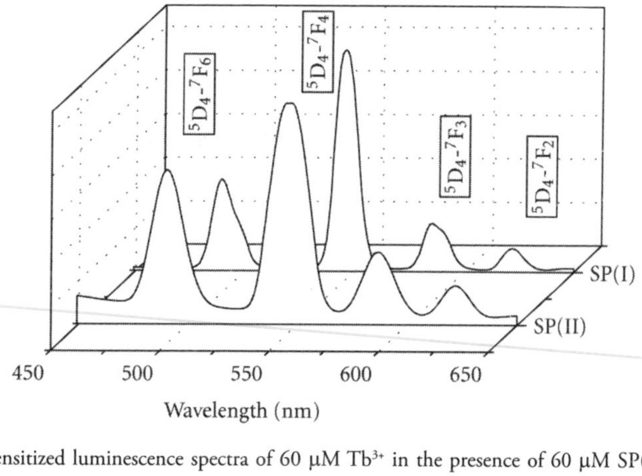

Figure 3: The sensitized luminescence spectra of 60 μM Tb^{3+} in the presence of 60 μM SP(I,II) in DMF at room temperature.

Conclusion

The substituent nature and the presence of lanthanide metal ions in solution of recently synthesized photochromic spirobenzopyrans of benzoxazine series induced significant changes in its photochromic parameters. It could be generally concluded that the presence of lanthanide ions significantly enhances the rate constants of both color development of the light-induced formation of the open merocyanine-like quinonoid species and the rate constant of its thermal bleaching. In most cases, colorability is enhanced. Moreover, strong characteristic sensitized luminescence of the lanthanide ions was observed due to efficient population of its emissive states via energy transfer from the open forms of the spirobenzoxazines.

Acknowledgement

Russian Foundation for Basic Research supported the Russian team (Grant 07-03-234).

References

1. F. Ebisawa, M. Hoshino, and K. Sukegawa, "Self-holding photochromic polymer Mach—Zehnder optical switch," Applied Physics Letters, vol. 65, no. 23, pp. 2919–2921, 1994.
2. M. Seibold, H. Port, and H. C. Wolf, "Fulgides as light switches for intra-supermolecular energy transfer," Molecular Crystals and Liquid Crystals, vol. 283, no. 1, pp. 75–80, 1996.
3. M. Hamano and M. Irie, "Rewritable near-field optical recording on photochromic thin films," Japanese Journal of Applied Physics, vol. 35, no. 3, pp. 1764–1767, 1996.
4. M. Irie and M. Mohri, "Thermally irreversible photochromic systems. Reversible photocyclization of diarylethene derivatives," Journal of Organic Chemistry, vol. 53, no. 4, pp. 803–808, 1988.
5. R. Wortmann, P. M. Lundquist, R. J. Twieg, et al., "A novel sensitized photochromic organic glass for holographic optical storage," Applied Physics Letters, vol. 69, no. 12, pp. 1657–1659, 1996.
6. R. M. Tarkka, M. E. Talbot, D. J. Brady, and G. B. Schuster, "Holographic storage in a near-ir sensitive photochromic dye," Optics Communications, vol. 109, no. 1-2, pp. 54–58, 1994.
7. J. P. Hagen, I. Becerra, D. Drakulich, and R. O. Dillon, "Effect of antenna porphyrins and phthalocyanines on the photochromism of benzospiropyrans in poly(methyl methacrylate) films," Thin Solid Films, vol. 398-399, no. 1, pp. 104–109, 2001.

8. R. C. Bertelson, "Photochromic processes involving heterolytic cleavage," in Photochromism, G. H. Brown, Ed.G. H. Brown, Ed., vol. 3 of Techniques of Chemistry, pp. 45–431, John Wiley & Sons, New York, NY, USA, 1971.

9. R. Guglielmetti, "Spiropyrans have been extensively studied," in Photochromism: Molecules and Systems, H. Dürr and H. Bouas-Laurent, Eds., vol. 40 of Studies in Organic Chemistry, pp. 314–466, Elsevier, Amsterdam, The Netherlands, 1990.

10. J. B. Flannery, Jr., "Photo- and thermochromic transients from substituted 1′,3′,3′-trimethylindolinobenzospiropyrans," Journal of the American Chemical Society, vol. 90, no. 21, pp. 5660–5671, 1968.

11. T. Bercovici, R. Heiligman-Rim, and E. Fischer, "Photochromism in spiropyrans. Part VI trimethylindolino-benzospiropyran and its derivatives," Molecular Photochemistry, vol. 1, pp. 23–55, 1969.

12. V. I. Minkin, "Photo-, thermo-, solvato-, and electrochromic spiroheterocyclic compounds," Chemical Reviews, vol. 104, no. 5, pp. 2751–2776, 2004.

13. A. V. Chernyshev, N. A. Voloshin, I. M. Raskita, A. V. Metelitsa, and V. I. Minkin, "Photo- and ionochromism of 5′-(4,5-diphenyl-1,3-oxazol-2-yl) substituted spiro[indoline-naphthopyrans]," Journal of Photochemistry and Photobiology A, vol. 184, no. 3, pp. 289–297, 2006.

14. T. Deligeorgiev, S. Minkovska, B. Jeliazkova, and S. Rakovsky, "Synthesis of photochromic chelating spironaphthoxazines," Dyes and Pigments, vol. 53, no. 2, pp. 101–108, 2002.

15. S. Minkovska, B. Jeliazkova, E. Borisova, L. Avramov, and T. Deligeorgiev, "Substituent and solvent effect on the photochromic properties of a series of spiroindolinonaphthooxazines," Journal of Photochemistry and Photobiology A, vol. 163, no. 1-2, pp. 121–126, 2004.

16. Yu. S. Alekseenko, B. Lukyanov, A. N. Utenyshev, et al., "Photo-and thermochromic spiranes. 24. Novel photochromic spiropyrans from 2,4-dihydroxyisophthalaldehyde," Chemistry of Heterocyclic Compounds, vol. 42, no. 6, pp. 803–812, 2006.

17. B. Lukyanov and M. B. Lukyanova, "Spiropyrans: synthesis, properties, and application," Chemistry of Heterocyclic Compounds, vol. 41, no. 3, pp. 281–311, 2005.

18. G. Favaro, V. Malatesta, U. Mazzucato, G. Ottavi, and A. Romani, "Thermally reversible photoconversion of spiroindoline-naphtho-oxazines to photomerocyanines: a photochemical and kinetic study," Journal of Photochemistry and Photobiology A, vol. 87, no. 3, pp. 235–241, 1995.

19. S. Kawauchi, H. Yoshida, N. Yamashina, M. Ohira, S. Saeda, and M. Irie, "A new photochromic spiro[3H-1,4-oxazine]," Bulletin of the Chemical Society of Japan, vol. 63, no. 1, pp. 267–268, 1990.
20. J.-L. Pozzo, A. Samat, R. Guglielmetti, and D. de Keukeleire, "Solvatochromic and photochromic characteristics of new 1,3-dihydrospiro[2H-indole-2,2'-[2H]-bipyrido[3,2-f][2,3-h][1,4]benzoxazines]," Journal of the Chemical Society, Perkin Transactions 2, no. 7, pp. 1327–1332, 1993.
21. V. Pimienta, C. Frouté, M. H. Deniel, D. Lavabre, R. Guglielmetti, and J. C. Micheau, "Kinetic modelling of the photochromism and photodegradation of a spiro[indoline-naphthoxazine]," Journal of Photochemistry and Photobiology A, vol. 122, no. 3, pp. 199–204, 1999.
22. B. B. Safoklov, B. Lukyanov, A. O. Bulanov, et al., "Photo- and thermochromic spiropyrans. 21. 2,8″-Formyl-3,6″-dimethyl-4-oxospiro(3,4-dihydro-2H-1,3-benzoxazine-2,2″-[2H]chromene) possessing photochromic properties in the solid phase," Russian Chemical Bulletin, vol. 51, no. 3, pp. 462–466, 2002.
23. K. A. Chibisov and H. Görner, "Complexes of spiropyran-derived merocyanines with metal ions: relaxation kinetics, photochemistry and solvent effects," Chemical Physics, vol. 237, no. 3, pp. 425–442, 1998.

Hydriding Reaction of $LaNi_5$: Correlations between Thermodynamic States and Sorption Kinetics during Activation

P. Millet, C. Lebouin, R. Ngameni, A. Ranjbari and M. Guymont

ABSTRACT

This research work concerns the hydriding reaction of $LaNi_5$ during the first hydriding cycles (activation process). Step-by-step sorption isotherms ($\Delta[H/M] \approx 0.03$) were measured at 298 K, in the composition range $0<H/M<2.0$, at the beginning (first hydriding cycle, where hysteresis is maximum) and at the end (tenth hydriding cycle, where hysteresis is minimum) of the activation process, offering the possibility to correlate thermodynamic *states (pressure-composition data points) to sorption kinetics. Using pneumatochemical impedance spectroscopy (PIS), experimental impedance diagrams were obtained for each data point of the isotherms. Microscopic rate parameters such as surface resistance and hydrogen diffusion coefficient were obtained as a function of composition, by fitting appropriate model equations to experimental impedances. It is found that the high-frequency pneumatochemical resistance significantly decreases during activation. This is correlated with the surface increase of the solid-gas interface area. The hydrogen diffusion coefficient is found to be larger at the beginning of the activation process and lower on a fully activated sample.*

Introduction

Metal-hydride systems are increasingly studied for hydrogen-storage applications. In particular, many intermetallic compounds (IMCs) such as the Haucke (AB5) phases offer several interesting features, in terms of specific energy, volumetric energy, and kinetics. Practical applications require the determination of both thermodynamic and kinetic properties which are, in general, measured separately. Pneumatochemical impedance spectroscopy (PIS) [1] now offers the possibility of analyzing hydriding/dehydriding reaction mechanisms, raw kinetic data being measured using a classical Sieverts experimental setup [2]. Using PIS, kinetic information is not only a measure of a global reaction rate (limited by any rate-determining step): the different steps involved in the sorption reaction are put into evidence and the values of individual rate parameters are inferred. Results obtained so far on various IMCs indicate that most hydriding mechanisms involve two microscopic reaction steps: (i) the surface chemisorption of molecular hydrogen which is dissociated into two surface ad-atoms; (ii) the diffusion-controlled transport of hydrogen atoms to bulk regions where they can either accumulate (in solid-solution domains) or precipitate to form metal hydrides (in two-phase domains). PIS is therefore an in situ spectroscopic tool which can be used to analyze surface and bulk processes in solid-gas systems.

Activation is the term referring to the first hydriding cycles of a brittle hydrogen absorbing IMC, during which the sample is gradually pulverized into a powder of fine particles. This pulverization is induced by the precipitation of a hydride phase of larger lattice parameters than the host IMC. This in turn generates internal strain and stresses and leads to the formation of a few micrometers thick particles, with large concentration of defects. During activation, thermodynamic properties (in particular pressure-composition equilibrium isotherms) and sorption kinetics are significantly modified. Whereas it is known from the literature that the absorption pressure plateau significantly decreases during the first hydriding cycles [3]; it is still unclear how individual rate parameters (surface resistance and hydrogen diffusion coefficient) are related to the thermodynamic states of the sample. The purpose of the work presented here is therefore to correlate thermodynamic states and sorption kinetics measured during the activation of LaNi5.

Experimental Section

The basic principles and details about pneumatochemical impedance spectroscopy have been previously reported [1]. A thorough description of the experimental setup and experimental requirements for measuring pneumatochemical impedances is described in [4]. In the present work, 2.9 g of $LaNi_5$ powder (Alfa Aesar GmbH, Karlsruhe, Germany) were placed in the reaction chamber of the PIS setup. Because the sample is prone to surface oxidation [5], it was loaded in an inert atmosphere of argon using a glove box. The setup can be gas-purged down to secondary vacuum levels (1×10^{-6} mbar) using a turbomolecular pumping station (BOC Edwards Co., Crawley, UK) to remove any residual impurity. Alphagaz grade 1 H_2 (Air Liquide Co., Paris, France) was used in the experiments.

Results and Discussion

Experimental Results

Two {pressure-composition} isotherms, measured at 298 K on $LaNi_5$-H_2 (g) up to the composition point $H/M \approx 2$, are plotted in Figure 1(a) (the maximum composition at 298 K, not represented here, is $H/M \approx 5.8$). The first isotherm was recorded during the first hydriding cycle and the second one was recorded at the end of the activation process, during the 10th hydriding cycle. At low-hydrogen content ($0.0 < H/M < 0.5$), a solid solution of hydrogen (α-$LaNi_5$) is formed. As the hydrogen content increases, a nonstoechiometric hydride phase ($LaNi_5H_{6-x}$ or β-$LaNi_5$ phase) begins to be formed. The system becomes monovariant and a pressure plateau develops. This is the two-phase domain which extends over the $0.5 < H/M < 5.8$ composition range. At higher hydrogen contents, a solid solution of hydrogen in β-$LaNi_5$ is formed. In this research work, experiments plotted in Figure 1 were intentionally stopped at $H/M \approx 2.0$ to avoid lengthy and unnecessary measurements over the whole composition range.

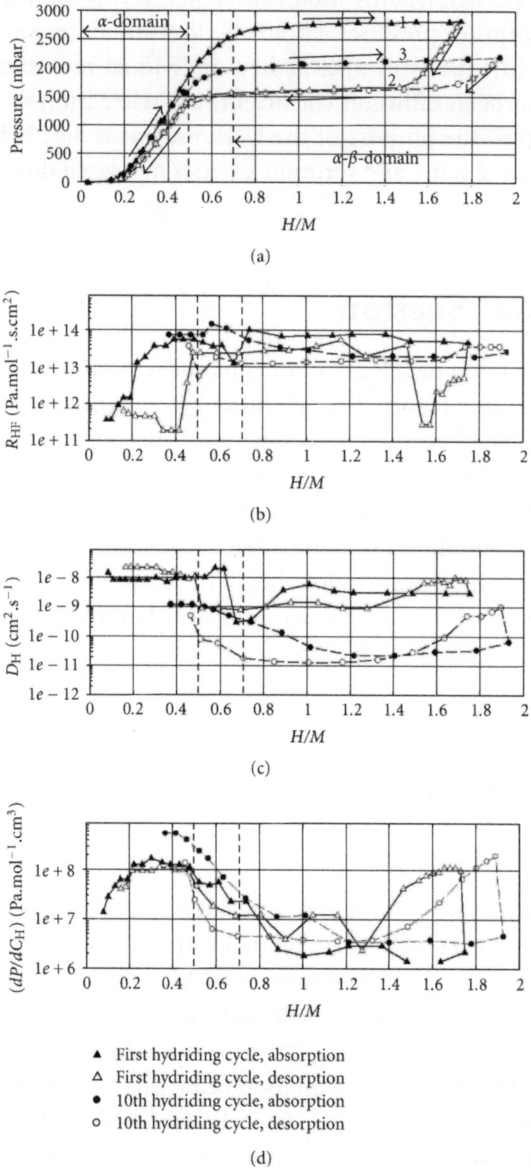

Figure 1: (a) Equilibrium pressure; (b) R_{HF}; (c) D_H; (d) dP/dC_H as a function of composition H/M measured on $LaNi_5$ at 298 K. (Data points labeled 1, 2, 3 in (a) refer to the impedance diagrams of Figure 2.)

The main characteristic associated with metal-hydride systems is the presence of a large hysteresis. The difference between absorption and desorption plateau pressures is significantly large (ca. 1400 mbar) for the first isotherm (Figure 1(a)).

The amplitude of hysteresis decreases during activation, and a final value of ca. 400 mbar, observed during the 10th hydriding cycle (Figure 1(a)), is typical of an activated material. Hysteresis also exists at phase boundaries, the composition at which the first hydride nucleus is formed during absorption being different from the composition at which the last hydride nucleus disappears during desorption (Figure 1(a)).

Each data point of Figure 1(a) corresponds to one gas transfer experiment during which the transient pressure measured in the reaction chamber and the transient hydrogen mass flow entering the reaction chamber were synchronously sampled, yielding one impedance diagram per experiment. Typical impedance diagrams measured during activation in two-phase domains are plotted in Figure 2. Impedance spectra numbers (1, 2, 3) of Figure 2 are referring to the gas transfer experiments (1, 2, 3) of Figure 1(a). They are expressed in Pa•mol−1•s and not in Pa•mol−1•s•cm2, that is, they are not corrected for the solid-gas interface area. As analyzed elsewhere [1], in two-phase domains, a typical semicircle is observed in the high-frequency (HF) domain along the real axis (see, e.g., diagram 1 in Figure 2). This is the signature of the parallel connection of a capacitance (associated with the dead volume of the reaction chamber) and a resistance in Pa•mol−1•s. In single-phase domains (SPDs), the diameter of this HF semicircle is equal to the value of the surface resistance Rs (the chemisorption step of the sorption process) but in two-phase domains (TPDs), this is the sum (Rs+RPT), RPT being the resistance associated with the phase transformation process [6]. The HF semicircle is then followed at lower frequencies by a characteristic medium frequency (MF) region related to hydrogen diffusion and ends up at low frequencies (LF) in a capacitive shape (a semi-infinite line along the imaginary axis).

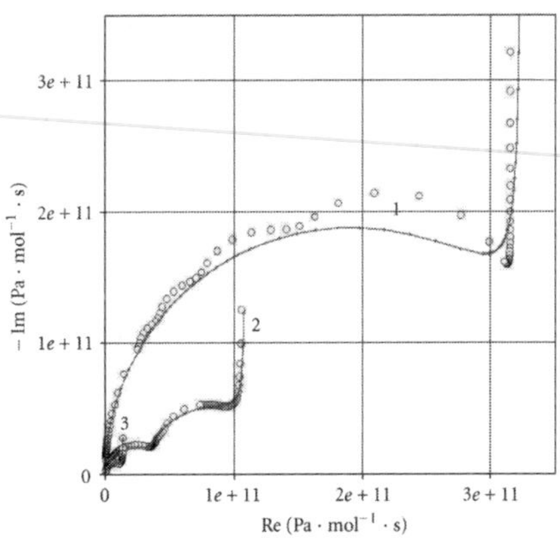

Figure 2: Experimental (o) and model (–) impedance diagrams. 1, 2, 3 refer to data points of Figure 1(a). (1) First hydriding cycle, absorption, H/M ≈ 1.5, (2) First hydriding cycle, desorption, H/M ≈ 1.4, (3) 10th hydriding cycle, absorption, H/M ≈ 1.5.

Each experimental impedance diagram of Figure 1(a) was fitted using model equations applicable to single- and two-phase domains, as described in [6]. From the fits, values of the HF resistance (RHF), the hydrogen diffusion coefficient (DH), and the slope of the isotherm at the measurement point (dP/dCH) were obtained. All results are plotted as a function of the composition H/M in Figures 1(b), 1(c), and 1(d), respectively. In Figure 1(b), the true solid-gas interface area has been taken into account and RHF values are expressed in Pa•mol−1•s•cm2.

Shape of Impedance Diagrams during Activation

The first impedance diagram of Figure 2 (curve 1) was measured on the unactivated sample, on the pressure plateau in the two-phase domain (data point 1 in Figure 1(a)). A large HF resistance of ca.4×10^{11} n $Pa \bullet mol^{-1} \bullet s$ is observed. Once the scan is reversed, the first impedance diagram measured on the desorption pressure plateau (point 2 in Figure 1(a)) is also plotted in Figure 2 (curve 2). An overall lower HF resistance is measured but two consecutive semicircles are observed, suggesting a bimodal distribution of particle size. In other experiments (not reported here), more than two HF semicircles were observed. It can be concluded that fragmentation of initial particles leads to particles of heterogeneous sizes. Finally, the impedance diagram measured in two-phase domains on the activated sample (point 3 in Figure 1(a)) is also plotted in Figure 2 (curve 3). The impedance is significantly lower for each frequency, indicating that the sorption kinetics will be higher on the activated sample.

High-Frequency Resistance

It is observed from the whole set of experimental impedance diagrams (not represented here) that the high-frequency resistance R_{HF} (expressed in $Pa \bullet mol^{-1} \bullet s$) decreases exponentially with the number of hydriding cycles, down to a constant value at the end of the activation process. This does not mean that the kinetics of the surface step (in SPDs) or the phase transformation process (in TPDs) are varying during activation: mostly HF resistances are a function of the operating temperature which is constant here. The reduction of R_{HF} must be correlated with the increasing solid-gas (and interphase) interface areas resulting from the sample pulverization. To take into account the modification of the solid-gas interface area during activation, R_{HF} values plotted in Figure 1(b) are expressed in $Pa \bullet mol^{-1} \bullet s \bullet cm^2$. Real surfaces areas were estimated from mean particle size values obtained from microscope observations. Assuming that particles are spherical, taking a mean particle radius of 50 μm for the first hydriding cycle and a density of 8.2 for $LaNi_5$ [7], a total surface of 212 cm^2 was used. Taking a mean particle radius of 5 n µm for the tenth hydriding cycle, a total surface of 2120 cm^2 was used. When HF resistances are expressed per cm^2 of solid-gas interface, they are mostly constant in TPDs (Figure 1(b)), although their values spread over a decade of units (this comes from experimental uncertainties and the estimation of the true interface area). In SPDs, HF resistances are lower than those in TPDs where R_{PT} values (the resistance associated with the phase transformation process) are included, as discussed in [6]. From the data of Figure 1(b), R_{PT} is found to remain almost constant during activation.

From these results, a limitation of PIS analysis should be pointed out (especially when powdered samples are used, as LaNi5 considered here): the mean particle radius (at the measurement point) must be known to normalize impedance diagrams. This is because different triplet values {r;D;dP/dC} lead to similar impedance diagrams. If not, it is difficult to analyze quantitatively the hydriding kinetics.

Hydrogen Diffusion Coefficient

Concerning hydrogen diffusion coefficients, the most striking result is that values are spread over several (three) decades of values. General trends are that values measured on unactivated samples are larger than those measured on fully activated samples and that values measured in solid-solution domains are larger than those measured in two-phase domains. At 298 K, D_H values measured on an unactivated sample are within the range 1×10^{-9}–1×10^{-8} $cm^2 \bullet s^{-1}$, whereas those measured on an activated sample are spread over the 1×10^{-11}–1×10^{-10} $cm^2 \bullet s^{-1}$ range. It is concluded that large concentrations of defects (point defects, dislocations), appearing in activated samples because of the precipitation of the β-$LaNi^5H$ hydride phase, contribute to a reduction of the kinetics of the diffusion step but a relationship between kinetics and microstructure still needs to be established to clarify the point. Anyway, concerning the overall kinetics, this effect is largely counterbalanced by the large increase of the solid-gas interface area.

Slope of the Isotherm

The slopes of the isotherms obtained from the automated fitting procedure are plotted in Figure 1(d) as a function of hydrogen composition. Although a certain discrepancy is observed, there is a qualitative agreement with the experimental isotherms of Figure 1(a), the slope being larger in solid-solution domains than in two-phase domains, except at the foot of the isotherms where hydrogen trapping effects prevail. From a quantitative viewpoint, care must be taken when comparing the data of Figure 1(d) to the isotherms of Figure 1(a). This is because in

Figure 1(a), the hydrogen content is expressed in H/M units whereas in Figure 1(d), the hydrogen content is expressed in concentration (CH in mol • cm^{-3}).

Conclusions

The hydriding reaction of $LaNi_5$ has been analyzed during the activation process using pneumatochemical impedance spectroscopy. Experimental impedance diagrams have been measured during the first and the tenth hydriding cycle. Using model equations, individual rate parameters (surface resistance and hydrogen diffusion coefficient) have been obtained as a function of composition H/M. Impedance diagrams show that the overall absorption kinetics largely increases during activation. This is mostly due to the increase of the solid-gas interface areas. When the solid-gas surface area is taken into account, it is shown that the high-frequency resistance related to surface hydrogen chemisorption and hydride precipitation remains mostly constant. To a lesser extent, the hydrogen diffusion coefficient decreases during activation. This is attributed to the formation of defects. Results obtained along hysteresis loops will be presented in a subsequent communication.

Acknowledgements

Financial support from the French Agence Nationale de la Recherche, within the frame of the Plan d'Action National sur l'Hydrogène, EolHy Project (ANR-06-PANH—008-06), is acknowledged.

References

1. P. Millet, "Pneumato-chemical impedance spectroscopy. 1. Principles," Journal of Physical Chemistry B, vol. 109, no. 50, pp. 24016–24024, 2005.
2. A. Sieverts, "Die Aufnahme von Gasen durch Metalle," Zeitschrift für Metallkunde, vol. 21, pp. 37–46, 1929.
3. T. Yamamoto, H. Inui, and M. Yamaguchi, "Effects of lattice defects on hydrogen absorption-desorption pressures in LaNi5," Materials Science and Engineering A, vol. 329–331, pp. 367–371, 2002.
4. P. Millet, C. Decaux, R. Ngameni, and M. Guymont, "Experimental requirements for measuring pneumato-chemical impedances," Review of Scientific Instruments, vol. 78, no. 12, Article ID 123902, 11 pages, 2007.

5. P. Millet, "Pneumato-chemical impedance spectroscopy. 2. Dynamics of hydrogen sorption by metals," Journal of Physical Chemistry B, vol. 109, no. 50, pp. 24025–24030, 2005.
6. P. Millet, C. Decaux, R. Ngameni, and M. Guymont, "Fourier-domain analysis of hydriding kinetics using pneumato-chemical impedance spectroscopy," Research Letter in Physical Chemistry, vol. 2007, Article ID 96251, 5 pages, 2007.
7. J.-M. Joubert, R. Černý, M. Latroche, et al., "A structural study of the homogeneity domain of LaNi5," Journal of Solid State Chemistry, vol. 166, no. 1, pp. 1–6, 2002.

Hollow Disc and Sphere-Shaped Particles from Red Blood Cell Templates

Preston B. Landon, Jose J. Gutierrez, Sara A. Alvarado, Sujatha Peela, Srinivasan Ramachandran, Fernando Teran Arce and Ratnesh Lal

ABSTRACT

Colloidal gold particles with uniform size distributions were fabricated utilizing human red blood cells (RBCs) as templates. The gold shells were charged with a metal chelating agent to prevent flocculation. The procedure described here allows control over the shape of the colloidal particles. Thus, it was possible to fabricate discs and spheres by controlling the osmotic pressure.

Introduction

Colloidal particles with specific shape and function [1, 2] are currently a subject of intense investigation for the development of pigments [3], microcapsules [3,

4], guided drug carriers [5, 6], synthetic opals [7], transparent metals [8], photonic crystals [7, 8], thin-film transistors, organic light emitting diodes [9], solar cells [10, 11], and tunable lasers [12].

Large populations of nonspherical colloidal particles with uniform size distributions have many potential applications. Nonetheless, there are only a few techniques for the creation of nonspherical particles since their formation has been proven difficult. In recent years there has been a considerable effort in the formation of colloidal nanorods of various materials [13, 14]. Another example of nonspherical particles is the formation of oval-shaped polystyrene by applying mechanical force to spherical particles in the solid phase [15].

On the one hand, it is difficult to create uniform populations of nonspherical colloids in the laboratory. On the other hand, nature has been extremely efficient at this task. Moreover, unique shapes of colloidal particles are found throughout nature.

Human blood cells have a biconcave disc shape with an average diameter of 8.1 μm. They are roughly 2 μm thick at the outer edges and 1-μm thick in their center [16]. In a typical person, red blood cells (RBCs) have a coefficient of variation of approximately 11% [16]. Thus, the size distribution in this colloidal population can be considered relatively uniform. The outer surface of a red blood cell is composed of a phospholipid bilayer membrane [16] similar to that of most eukaryotic cells. However, unlike most cells, RBCs do not divide or reproduce themselves. They are fabricated as individual particles elsewhere within the body [16] and are, thus, truly colloidal in nature. The colloidal nature of RBCs and their unique shape motivated us to fabricate metallic colloidal particles using them as templates. Several potential commercial applications are envisioned [1–10].

Experimental Section

Materials

All chemicals were obtained from Aldrich Chemical Company and were used as received unless otherwise noted. Gold chloride was prepared as described elsewhere [17]. RBCs were obtained using a diabetic finger lancet. A boric acid and sodium borate isotonic contact lens saline solution was used in all experiments. Samples were centrifuged using a Hamilton Bell Van Guard V6500, which is a fixed speed blood centrifuge. All solutions were prepared with deionized water with a resistivity below 17 M Ω cm.

Colloidal Discs

Colloidal gold discs were prepared at room temperature as follows. Three drops of blood were dispersed in 1.5 mL of saline solution in 2 mL centrifuge tubes. 100 μL of 0.06 M aqueous sodium citrate (used as anticoagulant) followed by 100 μL of 0.006 M aqueous gold chloride were added at once. The mixture was shaken by hand and centrifuged for 60 seconds. The supernatant was decanted and the remaining sediment was quickly redispersed in pure saline solution. The coating process (i.e., addition of aqueous $AuCl_3$) was repeated two additional times. The gold coated RBCs were centrifuged and redispersed several times in 400 μL of 0.035 M aqueous 16-mercaptohexadecanoic acid for the purpose of purification.

Colloidal Gold Spheres

Six drops of blood were dispersed in 1.5 ml of saline solution. 120 μL of .06 M aqueous sodium citrate followed by 400 μL of 0.006 M aqueous gold chloride were added at once. After addition of gold chloride, the sample was immediately shaken and centrifuged for 60 seconds. The supernatant was decanted and the remaining sediment was quickly redispersed in pure saline solution. The coating process was repeated one additional time. The sediment was redispersed in 1.5 mL of saline solution and 300 μL of aqueous gold chloride. On the last coating cycle, the gold chloride remained in solution one minute before the addition of 400 μL of 0.035 M aqueous 16-mercaptohexadecanoic acid and 50 μL of 7.5 M aqueous sodium hydroxide. The particles were centrifuged and redispersed in deionized water containing 50 μL of aqueous sodium hydroxide. The spheres were centrifuged and redispersed several times in deionized water to purify the samples with 400 μL of the aqueous 16-mercaptohexadecanoic acid. The spheres were then centrifuged and redispersed two additional times and were allowed to sediment for 7 minutes before removing all the supernatant with a pipette, leaving only a small amount of sediment at the bottom. The purified spheres were redispersed in water and were allowed to selfassemble by sedimentation.

Sintering of the Metal Layers

The metal-coated discs were thermally sintered in a vacuum tube furnace as described: the gold-coated cells were heated at 80°C in air for 4 hours and then placed under vacuum. Once under vacuum, the temperature was increased by 10°C per hour for 4 hours. Temperature was held at 120°C for 5 hours and then raised to 500°C, at a rate of 50°C per hour. The samples were kept at 500°C for 10 hours and then cooled to room temperature at a rate of 50°C per hour.

Results and Discussion

Self Assembly and Sintering

Colloidal metal disc-shaped particles were self-assembled by sedimentation and then slowly dried in air (Figure 1(a)). Once the particles were dried, they were slowly sintered as described in the experimental section (Figure 1(b)).

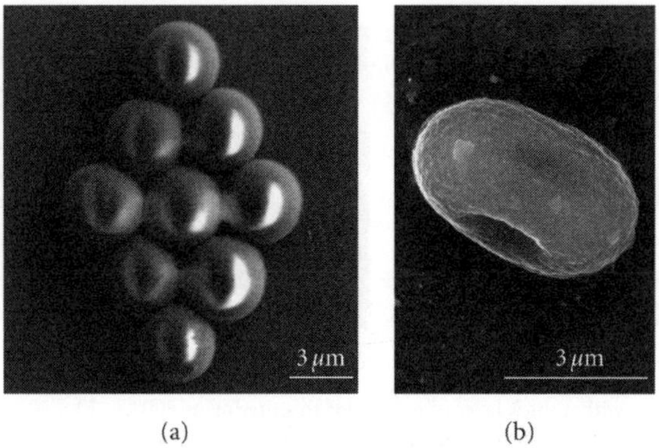

Figure 1: (a) Colloidal gold discs self-assembled on a silicon substrate prior to sintering. (b) A hollow colloidal gold replica of a red blood cell on a silicon substrate after sintering at 500°C.

Gold chloride is converted to metallic gold and chlorine gas at temperatures above 350°C [17]. The sintering process should therefore result in the thermal conversion of virtually all of the remaining gold chloride to metallic gold. This was verified by energy dispersive X-ray (EDX) (Table 1). Accordingly, chlorine was present in a ratio far below its composition in gold chloride. This was evident in both the atomic and weight composition.

Table 1: Elemental composition of gold chloride-coated red blood cells as determined by energy dispersive X-ray spectroscopy (EDX) after sintering.

Element	Weight %	Atomic %
Au	40.07	8.86
Cl	3.34	4.11

When particles with thick metal shells were sintered, the shells fused at their interfaces causing widening between the necking of the shells (Figure 2(a)). This suggests that there was excess gold chloride remaining in the metal shells before sintering allowing them to neck together during the thermal process, which was significantly below 1064°C, the melting temperature of gold. Roughening of the particles' surface was observed after sintering (Figure 2(a)). Interestingly, the gold-coated particles that were sintered have dodecahedral-shaped nodules visible on the surface of the particles, which were not present prior to sintering (Figure 2(b)). It would be interesting to investigate the formation and evolution of these dodecahedral aggregates on the particles surface during the sintering process, while the gold chloride converts to metallic gold.

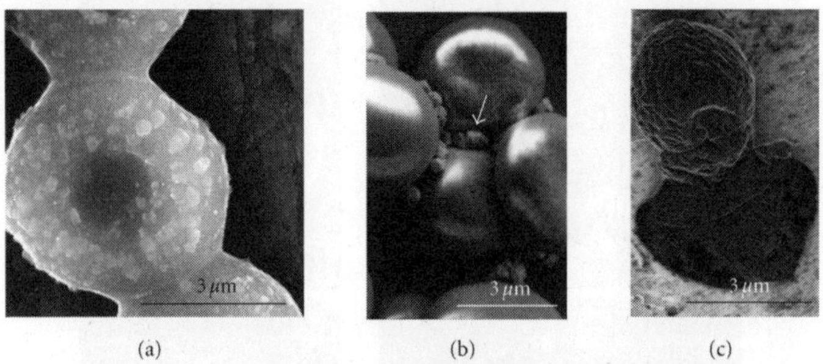

Figure 2: (a) Gold discs with a thick layer of gold prior to sintering at 500°C, (b) a 300–400 nm thick layer of gold deposited on the blood cells from repeated coatings of gold prior to sintering. The arrows indicate excess colloidal gold, (c) an SEM image of a red blood cell that bursts because the gold shell was too thin to contain it.

Properties of the Gold Shells

The coating process described in the experimental section outlines the optimum procedure to create the thinnest shell necessary to stabilize the particles. This allowed us to dry the samples and obtain SEM images without bursting (Figure 2(c)) or collapsing (Figure 3(a)) the shells. The thickness of this gold layer was found to be approximately 30–120 nm thick. This was determined by examining the remains of particles in the SEM that were shattered by thermal shock (Figure 3(a)).

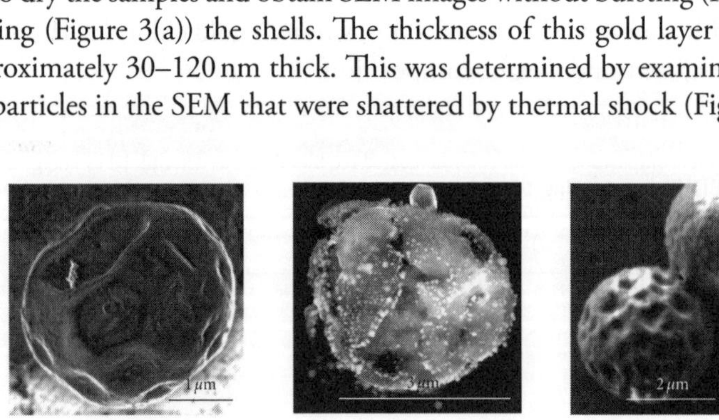

Figure 3: (a) Collapsed gold-coated RBC disc. (b) A thermally shattered hollow gold disc with a 50–120 nm thick layer of gold. (c) A doublet formed from two thinly coated spheres.

Once the cells were encapsulated with a layer of gold, they could be further coated by placing them in solutions that had much higher concentrations of gold chloride. The particles could also be left in these plating solutions for any desired amount of time since they were completely coated and indifferent to osmotic changes in solution (Figure 2(b)). This allowed the formation of much thicker gold shells and it resulted in reproducible control of the thickness of the gold shells.

Red Blood Cells

The cell membrane of an RBC is permeable to water, but only relatively permeable to ions [16]. Thus, the volume of the individual cells can quickly change due to flux of water across the membrane arising from a change in the osmotic pressure across the cells membrane [16, 18]. When the osmotic pressure drives water out of the interior of the cell, the volume of the cell contracts and the cell is said to be crenated. When the osmotic pressure drives water into the cell, its volume expands [19]. This allowed us to modify the shape.

The membrane of a red blood cell is relatively elastic and the volume of the cell can increase from 30% to 40% before bursting in a process called hemolysis [16, 19, 20]. When hemolysis occurred in our samples, it was evident because the samples turned blood red. This red color did not sediment when centrifuged during the coating process and was a very convenient indicator of how the coating process was proceeding. After hemolysis, the membrane of a red blood cell will often reclose or remain stable. These blood cells are called ghosts. Many interesting ghosts were found in our samples and were prevalent when the layer of gold was very thin or coated using a process where the gold layer was deposited too slowly.

Under normal physiological conditions the osmotic pressure across an RBC membrane is about 3.4 times stronger than the atmospheric pressure [19]. The osmotic sensitivity of the cells makes the coating process a challenging time-dependent task that is both pH and electrolyte sensitive. The coating process appears to be a time-sensitive one and only attempts using a quick coating process resulted in reproducible work. A failed coating attempt usually resulted in hemolysis of virtually all of the cells. The speed of the coating process was controlled by the concentration of gold chloride, which also affected the osmotic pressure. The increased osmotic pressure (resulting from excess solution) allowed the formation of spherical gold-coated RBCs (Figure 3(c)).

Colloidal Stabilization and Mechanical Properties

In order to stabilize the colloidal particles, their surfaces needed to be electrostatically charged. This was achieved using the following multistep process. First, the particles were charged using 16-mercaptohexadecanoic acid (a metal chelating agent) [21]. Addition of the chelating agent resulted in the removal of small unwanted gold particles (Figure 2(b)) during the rinsing process described in the experimental section. Thus, they remained dispersed after centrifuging and were decanted during the purification procedure. This also greatly reduced the number of doublet spheres found in the samples (Figure 3(b)). This means that the surfaces of the gold-coated cells must remain relatively malleable after the coating process. This malleability was observed in some of the SEM images of the gold-coated cells prior to sintering (Figure 4(a)). For comparison, an SEM image of dried and uncoated red blood cells is included as a control (Figure 4(c)).

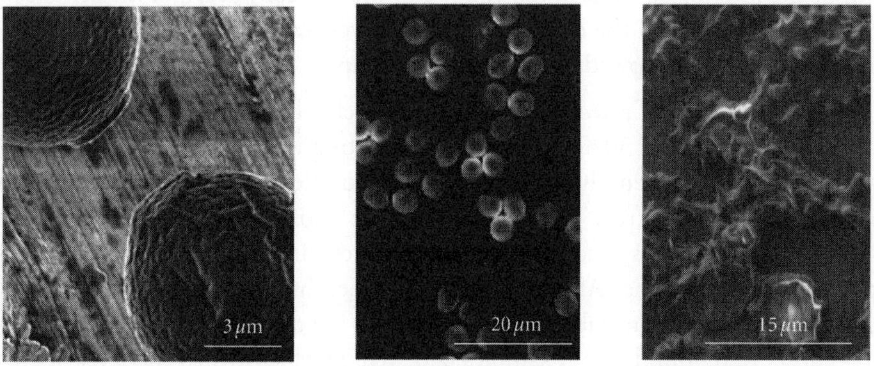

(a) (b) (c)

Figure 4: (a) A previous doublet formed from two gold spheres that had been forced apart. The arrows indicate where the two spheres were joined. (b) A wide-field SEM image of colloidal disc-shaped particles. (c) An SEM image of dried untreated RBCs on a silicon substrate.

Conclusion

Hollow gold colloidal disc and sphere-shaped shells were fabricated by encapsulating human red blood cell templates in gold chloride shells. The thickness of the gold shell and the shape of the resulting particles were found to be controllable during the encapsulating process. Coated red blood cells were electrostatically charged, self-assembled and the residual gold chloride was converted to metallic gold by sintering under vacuum at 500°C.

Acknowledgements

We would like to thank the NIH/ NHLBI funded Research Training grant (T32 HL 007605) at the University of Chicago and the Welch Foundation (BG-0017) at the University of Texas Pan-America for support of this work.

References

1. F. Caruso, C. Schüler, and D. G. Kurth, "Core-shell particles and hollow shells containing metallo-supramolecular components," Chemistry of Materials, vol. 11, no. 11, pp. 3394–3399, 1999.
2. A. Dokoutchaev, J. Thomas James, S. C. Koene, S. Pathak, G. K. Surya Prakash, and M. E. Thompson, "Colloidal metal deposition onto functionalized polystyrene microspheres," Chemistry of Materials, vol. 11, no. 9, pp. 2389–2399, 1999.
3. R. A. Caruso, A. Susha, and F. Caruso, "Multilayered titania, silica, and Laponite nanoparticle coatings on polystyrene colloidal templates and resulting inorganic hollow sphere," Chemistry of Materials, vol. 13, no. 2, pp. 400–409, 2001.
4. B. Neu, A. Voigt, R. Mitlöhner, et al., "Biological cells as templates for hollow microcapsules," Journal of Microencapsulation, vol. 18, no. 3, pp. 385–395, 2001.
5. F. Caruso, M. Spasova, A. Susha, M. Giersig, and R. A. Caruso, "Magnetic nanocomposite particles and hollow spheres constructed by a sequential layering approach," Chemistry of Materials, vol. 13, no. 1, pp. 109–116, 2001.
6. W. Schütt, C. Grüttner, J. Teller, et al., "Biocompatible magnetic polymer carriers for in vivo radionuclide delivery," Artificial Organs, vol. 23, no. 1, pp. 98–103, 1999.
7. P. B. Landon and R. Glosser, "Self-assembly of spherical colloidal silica along the [100] direction of the FCC lattice and geometric control of crystallite formation," Journal of Colloidal and Interface Science, vol. 276, no. 1, pp. 92–96, 2004.
8. V. Kamaev, V. Kozhevnikov, Z. V. Vardeny, P. B. Landon, and A. A. Zakhidov, "Optical studies of metallodielectric photonic crystals: bismuth and gallium infiltrated opals," Journal of Applied Physics, vol. 95, no. 6, pp. 2947–2951, 2004.
9. M. N. Shkunov, Z. V. Vardeny, M. C. DeLong, R. C. Polson, A. A. Zakhidov, and R. H. Baughman, "Tunable, gap-state lasing in switchable directions for opal photonic crystals," Advanced Functional Materials, vol. 12, no. 1, pp. 21–26, 2002.

10. B. J. Schwartz, T.-Q. Nguyen, J. Wu, and S. H. Tolbert, "Interchain and intrachain exciton transport in conjugated polymers: ultrafast studies of energy migration in aligned MEH-PPV/mesoporous silica composites," Synthetic Metals, vol. 116, no. 1–3, pp. 35–40, 2001.
11. K. Landfester, R. Montenegro, U. Scherf, et al., "Semiconducting polymer nanospheres in aqueous dispersion prepared by a miniemulsion process," Advanced Materials, vol. 14, no. 9, pp. 651–655, 2002.
12. P. N. Bartltt, P. R. Birkin, M. A. Ghanem, and C.-S. Toh, "Electrochemical syntheses of highly ordered macroporous conducting polymers grown around self-assembled colloidal templates," Journal of Materials Chemistry, vol. 11, no. 3, pp. 849–853, 2001.
13. B. M. I. van der Zande, M. R. Böhmer, L. G. J. Fokkink, and C. Schönenberger, "Aqueous gold sols of rod-shaped particles," Journal of Physical Chemistry B, vol. 101, no. 6, pp. 852–854, 1997.
14. Y. Sun and Y. Xia, "Multiple-walled nanotubes made of metals," Advanced Materials, vol. 16, no. 3, pp. 264–268, 2004.
15. L. Yu, Y. Yin, Z.-Y. Li, and Y. Xia, "Colloidal crystals made of polystyrene spheroids: fabrication and structural/optical characterization," Langmuir, vol. 18, no. 20, pp. 7722–7727, 2002.
16. L. Sherwood, Human Physiology, Thomson Brooks/Cole, Belmont, Calif, USA, 6th edition, 2007.
17. P. B. Landon, J. J. Gutierrez, J. P. Ferraris, et al., "Inverse gold photonic crystals and conjugated polymer coated opals for functional materials," Physica B, vol. 338, no. 1–4, pp. 165–170, 2003.
18. S. R. Randel and M. H. Navidi, Chemistry, West Publishing, New York, NY, USA, 2nd edition, 1994.
19. R. H. Garrett and C. M. Grisham, Biochemistry, Sanders College Publishing, Orlando, Fla, USA, 5th edition, 1999.
20. B. Alberts, D. Bray, J. Lewis, M. Raff, K. Roberts, and J. D. Watson, Molecular Biology of the Cell, Garland Publishing, New York, NY, USA, 3rd edition, 1994.
21. A. El Kasmi, J. M. Wallace, E. F. Bowden, S. M. Binet, and R. J. Linderman, "Controlling interfacial electron-transfer kinetics of cytochrome c with mixed self-assembled monolayers," Journal of the American Chemical Society, vol. 120, no. 1, pp. 225–226, 1998.

Ultrasoft and High Magnetic Moment CoFe Films Directly Electrodeposited from a B-Reducer Contained Solution

Baoyu Zong, Guchang Han, Jinjun Qiu, Zaibing Guo, Li Wang, Wee-Kay Yeo and Bo Liu

ABSTRACT

A methodology to fabricate ultrasoft CoFe nano-/microfilms directly via electrodeposition from a semineutral iron sulfate solution is demonstrated. Using boron-reducer as the additive, the CoFe films become very soft with high magnetic moment. Typically, the film coercivity in the easy and hard axes is 6.5 and 2.5 Oersted, respectively, with a saturation polarization up to an average of 2.45 Tesla. Despite the softness, these shining and smooth films still display a high-anisotropic field of ~45 Oersted with permeability up to 10^4. *This kind of films can potentially be used in current and future magnetic recording systems as well as microelectronic and biotechnological devices.*

Introduction

The ever expanding demand of the world market leads to magnetic recording data storage devices advancing toward much smaller exterior dimension and higher capacity [1, 2]. In order to achieve very high capacity and fast recording data storage in a miniature device, an ultrasoft and high magnetic moment material is required for producing high-saturation flux density (B_s), so that the necessary flux density can be preserved on reducing device dimensions, while simultaneously achieving a low coercivity (H_c) to match the hard magnetic media with high H_c, track density, and linear density [3]. Soft magnetic films with high moment are also widely used in modern electromagnetic devices, such as high-frequency field-amplifying components, versatile communication tools, and magnetic shielding materials in tuners [4]. Although numerous soft or high-Bs magnetic films have been achieved nowadays via sputtering, evaporation, and casting, most of them cannot be applied to device fabrication due to the following reasons: low-deposition rate (usually < 8 Å/S) and high-internal stress films from sputtering [5], overly thick films (commonly > 1 mm) from casting, and coarse-grained films from evaporation [6]. In addition, most of the reported materials have limited B_s of ≤ 2.1T (e.g., NiFe, FeCoNi films) [3]. Despite CoFe alloy possessing the highest B_s of about 2.45T theoretically, the CoFe nano/microfilms prepared always encounter two vital issues for application: poor magnetic properties [such as poor anisotropy, high value of H_c (always > 4 Oerted (O_e))] and poor mechanical properties (such as rough surface, and cracking film) [7]. Electrodeposition is a fast and simple method to achieve various thicknesses of soft Fe-based materials with lower stress [8, 9]. However, preventing Fe^{2+} from oxidization into Fe^{3+} (which decreases B_s), and minimizing the H_c and roughness of the metal films are challenges in electrodeposition [3, 7]. So far, only the softest magnetic CoFe film, which possesses an easy axis H_c (H_{ce}) and hard axis H_c (H_{ch}) of 15 and 4.8 O_e, respectively, has been reported [7].

Herein, we demonstrate a novel approach to the fabrication of ultrasoft magnetic CoFe films via electrodeposition from a sulfate salt-based solution containing dimethylamine borane [(CH3)2NHBH3] (B-reducer, Bayer AG, Germany). In comparison with commonly used strong acidic solutions (corroding devices) which contain S-additives (saccharin, Na Lauryl sulfate), the solution containing B-reducer is a semineutral medium with larger process latitude. For instance, the pH range (3.5–5.1) is much wider than that of the common plating solutions containing S-additives (pH=2.3–2.5 n or 2.5–3.0) [7, 9]. Furthermore, the plating solution does not require a salt bridge to protect Fe2+ from oxidization.

The thickness of the prepared films measured through atomic force microscopy (AFM) ranges from tens of nanometers to micrometers. The magnetic characterizations performed by using vibrating sample magnetometer (VSM) and superconducting quantum interference device (SQUID) show that the films are very soft magnets with good anisotropy and high magnetic moment properties.

Method and Results

The CoFe films were electrodeposited on Si(100) wafers. A seed layer, for example, Au, Cu, or CoFe, with thickness of 20–30 nm was sputtered onto each wafer surface as an electrical conducting layer for the electrodeposition. The Fe_xCo_{100-x} n (x=55–68) films were fabricated at the temperature of 40±2°C from a solution of 0.07 mol/L $CoSO_4 \cdot 7H_2O$, 0.10 mol/L $FeSO_4 \cdot 7H_2O$, 0.5 mol/L NH_4Cl, 0.5 mol/L H_3BO_3, and 2.8–3.2 g/L B-reducer additive. The electroplating system used was a Paddle cell with pulse DC power. During the electrodeposition of CoFe film, a magnetic field of 280 Oe was applied parallel to the substrate surface. The atomic ratio of Co to Fe in the electrodeposited films was determined by energy dispersive X-ray (EDX, JSM-6340F, JEOL Asia). While measuring H_c with the VSM possessing Helmoltz coils, high resolution of 0.01 Oe was obtained through the field control mode. The maximum relative permeability (μm) was calculated from easy axis M-H loops according to the formula μm=B(T)/H(T)≈4π magnetic moment (M, emu)/[magnetic field (H,Oe) × film volume (cm^3)], where (M,H) refers to the turning point in the lower branch of the easy axis M-H loop. It is noted that the effect of film sample shape (normally, 1.20 cm × 1.20 cm × ~0.6 μm) can be neglected due to the length and width being much larger than the thickness [10, 11]. The Bs values were calculated from the formula $B_s(T)=\mu_0 M_s=4\pi\times10^4 M$ (emu)/film volume (cm^3), where M refers to the saturation moment in the M-H loops. The B_s result is the average value obtained from a series of CoFe films having different thickness (80 nm -2.5 n μm). The dimension of the films was measured on ADM-60 Micro-Dicer. The total error of Bs measurement is 0.01T.

Table 1 shows that the B-reducer dramatically induces the decrease of Hc and the increase of Bs with its addition increasing from 0 to 100 mL/L. It is due to the fact that despite the absence of salt bridge (leading to a lower efficiency of plating current) to prevent Fe2+ from oxidization in the solution, the B-reducer can protect the as-synthesized CoFe film effectively from oxidization since B-reducer, acting as a reductant molecule, can reduce Fe3+ (produced during plating) back into Fe2+:

$$\begin{aligned} & Fe^{2+} - e \xrightarrow{O_2 \ or \ at \ node} Fe^{3+} \left(by - reaction\right) \\ & Fe^{3+} + e \xrightarrow{B-reducer} Fe^{2+}. \end{aligned} \tag{1}$$

Table 1: Property comparison of CoFe films (without annealing).

Property	No additive	B-reducer (1.5 g/L)	B-reducer (3 g/L)	S-additive[7]
H_{ce} (Oe)	63	6.6	6.5	13
H_{ch} (Oe)	53	5.8	2.5	4.8
H_k (Oe)	78	26	45	20.2
μ_m	7.8×10^1	4.5×10^4	5.8×10^4	—
B_s (T)	1.42	2.31	2.45	2.45

Despite the addition, the B-reducer did not lead to too much boron doping into the CoFe film. Hence, there is a very low content of oxygen and boron present in the deposited CoFe films, in which the low oxygen content cannot be detected by EDX, while the boron content is less than 0.8% after XPS analyzing. For the Co40Fe60 film deposited from a current density of 6.0 mA/cm2, with the concentration of B-reducer increasing from 0.0 to 3.0 g/L in the electroplating solution, the Hce and Hch values decrease from 63 and 53 to 6.5 and 2.5 Oe, respectively, while Bs increases from 1.4 to 2.45T in average. On the other hand, the permeability (μm) of CoFe film also increases from the order of 101 to 104 after the addition of B-reducer. Thus, the film becomes very soft due to the effect of B-reducer. Figure 1 shows the magnetic moment–applied field curves (M-H loops) of the CoFe films from the solutions containing different concentrations of B-reducer. It appears that the loops along the easy and hard axes change from being similar [such as in Figure 1(a)] to distinctly different [shown in Figure 1(c)]. Hence, the ultrasoft and high magnetic moment of Co40Fe60 film possess a much larger anisotropy compared to the softest CoFe films reported to date [7]. Further investigations revealed that although similar effects of B-reducer had also been observed for other components of CoFe films by varying the plating current density, only the films with formulas from Co35Fe65 to Co43Fe57 could achieve a high Bs of ≥2.4T, and such films were electrodeposited at a current density ranging between 4.8 and 7.2 mA/cm2.

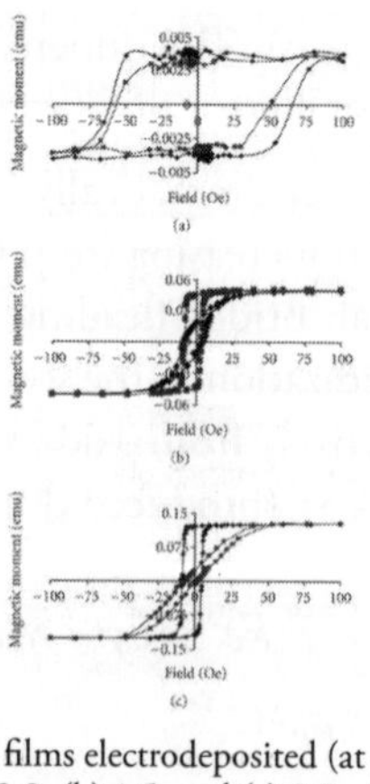

Figure 1: The M-H loops of the $Co_{40}Fe_{60}$ films electrodeposited (at a current density of ~6.0 mA/cm²) from the solutions which contain B-reducer of (a) 0.0, (b) 1.5, and (c) 3.0 g/L, respectively.

More measurements were carried out to identify the mechanism for the effect of B-reducer on the properties of CoFe films. In Figure 2, the AFM and FESEM images show the surface variations of the deposited Co40Fe60 films with the increase of B-reducer concentration from 0 to 1.5, then to 3.0 g/L. The sizes of CoFe nanoparticles become much smaller (roughly from 80 to 10–30 nm) with the increase of B-reducer concentration in the electroplating solution. Thus, for the films, Hc drops and μm increases drastically since the average magnetocrystalline anisotropy and the exchange coupling range become lower and wider, respectively, with the smaller ferromagnetic nanoparticles [10, 11]. In addition, X-ray diffraction (XRD) analysis (recorded from a powder sample) in Figure 3 shows that the added B-reducer eliminates the foreign Fe2O3-phase (viz. protects Fe2+ from oxidization), which leads to a texture-structural change from CoFe(110) and Fe2O3(209) polycrystalline to CoFe(110) single-crystalline for the Co40Fe60 films. As a result, the film possesses very low Hc [11] and high theoretical Bs value (~2.45T) [7]. Further investigation results revealed that too much of B-reducer (>10 g/L) led to an unstable plating solution and a high content of boron doping (>2%) in the CoFe films. The films possess an amorphous texture, a decreased Bs, and poor anisotropy.

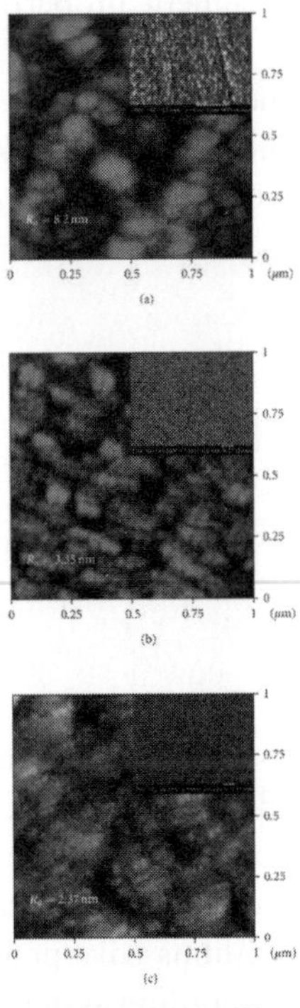

Figure 2: The AFM and FESEM (inset) images of the $Co_{40}Fe_{60}$ films electrodeposited (at a current density of ~6 mA/cm^2) from the solutions which contain B-reducer of (a) 0, (b) 1.5, and (c) 3.0 g/L, respectively. The scale bar in the inset (a)–(c) is 1 μm, 100, and 100 nm, respectively.

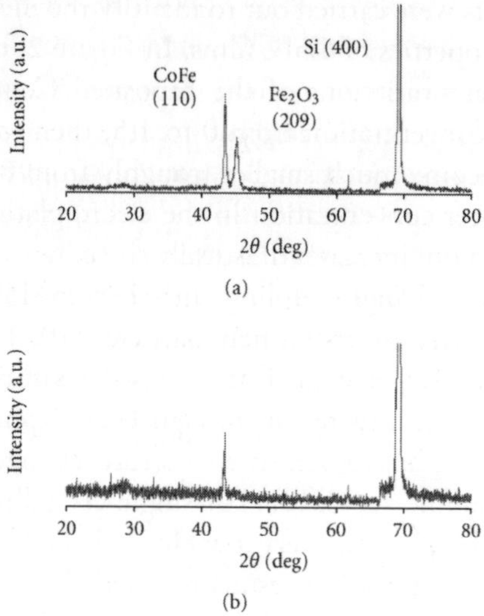

Figure 3: The 2θ XRD results of ~0.6 μm Co40Fe60 films on Si(100) wafer. (a) and (b) stand for without adding and adding B-reducer, respectively.

Apart from having excellent magnetic properties, our characterizations shown in Table 2 indicate that the CoFe films also displayed good mechanical and other properties, such as very low magnetostriction (λs) and surface roughness (Ra, also depicted in Figure 1), and absence of microcracking defects, which is a critical problem faced by current CoFe electrodeposition [7, 12]. These properties also contribute to low Hc of the films [10].

Table 2: Properties of soft $Co_{40}Fe_{60}$ films.

Film	H_{ce} (Oe)	H_{ch} (Oe)	B_s (T)	λ_s ($\times 10^{-6}$)	Surface	R_a (nm)
Before anneal	6.5	2.5	2.45	9.2	Shining	2.4
After anneal	5.6	1.8	2.43	3.4	Shining	2.2

Note: the annealing was conducted at 200°C for 4 hours.

Conclusion

The two challenging issues (poor magnetic and mechanical properties), encountered in the preparation of high-B_s magnetic CoFe films [7], can simply be solved by using boron reducer in electrodeposition. The boron reducer can greatly improve the softness, magnetic moment, mechanical, and other properties of the CoFe films. Thus, the as-prepared films have potential applications in ultrahigh density and frequency magnetic recording system, biotechnological, and microelectronic devices [2, 4].

Acknowledgement

The authors would like to thank Mr. J. F.Chong for XRD measurement.

References

1. S. R. Brankovic, X. Yang, T. J. Klemmer, and M. Seigler, "Pulse electrodeposition of 2.4T Co37Fe63 alloys at nanoscale for magnetic recording application," IEEE Transactions on Magnetics, vol. 42, no. 2, pp. 132–139, 2006.
2. P. C. Andricacos and N. Robertson, "Future directions in electroplated materials for thin-film recording heads," IBM Journal of Research and Development, vol. 42, no. 5, pp. 671–680, 1998.
3. T. Osaka, M. Takai, K. Hayashi, K. Ohashi, M. Saito, and K. Yamada, "A soft magnetic CoNiFe film with high saturation magnetic flux density and low coercivity," Nature, vol. 392, no. 6678, pp. 796–798, 1998.
4. Ultra-soft magnetic films for high-frequency inductors (GWN.4561), http://www.stw.nl/Projecten/G/gwn/gwn4561.htm.
5. S. X. Wang, N. X. Sun, M. Yamaguchi, and S. Yabukami, "Sandwich films: properties of a new soft magnetic material," Nature, vol. 407, no. 6801, pp. 150–151, 2000.
6. W. Yu, J. A. Bain, Y. Peng, and D. E. Laughlin, "Magnetization reduction due to oxygen contamination of bias sputtered Fe35Co65 thin films," IEEE Transactions on Magnetics, vol. 38, no. 5, pp. 3030–3032, 2002.
7. E. I. Cooper, C. Bonhôte, J. Heidmann, et al., "Recent developments in high-moment electroplated materials for recording heads," IBM Journal of Research and Development, vol. 49, no. 1, pp. 103–126, 2005.
8. B. Zong, Y. Wu, G. Han, et al., "Synthesis of iron oxide nanostructures by annealing electrodeposited Fe-based films," Chemistry of Materials, vol. 17, no. 6, pp. 1515–1520, 2005.
9. M. Kawasaki and Y. Kanada, "Soft magnetic film having high saturation magnetic flux density, thin-film magnetic head using the same, and manufacturing method of the same," US patent no. 6765757, August 2002.
10. G. Herzer, "Soft magnetic nanocrystalline materials," Scripta Metallurgica et Materialia, vol. 33, no. 10-11, pp. 1741–1756, 1995.
11. Q. Zeng, I. Baker, V. McCreary, and Z. Yan, "Soft ferromagnetism in nanostructured mechanical alloying FeCo-based powders," Journal of Magnetism and Magnetic Materials, vol. 318, no. 1-2, pp. 28–38, 2007.

12. F. Lallemand, L. Ricq, E. Deschaseaux, L. De Vettor, and P. Berçot, "Electrodeposition of cobalt-iron alloys in pulsed current from electrolytes containing organic additives," Surface & Coatings Technology, vol. 197, no. 1, pp. 10–17, 2005.

Determination of Differential Enthalpy and Isotherm by Adsorption Calorimetry

V. Garcia-Cuello, J. C. Moreno-Piraján, L. Giraldo-Gutiérrez, K. Sapag and G. Zgrablich

ABSTRACT

An adsorption microcalorimeter for the simultaneous determination of the differential heat of adsorption and the adsorption isotherm for gas-solid systems are designed, built, and tested. For this purpose, a Calvet heat-conducting microcalorimeter is developed and is connected to a gas volumetric unit built in stainless steel to record adsorption isotherms. The microcalorimeter is electrically calibrated to establish its sensitivity and reproducibility, obtaining $K=154.34\pm0.23\ WV^{-1}$. The adsorption microcalorimeter is used to obtain adsorption isotherms and the corresponding differential heats for the adsorption of CO_2 on a reference solid, such as a NaZSM-5 type zeolite. Results for the behavior of this system are compared with those obtained with commercial equipment and with other studies in the literature.

Introduction

It is widely accepted that the knowledge of adsorption heats is vital in the description of gas-solid interaction. This is particularly useful when adsorption heat measurements are combined with simultaneous measurement of the adsorption isotherm. These measurements obviously may provide information about the energetic of surface processes. In some simple cases, even information on the structure of the surface itself, like for example the energetic topography, can be retrieved from adsorption heats and isotherms [1, 2]. Chemisorption and catalyzed reactions, like any chemical reaction, are associated with changes of enthalpy and can therefore be studied by means of calorimeters. Many calorimeters, operating on different principles, have indeed been used for this purpose [3–5]. Adsorption calorimeters are particularly convenient for these studies [4]. They offer a number of advantages which will be illustrated by means of selected examples.

Adsorption calorimetry, preferably in association with other physicochemical or physical techniques, may be used to describe the surface properties of a solid.

Information on the binding energy, deduced from calorimetric data, is needed to achieve a theoretical description of the adsorbate-adsorbent bond. It has been shown, for instance, that, in the case of the adsorption of hydrogen on nickel-copper alloys, a correlation between heats of adsorption and surface magnetic properties can be found. The correlation indicates that the energy of the bond between adsorbed hydrogen and nickel atoms is regulated by the electron density of states, near the Fermi level, for the metal surface [6–8].

In these works, we present the design, construction, and test of an adsorption microcalorimeter capable of measuring simultaneously adsorption isotherms and heats.

Experimental

Description of the New Microcalorimeter

Figure 1 shows a complete view of the adsorption calorimeter built here, which is not very common and has not been considered in the literature.

A detailed general view of the equipment calorimeter is shown in Figure 1. The diagram shows microcalorimeter with the calorimetric cells made of stainless steel (sample and reference), which are embedded inside a large block (also divided in two parts) in stainless steel, which acts as deposit of the thermostatic liquid. Due to its thermal diffusion coefficient, this set allows the rapid heat conduction towards the surrounding of the calorimeter. The whole set is placed inside a nylon block to isolate it from the surroundings and to allow the rapid stabilization of the temperature. The thermal effects are sensed through ten thermopiles and trademark Melcor Corporation, NJ, USA, connected in series to increase the sensitivity of the microcalorimeter. The microcalorimeter designed in this work connected to the adsorption system constructed specially for this equipment in stainless steel to allow the simultaneous measurement of the heat of adsorption and the isotherm. The connection is through two pressure transducers, one in the range of high pressure (1000 Torr), and the other in the range of low pressure (10 Torr) (see Figure 1).

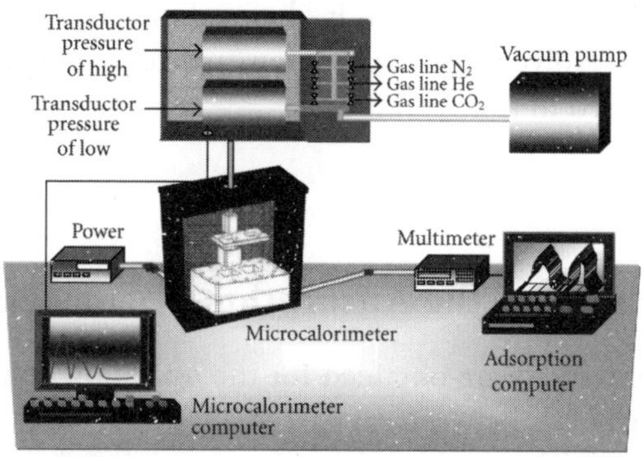

Figure 1: General scheme of station for the simultaneous measurement of isotherm and heat of adsorption.

Electric Calibration of the Adsorption Microcalorimeter

In order to establish the correct functioning of the microcalorimeter, which is then connected to the volumetric adsorption unit, the sensitivity is evaluated determining the calorimeter constant. The calibration constant reports the voltage generated by the calorimeter when a heat flow is emitted from inside the microcalorimetric cell. There are two methods to determine the calibration constant K: by application of electric power and by the stationary method [9, 10].

Description of the Unit for Simultaneous Measurement of Isotherm and Adsorption Heat

Heats of adsorption have been measured at 273 K by means of the adsorption microcalorimeter and by contacting the solid with small successive doses of the adsorptive. This allows the evolution of the interaction energy along with the coverage to be measured. In the system, an ultra-high vacuum pump (Pfeiffer Vacuum Ref. TSH 071E) is previously connected to an oil rotary pump which initially allows to have a previous vacuum in all the system. Once the system has a pressure of about 10–3 Torr, the ultra-high vacuum pump starts working and is kept functioning until the pressure reaches at least 10–5 Torr. This part of the station is also composed of a joint built all in one unit, all in stainless steel, which was previously calibrated and was specially designed to obtain precise and accurate measurements. This is a novel contribution to the research equipments normally used in this type of measurements where this part consists of equipment constructed in glass with the problems associated with it. The cell containing the sample is also shown as well as the pirani pressure transductor which is connected to a computer through an interface RS-232.

The differential heat of adsorption is obtained directly from the calorimetric, measuring the heat evolved, as small increments of adsorbate are added. This method is the one used in this work.

Results

Electric Calibration of the Adsorption Microcalorimeter

The calibration constants were obtained for the operation conditions of the microcalorimeter. Constants between range 134.11 ± 0.19 WV^{-1} to 156.67 ± 0.23 WV^{-1} are determined. These values show the sensitivity of the microcalorimeter built here, which is higher than that of equipments reported in literature and even of those built in our laboratory previously. This constitutes a significant contribution to the construction of this type of instruments. Values by the method of state stationary condition were obtained and were of the order same and magnitude.

Isotherms and Differential Heats of Adsorption

Table 1 reports the characterization results obtained with the equipment built here for the probe sample, type NaZSM-5 zeolite, previously characterized in an Autosorb Quantachrome 3B equipment. The superficial characteristics and microporosities obtained with the two equipments are compared.

Table 1: Comparison of principal superficial characteristics of NaZSM-5 zeolite.

NaZSM-5 ZEOLITE	Quantachrome 3B™	Microcalorimeter built in this work
$S_{\mu p}$-DRK-method (m^2/g)	285	296
$V_{\mu p}$-DRK-method (cm^3/g)	0.24	0.26

These values are evaluated from the adsorption of CO2 at 298 K. The results show a very good agreement between the commercial equipment and the microcalorimeter built here, reinforcing the excellent functioning of this equipment. Figure 2 shows the adsorption isotherm of CO2 at 273 K obtained for the zeolite analyzed in this investigation. This isotherm was reproduced also on the commercial equipment with a good concordance, reinforcing again the satisfactory behavior of our apparatus.

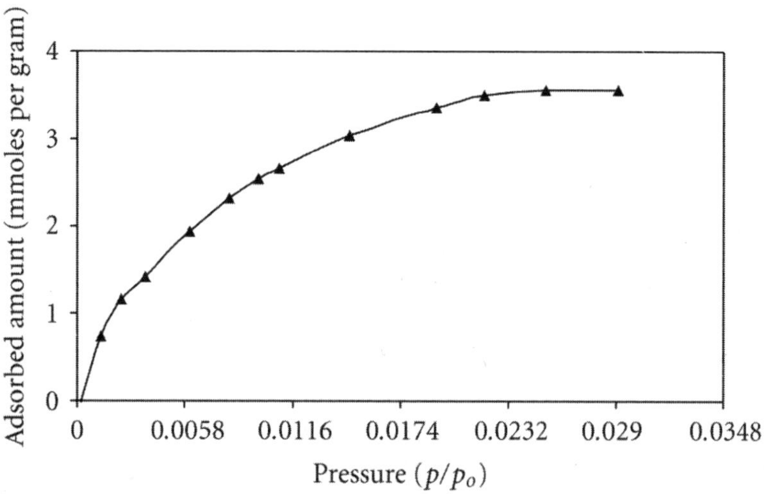

Figure 2: Adsorption isotherm for zeolite in CO_2 at 273 K.

It is interesting to analyze jointly the data obtained from the adsorption isotherm (see Figure 2) and those for the differential heat of adsorption (see Figure 3). In Na exchanged ZSM-5 zeolites, Na cations neutralize the acidity of the zeolite and develop the basicity for adsorbing acidic CO2. Thus, NaZSM-5 provides two kinds of adsorption sites for CO2: stronger sites around a Na cation (which saturates rapidly) and weaker sites on the pore walls [11]. The steep increase of the adsorption isotherm and the high value of qd at low pressure (p/po < 0.01) in Figures 2 and 3 reveal in a clear way the presence of these strong sites, which become rapidly saturated. Note also that the steep decrease in qd at very low pressure from 48 to 46 kJ/mol is indicating that the adsorption strength of these sites is not uniform (indication of energetic heterogeneity). After the strong sites become saturated (p/po > 0.01), qd steps down to a lower and almost constant value, corresponding to adsorption on the zeolite walls, and simultaneously the amount adsorbed increases more slowly in the isotherm. This is in [11] and shows the potentiality of the microcalorimetric station.

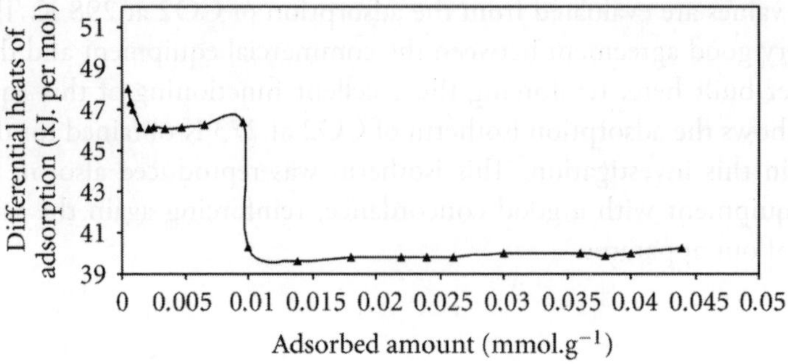

Figure 3: Differential heats of adsorption for CO_2 on NAZSM-5 zeolite.

Conclusions

A modern adsorption microcalorimeter was built for the simultaneous measurement of isotherms and adsorption heats, establishing its correct functioning through adequate calibration of both the calorimeter part and the volumetric equipment of the adsorption part. For this purpose, the microcalorimeter calibration constant was found with values that go from 134.11 ± 0.19 WV^{-1} to 156.67 ± 0.23 WV^{-1}. The adsorption isotherm was determined for a type NAZSM-5 zeolite as a reference solid to establish the correct functioning of the equipment. Micropore volume and superficial area were determined to be 0.20 cm^3/g and 296 m^2/g, respectively. These results agree very well with those obtained with commercial equipment. Finally, the differential heats of adsorption, for the same solid, were measured. The analysis of results gives valuable information about the studied CO_2/NaZSM-5 system, which is in concordance with other studies in the literature.

Acknowledgements

The authors thank the Departments of Chemistry of National University of Colombia, University of the Andes (Colombia), and Universidad Nacional de San Luis (Argentina), and the Master Agreement established between these institutions. Special gratitude is due to Fondo Especial de Investigaciones de la Facultad de Ciencias de la Universidad de Los Andes (Colombia) for its partial financing. Dr. Diana Azevedo is also kindly acknowledged for profiting discussions on the characteristics of the CO_2/NaZSM-5 system. One of the authors, G. Zgrablich, thanks CAPES (Brazil) for a Visiting Professor Fellowship at UFC.

References

1. W. Rudzinski, W. A. Steele, and G. Zgrablich, Eds., Equilibria and Dynamics of Gas Adsorption on Heterogeneous Solid Surfaces, W. Rudzinski, W. A. Steele, and G. Zgrablich, Eds., Elsevier, Amsterdam, The Netherlands, 1997.
2. F. Bulnes, A. J. Ramirez-Pastor, and G. Zgrablich, "Scaling behavior of adsorption on patchwise bivariate surfaces revisited," Langmuir, vol. 23, no. 3, pp. 1264–1269, 2007.
3. P. C. Gravelle, "Calorimetry in adsorption and heterogeneous catalysis studies," Catalysis Reviews, vol. 16, no. 1, pp. 37–110, 1977.
4. P. C. Gravelle, "Heat-flow microcalorimetry and its application to heterogeneous catalysis," Advances in Catalysis, vol. 22, pp. 191–263, 1972.
5. A. Auroux, J. C. Vedrine, and P. C. Gravelle, in Adsorption at the Gas-Solid and Liquid-Solid Interface, J. Rouquerol and K.S.W. Sing (Eds), pp. 305–322, Elsevier, Amsterdam, The Netherlands, 1982.
6. J. J. Prinsloo and P. C. Gravelle, "Volumetric and calorimetric study of the adsorption of hydrogen, at 296 K, on silica-supported nickel and nickel-copper catalysts," Journal of the Chemical Society, Faraday Transactions 1, vol. 76, pp. 2221–2228, 1980.
7. A. Auroux and P. C. Gravelle, "Comparative study of the bond energy of oxygen at the surface of supported silver catalysts and of the activity of these catalysis for ethylene epoxidation," Thermochimica Acta, vol. 47, no. 3, pp. 333–341, 1981.
8. P. C. Gravelle and S. J. Teichner, "Carbon monoxide oxidation and related reactions on a highly divided nickel oxide," Advances in Catalysis, vol. 20, pp. 167–266, 1969.
9. J. C. Moreno, Diseño, construcción, calibración y aplicación de un Microcalorímetro de conducción de calor, Ph.D. Thesis, National University of Colombia, Bogotá, Colombia, 1996.
10. L. Giraldo, Diseño de un Microcalorímetro de flujo tipo gemelo y su aplicación el estudio de las interacciones alcoholes en solución acuosa, Ph.D. Thesis, National University of Colombia, Bogotá, Colombia, 1996.
11. S. K. Wirawan and D. Creaser, "CO2 adsorption on silicalite-1 and cation exchanged ZSM-5 zeolites using a step change response method," Microporous and Mesoporous Materials, vol. 91, no. 1–3, pp. 196–205, 2006.

Comparison of the Molecular Dynamics of C_{70} in the Solid and Liquid Phases

R. M. Hughes, P. Mutzenhardt and A. A. Rodriguez

ABSTRACT

A previous study of C_{70} in deuterated benzenes generated evidence suggesting C_{70} exhibited unique reorientational behavior depending on its environment. We present a comparison of the dynamic behavior of this fullerene, in the solid and solution phases, to explore any unique features between these two phases. The effective correlation times, τ_C^{eff}, of C_{70} in the solid state are 2 to 3 times longer than in solution. In the solid state, a noticeable decrease in all the carbons' correlation times is seen between 293 K to 303 K; suggesting a transition from isotropic to anisotropic reorientational behavior at this temperature change. Although C_{70} in solution experiences van der Waals type interactions, these interactions are not strong enough to slow the solution-state motion below what is observed in the solid state. All observed differences in the *diffusion constants, D_X and D_Z, in solution are smaller than in the solid state suggesting a lower energy of activation between these two modes of reorientation in the liquid phase. A small-step diffusion "like" condition appears to be thermally generated in the solid phase at 323 K.*

Introduction

Molecular reorientational dynamics are sensitive to several factors including temperature, pressure, free volumes, solute-solvent interactions, and, henceforth, are commonly employed to probe a solute's rotational energetics or the effects of the immediate environment on the rotational motion. While fullerene C_{60}'s molecular dynamics have been studied extensively, its close family member C_{70} has received limited attention [1–10]. Early solid-state NMR measurements indicated that C_{70}'s reorientational motion was anisotropic near room temperature [9]. Later differential scanning calorimetry measurements on solid C_{70} indicated that C_{70}'s dynamic behavior dependent on the thermally stable solid-phase structure (e.g., fcc, rhomobohedral, or monoclinic phase) at a given temperature while zero-field measurements suggested C_{70} molecules rotated isotropically [11]. In a later study, Tycko and coworkers found that solid C_{70} molecules reorientational motion was anisotropic between 223–330 K (monoclinic and rhombohedral phases) but became isotropic at temperatures beyond 330 K (face-centered phase) [10]. In terms of molecular dynamics in solution, our recent investigation into the molecular dynamics of C_{70} in several deuterated benzenes revealed evidence suggesting C_{70}'s reorientational motion oscillated between anisotropic and isotropic behavior depending on the solvent and the temperature [12]. In order to explore any unique features of the dynamic behavior of C_{70} in the solid and liquid phase, we present a comparison of our solution measurements to values obtained in the solid state [10].

Theoretical Background

^{13}C Spin-Lattice Relaxation

^{13}C spin-lattice relaxation in fullerenes, both in the solid and liquid-phase, is known to be dominated by the chemical shift anisotropy, R_1^{CSA} (CSA). The spin-rotation mechanism, R_1^{SR} (SR), may also contribute in the liquid phase if the temperature is above 320 K [13–16]. Therefore, the overall ^{13}C spin-lattice relaxation rate in C_{70} can be expressed as sum of these two mechanisms [17]:

$$R_1 = \frac{1}{T_1} = R_1^{CSA} + R_1^{SR} \tag{1}$$

While it is possible for the spin-rotation mechanism to be present in the solid state, (1) can be reduced to R1=R1CSA at high magnetic fields and at moderate temperatures. Under extreme narrowing arguments, and assuming axial symmetry of the chemical shift tensor (CST), the CSA relaxation process is described by [18]

$$R_1^{CSA} = \left(\frac{2}{15}\right)\gamma_C^2 B_0^2 \Delta\sigma^2 \tau_C^{eff} \tag{2}$$

where γ_C is the carbon magnetogyric ratio, B_0 represents the field strength, $\Delta\sigma$ is the CSA, and τ_C^{eff} is the effective reorientational correlation time. The CSA is obtained from the three principal components (i.e., $\sigma_{zz} \geq \sigma_{yy} \geq \sigma_{xx}$) of the CST as defined by (3):

$$\Delta\sigma = \sigma_{zz} - 0.5\left(\sigma_{xx} + \sigma_{yy}\right) \tag{3}$$

Although solid-state NMR measurements can allow the determination of $\Delta\sigma$, the approach can be experimentally challenging and is primarily limited to chemical systems with high symmetry. In fact, Tycko and coworkers experienced these challenges when attempting to evaluate the CST for the carbons in C70 [10]. The relative high noise in their measurements only allowed for an average value of 200 ppm to be determined for each of the carbons in C70. Alternatively, $\Delta\sigma$ can be computed with a high degree of accuracy using standard computational methods [19–26]. The $\Delta\sigma$ values used in this work were calculated using a Gaussian software package [27], employing the B3LYP exchange-correlation energy density functional, the 6-31G* basis set [28, 29] and employing the gauge-independent atomic orbital (GIAO) method [30–33]. The calculated $\Delta\sigma$ values are given in Table 1. It is worth noting the similarity of $\Delta\sigma$ in C70 to the value in C60 (178 ppm) [15], suggesting that anisotropies in fullerenes fall within this range. Substitution of (2) into (1) gives

$$R_1 = \left(\frac{2}{15}\right)\gamma_C^2 B_0^2 (\Delta\sigma)^2 \tau_C^{eff} + R_1^{SR} \tag{4}$$

In the liquid state, a fit of the overall relaxation rate against two or more field strengths (B02) allows the separation of the CSA from the SR contribution which then leads to the determination of τCeff.

Table 1: Average values for the calculated tensor components, chemical shift anisotropies, and the chemical shift tensor (CST) orientation of carbons in C_{70}.

Carbon	σ_{xx} (ppm)	σ_{yy} (ppm)	σ_{zz} (ppm)	$\Delta\sigma$ (ppm)	θ (degrees)
1	−33.56	−0.97	162.06	179.32	13.94
2	−24.35	0.04	164.65	176.81	51.99
3	−32.24	3.04	165.95	180.55	62.15
4	−22.30	−2.62	167.19	179.65	72.88
5	−8.62	11.56	181.77	180.30	90.00

Reorientational Dynamics

Reorientation dynamics in liquids is described by either diffusion constants, D_i, or reorientational correlation times, τ_C, since these two parameters are closely correlated. D_i is the diffusion rate about a given molecular axis while τ_C is the time period required for the angular correlation function to decay to 1/e of its initial value [34, 35]. For symmetric-top molecules, such as C_{70}, two diffusion constants, D_Z and D_X, are usually required to characterize the overall motion. D_Z and D_X represent rotational diffusion about and of the top axis, respectively. The overall motion is now characterized by an effective reorientational correlation time, τ_C^{eff}, that, in the limit of small-step diffusion, is given by [36]

$$\tau_C^{eff} = \frac{0.25\left(3\cos^2\theta - 1\right)^2}{6D_X} + \frac{3\sin^2\theta\cos^2\theta}{5D_X + D_Z} + \frac{0.75\sin^4\theta}{2D_X + 4D_Z} \tag{5}$$

For CSA relaxation, θ is the orientation of the CST tensor relative to the molecular symmetry axis. In principle it is possible to determine DZ and DX for a symmetric-top molecule provided τCeff and θ values are known for different nuclei in the molecule. We employed the solution and solid-state experimental correlation times, along with the Gaussian generated θ values, in (5) to simultaneously solve an equation for each carbon and obtained the best-fit values for DZ and DX at each temperature. While (5) assumes a small-step diffusion process, one can employ this equation to characterize reorientational motion outside this limit provided the calculated values are viewed as rough or base-line estimates of the diffusion process.

Experimental Methods

Benzene-d_6, chlorobenzene-d_5, o-dichlorobenzene-d_4, and C_{70} were purchased from the Aldrich Chemical Company, USA, and were used as received. Solution and solid-state ^{13}C NMR spectra of C_{70} showed the 5 unique carbon resonances at approximately 151 ppm, 147 ppm, 148 ppm, 146 ppm, and 131 ppm. These carbon resonances correspond to the carbons labeled in Figure 1.

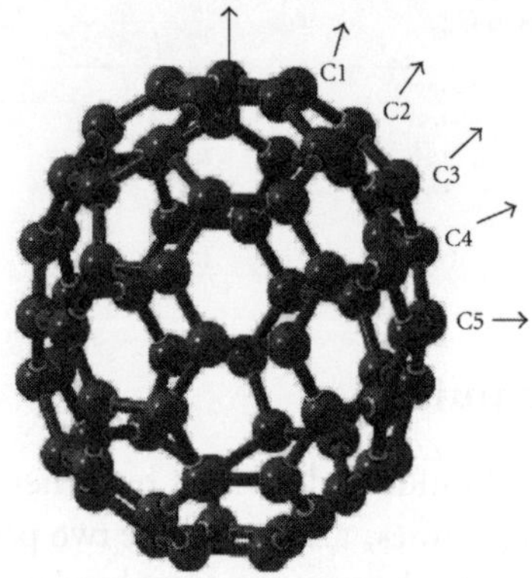

Figure 1: Carbon assignments in C_{70}. Carbon 1 (151 ppm), carbon 2 (146.5 ppm), carbon 3 (147.5 ppm), carbon 4 (145.5 ppm), and carbon 5 (131 ppm). The approximate carbon CSA tensor orientations (θ), relative to the molecular symmetry axis, are indicated by the various arrows.

In solution, resonances for carbons 1–4 were only used for the analysis since slight solvent peak interference and weak peak intensity of the carbon 5 in chlorobenzene-d5 prohibited the inclusion of this peak across all solvents. Optimum room temperature concentrations of C70 in the solvents were calculated from published data [37]. To eliminate the potential for C70 clustering, solutions were prepared with a very low mole fraction of 1.0×10–4. Samples were contained in 5 mm tubes, connected to a vacuum line and thoroughly degassed by several freeze-pump-thaw cycles to remove molecular oxygen. The tubes were then sealed under vacuum.

Solution 13C spin-lattice relaxation measurements were performed on instruments operating at 11.75 and 7.05 Tesla. Experiments were conducted at five different temperatures (283 K, 293 K, 303 K, 313 K, and 323 K). Temperature accuracy for these measurements is ±0.1 K. Relaxation times were obtained using the standard inversion-recovery pulse sequence as described in our earlier work [1–5]. Relaxation times in the various solvents are shown in Tables 2, 3, and 4.

Table 2: Experimental ^{13}C relaxation times for the different carbons at various temperatures for C_{70} in benzene-d_6. (Relaxation times measured at 11.7 Tesla.)

	Carbon 1	Carbon 2	Carbon 3	Carbon 4
T	T_1	T_1	T_1	T_1
(K)	(s)	(s)	(s)	(s)
283	15	27	27	28
293	25	28	28	29
303	29	30	31	30
313	24	24	25	23
323	18	17	17	15

Table 3: Experimental ^{13}C relaxation times for the different carbons at various temperatures for C_{70} in chlorobenzene-d_5. (Relaxation times measured at 11.7 Tesla.)

	Carbon 1	Carbon 2	Carbon 3	Carbon 4
T	T_1	T_1	T_1	T_1
(K)	(s)	(s)	(s)	(s)
283	18	21	27	23
293	22	27	28	25
303	26	29	31	28
313	29	39	37	33
323	25	29	35	29

Table 4: Experimental [3]C relaxation times for the different carbons at various temperatures for C_{70} in o-dichlorobenzene-d_4. (Relaxation times measured at 11.7 Tesla.)

	Carbon 1	Carbon 2	Carbon 3	Carbon 4
T	T_1	T_1	T_1	T_1
(K)	(s)	(s)	(s)	(s)
283	14	23	23	20
293	20	25	25	24
303	22	27	28	25
313	20	26	27	24
323	16	19	24	23

Solid-state relaxation times were obtained from the work of Tycko and co-workers and their temperature-dependent data, ranging from 223 to 342, was interpolated to the corresponding solution phase temperatures to permit the comparisons [10]. These relaxation times are given in Table 5.

Table 5: Experimental ^{13}C relaxation times for the different carbons at various temperatures for C_{70} in the solid state. (Values interpolated from [10].)

	Carbon 1	Carbon 2	Carbon 3	Carbon 4	Carbon 5
T	T_1	T_1	T_1	T_1	T_1
(K)	(s)	(s)	(s)	(s)	(s)
283	13	8	8	8	11
293	16	12	11	12	16
303	20	20	20	20	29
313	20	22	24	24	30
323	28	35	22	30	42

Discussion

The temperature behavior of the relaxation times of C_{70} in benzene-d_6, chlorobenzene-d_5, o-dichlorobenzene-d_4 and in the solid-state are illustrated in Figures 2, 3, 4, and 5; respectively. Tables 6, 7, 8, and 9 contain the reorientational correlation times of the various carbons, as well as the calculated diffusion coefficients, of C_{70} in the solid state, in benzene, chlorobenzene, and 1,2-dichlorobenzene, respectively.

Table 6: Experimental correlation times and predicted rotational diffusion constants for the different carbons at various temperatures for C_{70} in the solid state.

T (K)	Carbon 1 τ_C^{eff} (ps)	Carbon 2 τ_C^{eff} (ps)	Carbon 3 τ_C^{eff} (ps)	Carbon 4 τ_C^{eff} (ps)	Carbon 5 τ_C^{eff} (ps)	$D_X \times 10^{-9}$ (1/s)	$D_Z \times 10^{-9}$ (1/s)
283	45	75	72	73	52	2.2	2.5
293	37	52	50	51	36	3.4	3.6
303	29	30	29	29	20	5.4	7.5
313	29	27	24	24	19	5.5	9.9
323	20	17	26	19	13	8.0	10.5

Table 7: Experimental correlation times and predicted rotational diffusion constants for the different carbons at various temperatures for C_{70} in benzene-d_6.

T (K)	Carbon 1 τ_C^{eff} (ps)	Carbon 2 τ_C^{eff} (ps)	Carbon 3 τ_C^{eff} (ps)	Carbon 4 τ_C^{eff} (ps)	$D_X \times 10^{-10}$ (1/s)	$D_Z \times 10^{-10}$ (1/s)
283	25	14	13	13	0.6	2.9
293	14	12	12	12	1.2	1.7
303	11	11	10	10	1.5	1.9
313	8	8	7	7	2.0	2.7
323	7	7	5	6	2.4	3.5

Table 8: Experimental correlation times and predicted rotational diffusion constants for the different carbons at various temperatures for C_{70} in chlorobenzene-d_5.

T (K)	Carbon 1 τ_C^{eff} (ps)	Carbon 2 τ_C^{eff} (ps)	Carbon 3 τ_C^{eff} (ps)	Carbon 4 τ_C^{eff} (ps)	$D_X \times 10^{-10}$ (1/s)	$D_Z \times 10^{-10}$ (1/s)
283	20	17	13	15	0.8	1.6
293	13	10	12	13	1.4	1.6
303	9	9	9	10	1.8	1.8
313	7	8	6	7	2.3	2.6
323	6	6	4	5	2.6	3.9

Table 9: Experimental correlation times and predicted rotational diffusion constants for the different carbons at various temperatures for C_{70} in o-dichlorobenzene-d_4.

T (K)	Carbon 1 τ_C^{eff} (ps)	Carbon 2 τ_C^{eff} (ps)	Carbon 3 τ_C^{eff} (ps)	Carbon 4 τ_C^{eff} (ps)	$D_X \times 10^{-10}$ (1/s)	$D_Z \times 10^{-10}$ (1/s)
283	26.1	15.9	15.9	17.9	0.6	1.9
293	13.5	13.8	13.7	14.2	1.2	1.2
303	10.3	11.3	10.5	11.7	1.5	1.5
313	8.9	7.6	8.0	8.5	1.9	2.3
323	7.8	6.5	6.2	6.2	2.1	3.6

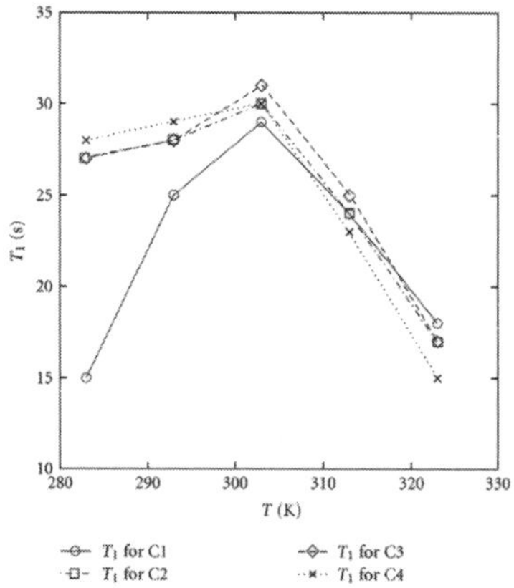

Figure 2: C13 relaxation times for C_{70} carbons at various temperatures in benzene-d_6.

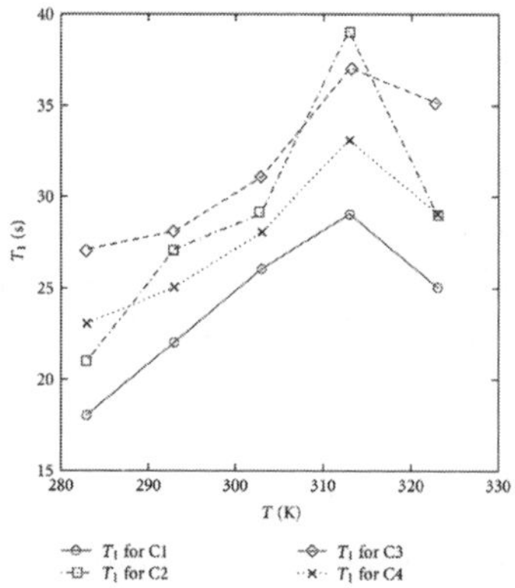

Figure 3: C13 relaxation times of C_{70} carbons at various temperatures in chlorobenzene-d_5.

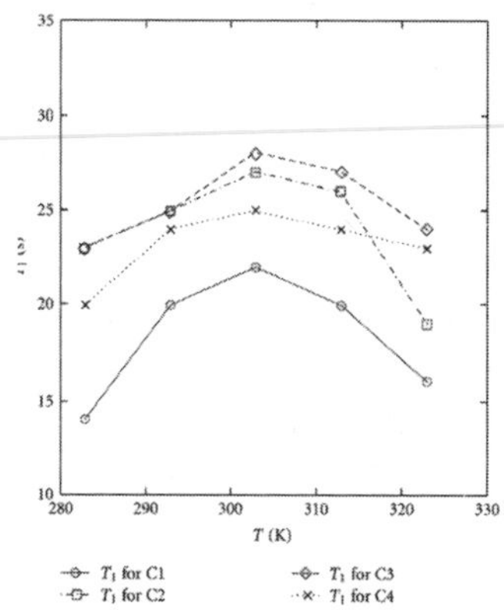

Figure 4: C13 relaxation times of C_{70} carbons at various temperatures in dichlorobenzene-d_4.

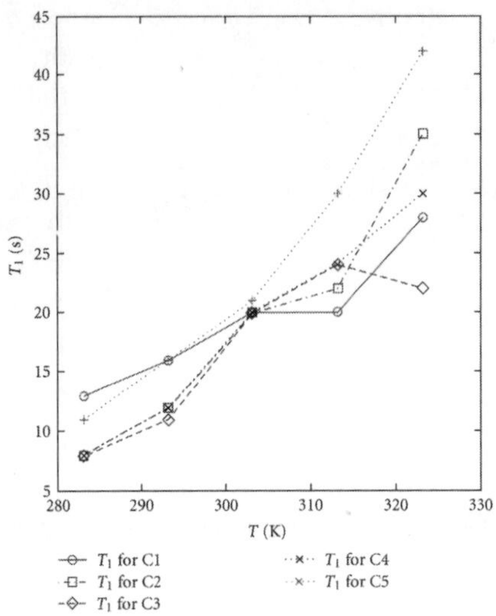

Figure 5: Relaxation times of carbons in C_{70} in the solid state.

The reorientational times for all carbons, both in the solid and in solution, are seen to systemically decrease with rising temperature indicating more rapid reorientational motion with escalating temperature. Unlike in C60, where faster rapid rotational motion is observed in the solid state than in solution, the τCeff values of C70 in the solid state are found to be from 2 to 3 times longer than in solution [14].

The slower rotational dynamics in solid C70, as compared to solid C60, can be attributed to differences in the orientational ordering of their lattice structures. Also of interest in the solid state is the noticeable decrease in all the carbons' correlation times in going from 293 K to 303 K. Noting the closeness of the two diffusion constants, DX and DZ, at the two lowest temperatures, which is indicative of a quasi-isotropic reorientation condition, the observed drastic drop in all the τCeff values suggest a definite transition from isotropic to anisotropic reorientational behavior at this temperature change. The enhanced difference between the two diffusion coefficients as temperature rises is particularly interesting since this correlates to the phase transition occurring from rhombohedral to face-centered cubic suggesting in increase in molecular disorder. We must however be cautious in overly interpreting this observation since diffusion values were calculated via (5) which assumes small-step diffusion and, as indicated by Tyco and coworkers, molecular reorientation in this phase is attributed to "thermally activated" orientational jumps rather than to step-wise diffusion. Nonetheless, the calculated diffusion values, and their differences, indicate preferred axial reorientation, becoming more pronounced with rising temperature.

As pointed out above, solution correlation times are significantly shorter than in the solid phase indicating faster reorientational motion in the liquid environment. Unlike the solid-state, C70 in solution is bathed in solvent molecules which, through van der Waals type interaction, experiences solute-solvent interactions which can affect the overall motion. However, these interactions do not appear to be strong enough to slow the solution-state motion below what is observed in the solid state. Additionally, at any common temperature, one observes that the differences between DX and DZ in solution are smaller than in the solid state suggesting that, although solvent displacement is needed in solution, the energetic difference between DX and DZ type motion in solution is lower than in the solid phase. It is interesting to note that diffusion values of solid-state C70 at 323K are comparable to values observed at 283K in solution suggesting that, at 323K, enough thermal energy is present to promote small-step diffusion "like" behavior in the solid phase.

A comparison of the diffusion coefficients, DZ and DX, for C70 in the various solvents generate some interesting observations. Since the value for DZ is slightly lower in the more viscous solvent of 1,2-dichlorobenzene-d4 than in benzene-d6 or chlorobenzene-d5, this observation suggests that the viscosity parameter is not the dominant factor giving rise to the observed spinning behavior of C70 in these solvents. The data suggest that, in chlorobenzene-d5 and 1,2-dichlorobenzene-d4, there is a balance between the strength of intermolecular forces and solvent structure which determines the reorientational behavior. Since the free volume is greater in 1,2-dichlorobenzene-d4, this suggests that the available free space is more important than intermolecular interactions in determining the spinning rate of C70 in these solvents. The tumbling motion, DX, is slowest in 1,2-dichlorobenzene-d4; consistent with the higher viscosity of this solvent. One must however be cautious of oversimplifying this observation since, as we saw for the other type of motion, other solvent-related factors are also present.

Conclusions

Reorientational times for all carbons, in solution and in the solid phase, decrease with rising temperature indicating faster rotational motion with escalating temperature. The effective correlation times, τ_C^{eff}, of C_{70} in the solid-state are 2 to 3 times longer than in solution. A comparison of the rotational dynamics of solids C_{60} and C_{70} indicates that C_{70} reorients slower which can be attributed to differences in the orientational ordering of their lattice structures. A noticeable decrease in all the carbons' correlation times is observed between 293 K to 303 K; suggesting a definite transition from isotropic to anisotropic reorientational behavior at this temperature change. Solution correlation times are seen to be significantly shorter than in the solid phase indicating faster reorientational motion in the liquid environment. Although C_{70} in solution experiences van der Waals type interactions, these interactions do not appear strong enough to slow the solution-state motion below what is observed in the solid state. All observed differences in the diffusion constants, D_X and D_Z, in solution are smaller than in the solid state suggesting a lower energy of activation between these two modes of reorientation in the liquid phase. A small-step diffusion "like" condition appears to be thermally generated in the solid phase at 323 K.

Acknowledgement

The authors are grateful to the National Science Foundation for support of this project under Grant CH-9707163.

References

1. V. K. Jones and A. A. Rodriguez, "Analysis of the reorientational motion of C60 in toluene-d8 at 303 K," Chemical Physics Letters, vol. 198, no. 3-4, pp. 373–378, 1992.
2. X. Shang, L. A. Fisher, and A. A. Rodriguez, "C13 spin-lattice relaxation and molecular dynamics of C60 in 1,2-dichlorobenzene-d4," The Journal of Physical Chemistry, vol. 100, no. 11, pp. 4361–4364, 1996.
3. J. A. Jones and A. A. Rodriguez, "Oxygen and solvent intermolecular effects on the C13 spin-lattice relaxation process of C60 in deuterated toluene at 303 K," Chemical Physics Letters, vol. 230, no. 1-2, pp. 160–164, 1994.
4. X. Shang and A. A. Rodriguez, "Anomalous relaxation and molecular dynamics of buckminsterfullerene in carbon disulfide," The Journal of Physical Chemistry A, vol. 101, no. 2, pp. 103–106, 1997.
5. N. H. Martin, M. H. Issa, R. A. McIntyre, and A. A. Rodriguez, "Field-dependent relaxation and molecular reorientation of C60 in chlorobenzene," The Journal of Physical Chemistry A, vol. 104, no. 48, pp. 11278–11281, 2000.
6. A. D. Irwin, R. A. Assink, C. C. Henderson, and P. A. Cahill, "Study of the reorientational motion of C60H2 in toluene-d8 by proton NMR," The Journal of Physical Chemistry, vol. 98, no. 46, pp. 11832–11834, 1994.
7. I. V. Rubtsov, D. V. Khudiakov, V. A. Nadtochenko, A. S. Lobach, and A. P. Moravskii, "Rotational reorientation dynamics of C60 in various solvents. Picosecond transient grating experiments," Chemical Physics Letters, vol. 229, no. 4-5, pp. 517–523, 1994.

8. R. M. Hughes, P. Mutzenhardt, L. Bartolotti, and A. A. Rodriguez, "Analysis of the spin-lattice relaxation rate and reorientational dynamics of fullerene C70 in chlorobenzene-d5," Journal of Molecular Liquids, vol. 116, no. 3, pp. 139–146, 2005.
9. R. Tycko, R. C. Haddon, G. Dabbagh, S. H. Glarum, D. C. Douglass, and A. M. Mujsce, "Solid-state magnetic resonance spectroscopy of fullerenes," The Journal of Physical Chemistry, vol. 95, no. 2, pp. 518–520, 1991.
10. R. Tycko, G. Dabbagh, G. B. M. Vaughan, et al., "Molecular orientational dynamics in solid C70: investigation by one- and two-dimensional magic angle spinning nuclear magnetic resonance," The Journal of Chemical Physics, vol. 99, no. 10, pp. 7554–7564, 1993.
11. M. Sprik, A. Cheng, and M. L. Klein, "Modeling the orientational ordering transition in solid C60," The Journal of Physical Chemistry, vol. 96, no. 5, pp. 2027–2029, 1992.
12. R. M. Hughes, P. Mutzenhardt, L. Bartolotti, and A. A. Rodriguez, "Experimental and theoretical analysis of the reorientational dynamics of fullerene C70 in various aromatic solvents," The Journal of Physical Chemistry A, vol. 112, no. 18, pp. 4186–4193, 2008.
13. C. S. Yannoni, R. D. Johnson, G. Meijer, D. S. Bethune, and J. R. Salem, "C13 NMR study of the C60 cluster in the solid state: molecular motion and carbon chemical shift anisotropy," The Journal of Physical Chemistry, vol. 95, no. 1, pp. 9–10, 1991.
14. J. H. Walton, A. K. Kamasa-Quashie, J. M. Joers, and T. Gullion, "Spin-rotation relaxation in C60," Chemical Physics Letters, vol. 203, no. 2-3, pp. 237–242, 1993.
15. R. D. Johnson, C. S. Yannoni, H. C. Dorn, J. R. Salem, and D. S. Bethune, "C60 rotation in the solid state: dynamics of a faceted spherical top," Science, vol. 255, no. 5049, pp. 1235–1238, 1992.
16. H. C. Dorn, J. Gu, D. S. Bethune, R. D. Johnson, and C. S. Yannoni, "The nature of fullerene solution collisional dynamics. A C13 DNP and NMR study of the C60/C6D6/TEMPO system," Chemical Physics Letters, vol. 203, no. 5-6, pp. 549–554, 1993.
17. E. D. Becker, High Resolution NMR: Theory and Chemical Applications, Academic Press, New York, NY, USA, 2nd edition, 1980.
18. A. Abragam, The Principals of Nuclear Magnetism, Oxford University, New York, NY, USA, 1961.

19. O. Walker, P. Mutzenhardt, P. Tekely, and D. Canet, "Determination of carbon-13 chemical shielding tensor in the liquid state by combining NMR relaxation experiments and quantum chemical calculations," Journal of the American Chemical Society, vol. 124, no. 5, pp. 865–873, 2002.
20. D. L. Bryce and R. E. Wasylishen, "A M95O and C13 solid-state NMR and relativistic DFT investigation of mesitylenetricarbonylmolybdenum(0)—a typical transition metal piano-stool complex," Physical Chemistry Chemical Physics, vol. 4, no. 15, pp. 3591–3600, 2002.
21. J. C. Facelli, B. K. Nakagawa, A. M. Orendt, and R. J. Pugmire, "Cluster analysis of C13 chemical shift tensor principal values in polycyclic aromatic hydrocarbons," The Journal of Physical Chemistry A, vol. 105, no. 31, pp. 7468–7472, 2001.
22. J. N. Latosińska, "C13 CP/MAS NMR and DFT studies of thiazides," Journal of Molecular Structure, vol. 646, no. 1–3, pp. 211–225, 2003.
23. F. L. Colhoun, R. C. Armstrong, and G. C. Rutledge, "Chemical shift tensors for the aromatic carbons in polystyrene," Polymer, vol. 43, no. 2, pp. 609–614, 2002.
24. D. Stueber and D. M. Grant, "The C13 chemical shift tensor principal values and orientations in dialkyl carbonates and trithiocarbonates," Solid State Nuclear Magnetic Resonance, vol. 22, no. 4, pp. 439–457, 2002.
25. G. Zheng, J. Hu, X. Zhang, L. Shen, C. Ye, and G. A. Webb, "Quantum chemical calculation and experimental measurement of the C13 chemical shift tensors of vanillin and 3,4-dimethoxybenzaldehyde," Chemical Physics Letters, vol. 266, no. 5-6, pp. 533–536, 1997.
26. V. G. Malkin, O. L. Malkina, and D. R. Salahub, "Influence of intermolecular interactions on the C13 NMR shielding tensor in solid α-glycine," Journal of the American Chemical Society, vol. 117, no. 11, pp. 3294–3295, 1995.
27. M. J. Frisch, G. W. Trucks, H. B. Schlegel, et al., Gaussian 98, Revision A.6, Gaussian, Pittsburgh, Pa, USA, 1998.
28. A. D. Becke, "Density-functional thermochemistry—III. The role of exact exchange," The Journal of Chemical Physics, vol. 98, no. 7, pp. 5648–5652, 1993.
29. C. Lee, W. Yang, and R. G. Parr, "Development of the Colle-Salvetti correlation-energy formula into a functional of the electron density," Physical Review B, vol. 37, no. 2, pp. 785–789, 1988.
30. R. McWeeny, "Hartree-Fock theory with nonorthogonal basis functions," Physical Review, vol. 114, no. 6, pp. 1528–1529, 1959.

31. R. Ditchfield, "Self-consistent perturbation theory of diamagnetism," Molecular Physics, vol. 27, no. 4, pp. 789–807, 1974.

32. K. Wolińnski and A. J. Sadlej, "Self-consistent perturbation theory. Open-shell states in perturbation-dependent non-orthogonal basis sets," Molecular Physics, vol. 41, no. 6, pp. 1419–1430, 1980.

33. K. Wolinski, J. F. Hinton, and P. Pulay, "Efficient implementation of the gauge-independent atomic orbital method for NMR chemical shift calculations," Journal of the American Chemical Society, vol. 112, no. 23, pp. 8251–8260, 1990.

34. N. Bloembergen, E. M. Purcell, and R. V. Pound, "Nuclear magnetic relaxation," Nature, vol. 160, no. 4066, pp. 475–476, 1947.

35. N. Bloembergen, Nuclear Magnetic Resonance, W. A. Benjamin, New York , NY, USA, 1969.

36. W. T. Huntress, Advances in Magnetic Resonance. Vol. 4, Academic Press, New York , NY, USA, 1971.

37. M. V. Korobov and A. L. Smith, Fullerenes: Chemistry, Physics, and Technology, John Wiley & Sons, New York , NY, USA, 2000.

Caustic-Side Solvent Extraction Chemical and Physical Properties: Equilibrium Modeling of Distribution Behavior

Laetitia H. Delmau, Tamara J. Haverlock, Tatiana G. Levitskaia, Frederick V. Sloop, Jr. and Bruce A. Moyer

ABSTRACT

A multivariate mathematical model describing the extraction of cesium from different mixtures of sodium hydroxide, sodium nitrate, sodium chloride, and sodium nitrite containing potassium at variable concentrations has been established. It was determined based on the cesium, potassium, and sodium distribution ratios obtained with simple systems containing single salts. These experimental data were modeled to obtain the formation constants of complexes formed in the organic phase based on specified concentrations of components in both organic and aqueous phases. The model was applied to five different SRS waste simulants, and the corresponding cesium extraction results were predicted satisfactorily, thus validating the model.

Introduction

The solvent extraction process proposed and considered for cesium removal from the waste present at the Savannah River Site (SRS) is being investigated with respect to the behavior of system components under different conditions. A thorough understanding of the process is in part demonstrable by establishing a model that predicts the extraction of cesium based on the major components of the waste (or simulant). The ability to predict distribution behavior facilitates appropriate flowsheet design to accommodate changing feed composition and temperature. It also provides greater confidence in the robustness of the process overall. Finally, given the knowledge of the composition of any particular feed, a reliable model yields an immediate estimate of expected flowsheet performance for comparison with process data. The scope of this modeling study was directed toward predicting the cesium distribution ratios obtained with five different SRS simulants corresponding to five real-waste tanks. Chemical analyses of the tanks provided the concentrations of sodium, potassium, cesium, nitrate, and free hydroxide. When preparing the simulants, the total concentration of cations could be as high as 5.6 M. The nitrate and hydroxide concentrations measured in the tanks could not balance the cation concentration. The quantity of anion still not accounted for by these analyses was filled either with chloride or with nitrite anions. Based on the total composition of the SRS waste, these four anions and three cations were determined to be the main components. The model will include species of these ions, and corresponding formation constants will be determined by the sequential modeling of simple systems containing first one cation and one anion at the same time, then systematically increasing the number of components. A model representing the extraction of cesium from different media will then be established and cesium extraction behavior could be predicted by a simple input of the concentrations in the aqueous phase before extraction.

Experimental Program

Materials

Stock solutions of HNO_3, $NaNO_3$, $NaN0_2$, NaC1, and NaOH were prepared and all other concentrations prepared as a dilution of the stock. Sodium hydroxide was diluted from 50% wt/wt received from J. T. Baker, Lot 517045. All salts were dried at 110 °C for >18 hours and stored in a desiccator prior to solution preparation. Sodium chloride was received from EM Scientific, Lot 33131325; $NaN0_2$ was received as 99.5% from Aldrich, Lot 07012MS; $NaNO_3$, was received from J. T. Baker as reagent grade crystal, Lot M14156. Cesium nitrate was received from Alpha Aesar, 99.9% and dried prior to use. Potassium nitrate was received from EM Science. Sodium concentrations prepared were 5.6, 4.5, 2.25, 1.00, 0.50, 0.10, and 0.01 M. $CsNO_3$, and KNO_3, were added at 0.5 mM and 60 mM, respectively, directly to the sodium salts, effecting a slight dilution of the initial sodium in solution. Binary salt solutions at anion ratios of 0, 0.25, 0.50, 0.75, 0.90, and 1.0 and total sodium concentration of 4.5 or 5.6 M were also prepared with $CsNO_3$ added at 0.5 mM as well as with and without KNO_3 at 60 mM. Potassium extractions from KNO_3 solutions at concentrations of 1.0, 0.30, 0.10, 0.01 M were also performed. Measurements of cesium extraction from nitric acid involved pristine solvent that had not been preequilibrated with the corresponding solution of nitric acid without cesium. The organic phase consisted of washed solvent Cs-7SB/ Isopar® L, ORNL Lot# PVB-B000718-156W (7-28-2000). The radiotracers ^{22}Na and ^{137}Cs were obtained from Isotope Products, Burbank, CA.

General Contacting and Counting Procedure

The capped vials were mounted by clips on a disk that was rotated in a constant-temperature air box at 25.0 ± 0.5°C for 90 minutes. After the contacting period, the vials were centrifuged for 3 minutes at 3600 RPM and 25°C in a Sanyo MSE Mistral 2000R temperature-controlled centrifuge. A 300 µL aliquot of each phase was subsampled and counted using a Packard Cobra IT Auto-Gamma counter. All samples were counted for a period of 10minutes using a window of 580-750 keV.

Variable Temperature Experiment

A series of experiments to determine cesium distribution using ^{137}Cs tracer techniques was completed. The distribution of cesium in response to increasing concentrations of NaOH and $NaNO_3$ at two temperatures, 20°C and 35°C, was examined. Contacting experiments were carried out using an O/A of unity. All contacts were performed in duplicate. The aqueous phase consisted of 0.5 mM $CsNO_3$ and varied concentrations of either NaOH or $NaNO_3$ at 0.01, 0.1, 0.5, 1.0, 2.25, 4.5, and 5.6 M. ^{137}Cs tracer was introduced at 0.1 µCi/mL aqueous phase. The series of 0.5 mM $CsNO_3$ in NaOH or $NaN0_3$ solutions were prepared by aqueous dilutions. The contacts were carried out for a period of 90 minutes in 5.0 mL VWR Cat. No. 66008-400 clear polypropylene vials.

The capped vials were mounted by clips on a disk which was rotated in a constant-temperature air box at 20.2 ± 0.5 °C for 90 min or on a similar wheel located in a LabLine Imperial III Model 306M Incubator at 35.8 °C for the same period of time. After the 90 minute contacting period, the vials were centrifuged for 3 minutes at 3600 RPM and 22°C in a Sanyo MSE Mistral 2000R temperature-controlled centrifuge. A 300 µL aliquot of each phase was subsampled and counted.

Experiments with Calixarene-Free Solvent

In this experiment cesium extraction as a function of the solvent/modifier/trioctylamine (TOA) system was investigated. A contacting experiment was carried out, using an O/A volume ratio of unity, in which 1 mL of washed calix-free CSSX solvent was contacted at 25°C with 1 mL of an aqueous phase consisting of 0.5 mM $CsNO_3$ and variable concentrations of NaOH (0.01, 0.1, 0.5, 1.0, 2.25, 4.5, and 5.6 M).

The calix-free organic phase was prepared by adding Cs-7SB modifier (Lot no. BO00718-24DM) at 0.5 M, and trioctylamine (Lot no. B000718-105L) at 0.001M to Isopar®L (Lot no. 0306-10967). This solvent was then washed in Teflon® FEP labware using an O/A volume ratio of unity, twice with 0.1 M NaOH and 50 mM HN03, and three times with DDI water. The aqueous phases were made by appropriately diluting a 5.6 M working stock of NaOH and a 50 mM solution of CsNO3- I37Cs tracer was introduced at 0.1 µCi/mL aqueous phase. The contacts were camed out for a period of 90 minutes in 5.0 mL VWR Cat. No. 66008-400 clear polypropylene vials.

Ion-Chromatography Experiments

The solvent (Lot B000718-156W) was contacted with an equal volume of the appropriate salt solution in 2 mL polypropylene vials for 1 hour by rotation in a thermostated air box set at 25 ± 0.1 °C. All samples were centrifuged for 3 minutes at 3500 rpm to confirm complete phase disengagement. The organic phase was then contacted with a five to ten-fold volume of dilute HNO_3 (I mM) to strip the metal ions into the aqueous phase. Results were based on the first strip since the metal recovery was equal or greater than 98%. The strip solutions were analyzed with a Dionex Model DX.500 equipped with a GP40 pump and a CD20 conductivity detector. The cations sodium, potassium and cesium were separated and analyzed using a CS12A analytical column coupled with a CG12A guard column. The analysis used 20 mM H_2SO_4 eluent at 1 mL/min in an isocratic run of 20 minutes. Background conductivity was 0.2 μS using CSRS-Ultra suppressor in auto-regeneration mode set at 300 mA. A five-level external standardization for each metal, Na, K, and Cs, was used. Duplicates were run for each sample and were analyzed with ± 2 % error.

Program SXFIT

Description of the Program

The program SXFIT is a program that can model thermodynamics data based on the constituents of the systems and the species that are being formed. Although the program's ability to model different kinds of systems is almost limitless, we will describe its capability to handle distribution ratios of ions, since our interest here is to be able to predict cesium extraction behavior. Like the preceding codes SXLSQ[2], SXLSQA[3], and SXLSQ1[4], SXFIT[5] is a program written in FORTRAN that refines a series of given inputs based on the least-squares minimization of the difference between the observed and the calculated quantities. The main improvement of SXFIT over predecessor codes is the fact that an unlimited number of constituents can be input. The program then calls for the parameters that are used to calculate the activity effects occurring in the aqueous phase (Masson and Pitzer coefficients) and the organic phase (Solubility parameters). In addition, molecular weights and non-aqueous molar volumes of the different constituents need to be provided along with the dielectric constant and the solubility parameter of the diluent in the organic phase. All initial concentrations of constituents are entered in a data file. Finally, based on the knowledge of the extraction reactions that occur during the process, a few reasonable species (products of the extraction system) may also be supplied with their formation constants. The program then calculates all concentrations of all constituents at equilibrium and the distribution coefficients of the ion of interest. Depending on the differences between the observed and calculated values, the program will then adjust the formation constants of the input species until the best fit is obtained. Of course, this could easily become a simple curve-fitting exercise in which a large number of parameters are used to fit a smaller number of data points. However, the user must ensure that all the species and their relative formation constants are chemically reasonable. Usually, the preference will be given to a model that contains the lowest number of species for a given goodness of fit, represented in the program by the agreement factor. A perfect fit with an accurate experimental error on all the data points yields an agreement factor of 1. A value greater than 1 indicates a poorer fit or an underestimate of the experimental error, while a value between 0 and 1 indicates an overestimate of the experimental error.

Assumptions

The solvent used in this system comprises 0.01 M Calix[4]arene-bis(tert-octylbenzo crown-6) (BOBCalixC6), 0.5 M 1-(2,2.3,3-tetrafluoropropoxy)-3-(-4-sec-butylphenoxy)-2-propanol (Cs7-SB modifier), and 0.001 M trioctylamine (TOA) in Isopar® L. The concentrations of the modifier and of TOA are held constant. The concentration of the modifier is large enough to neglect the amount that is being complexed during the extraction of the cations. Therefore, it will not appear in any species of the model. Regarding TOA, we chose not to include it in the model. Its only influence occurs when the aqueous phase contains enough acid to convert TOA into its acidic form, which in turn increases the amount of nitrate in the organic phase. In this work, only two sets of data involved nitric acid, and it was found that TOA did not have any influence on the cesium or potassium distribution ratios. In future modeling, this restriction will need to be lifted to properly account for acid balance in scrubbing and stripping. Likewise, an accurate accounting of volume and concentration changes would benefit from knowledge of water transferred to and from the solvent; this was omitted from the present treatment.

Parameters Used

The program requires a series of input parameters, most of which are available in published handbooks and literature. Those parameters involved in the activity coefficients in the organic and aqueous phases can be refined by the program (Pitzer parameters, solubility parameters). However, for present purposes, the parameters were either calculated prior to any modeling or obtained from referenced sources and kept constant. The only parameters refined during the modeling process itself were the formation constants of the species in the organic phase.

Table 1. Molecular weights and non aqueous molar volumes of the constituents

Constituent	Molecular Weight (g/mol)	Non aqueous molar volume (cm^3/mol)*
Na^+	22.990	10
K^+	39.098	9
Cs^+	132.91	21.5
H^+	1.008	0
NO_3^-	62.005	29
Cl^-	35.450	18
NO_2^-	46.006	26
OH^-	17.008	18
BOBCalixC6	1149.53	500
Diluent (Isopar® L)	170	227
Water		18

*The values for the ions are based on their aqueous molar volumes V_0 presented in Table 2. The value for sodium is a personal communication from Charles F. Baes, Jr.

The values for the constituents presented in Table 1 are those called by the program and changeable by the users. The molecular weight of water is 18.015 g/mol. This value is a constant and non-changeable.

Table 2. Masson coefficients[6] of ions present in the system.

Constituent	V_θ	Sv
Na^+	-1.5	1.89
K^+	8.73	1.10
Cs^+	21.40	1.29
H^+	0	0
NO_3^-	29.33	0.543
Cl^-	18.12	0.83
NO_2^-	26.5	2.00
OH^-	-1.04	2.32

Table 3. Pitzer parameters[7] for the interactions between ions present in the system.

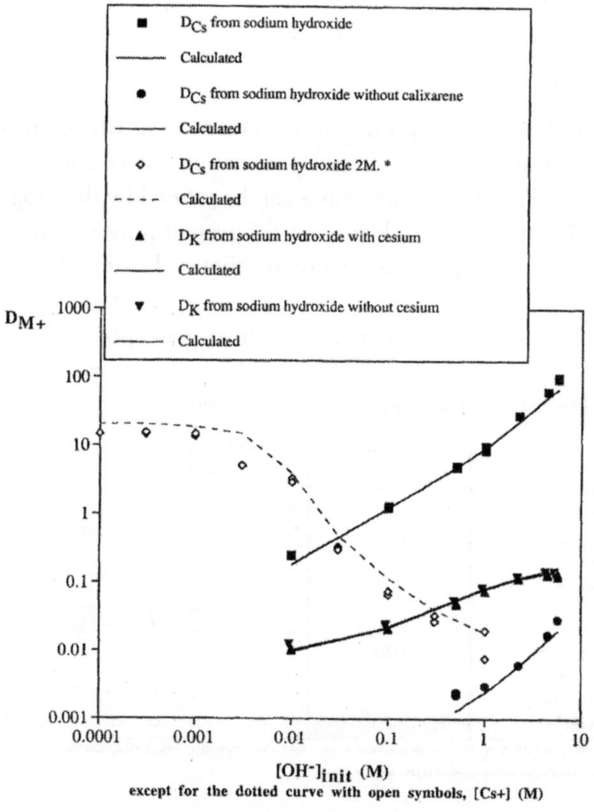

All ß2 values are set to 0. Parameter $\alpha 1= 2$, and $\alpha 2= 0$, since all the interactions are between two monocharged ions.

Regarding the activity coefficients in the organic phase, all the product species were assigned a similar value. The solubility parameter of the diluent (Isopar® L) and the extractant BOBCalixC6 were determined by group contribution calculations8 A best estimate of the values of the product species was made. They were kept constant as no reliable source for better values is available. In addition, the solubility parameters usually do not have a major impact on the determination of the product species formation constants. They avoid assuming ideality in the organic phase, but do not have a crucial effect on the final results, as the mole fraction of extracted species in the solvent is very small. The dielectric constant of the diluent equals 2.014; its solubility parameter is set to 18.40 J1/2cm-3/2. The solubility parameter of water is set to 51.13 J1/2cm-3/2.

The calixarene solubility parameter was estimated with the group contributions and determined to be 21 J1/2cm-3/2. A11 organic species formed in the organic phase were assigned a solubility parameter of 19.8 J1/2cm-3/2,which is also the solubility parameter of the modifier. Previous studies showed that at least one molecule of modifier was included in the complexes, and the solubility parameter is close enough to the value for the calixarene to avoid any major activity effect.

Results and Discussion

Tables with all experimental results used in this modeling can be found in the appendix. By way of brief explanation, we include here a description of the approach chosen to find the best model. The method used to model the first set of data is discussed in detail. Fewer details will be given for subsequent groups, as the modeling technique and approach remain the same.

Extraction Modeling From Nitrate Media

The first step was to model data that involved only one anion. Indeed, all the other sodium salts were spiked with potassium nitrate and/or cesium nitrate, and the corresponding amount of anion, however small, was taken into account.

Data on cesium extraction from cesium nitrate and nitric acid yielded the formation constants of CsNO3Calix (o) and (CsNO3)2Calix(o); the notation (o) refers to the organic phase. Addition of data on potassium extraction from potassium nitrate alone or mixed with cesium nitrate yielded the formation constant of KNO3Calix(o). Finally, addition of data on potassium and cesium extraction from sodium nitrate allowed us to calculate the formation constant of NaNO3Calix(o).

Table 4. Species and formation constants for the model derived for nitrate data

Species	Formation constant
$CsNO_3Calix(o)$	$Log_{10} K = 3.615$
$(CsNO_3)_2Calix(o)$	$Log_{10} K = 4.317$
$KNO_3Calix(o)$	$Log_{10} K = 1.387$
$NaNO_3Calix(o)$	$Log_{10} K = -0.943$

The fit of all the data is presented in Figures 1 and 2. The overall agreement factor is 2.7, with an assumption of a uniform 5% error on all the data points.

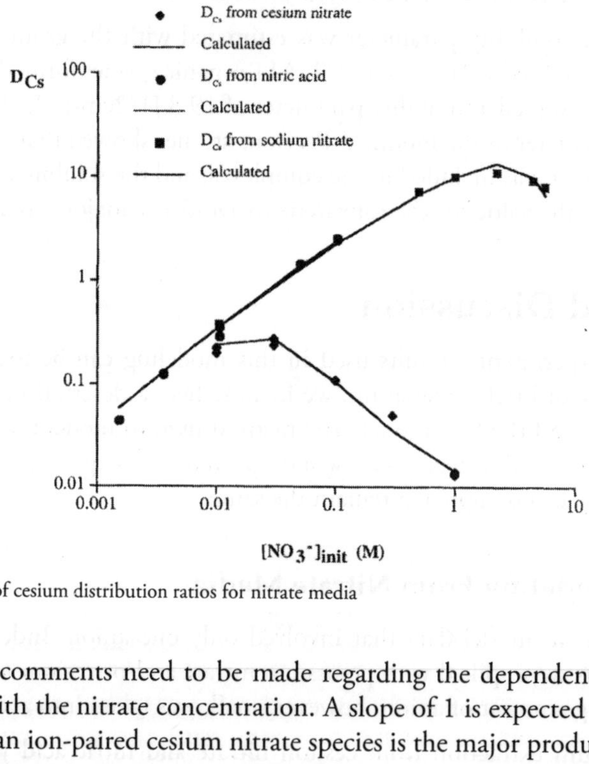

Figure 1. Fit of cesium distribution ratios for nitrate media

A few comments need to be made regarding the dependence of cesium extraction with the nitrate concentration. A slope of 1 is expected when a complex involving an ion-paired cesium nitrate species is the major product formed in the organic phase.

$$Cs^+ + NO_3^- + BOBCalixC6(o) \text{€} \; CsNO_3(BOBCalix)(o).$$

This is well-demonstrated with the nitric acid experiment. The other two depart from the previous statement as the calixarene is loaded when cesium is extracted from increasing concentrations of cesium nitrate. A similar phenomenon is observed when reaching high concentrations of sodium nitrate where loading and activity effects give a trend that shows the cesium distribution ratios reach a maximum and then decrease.

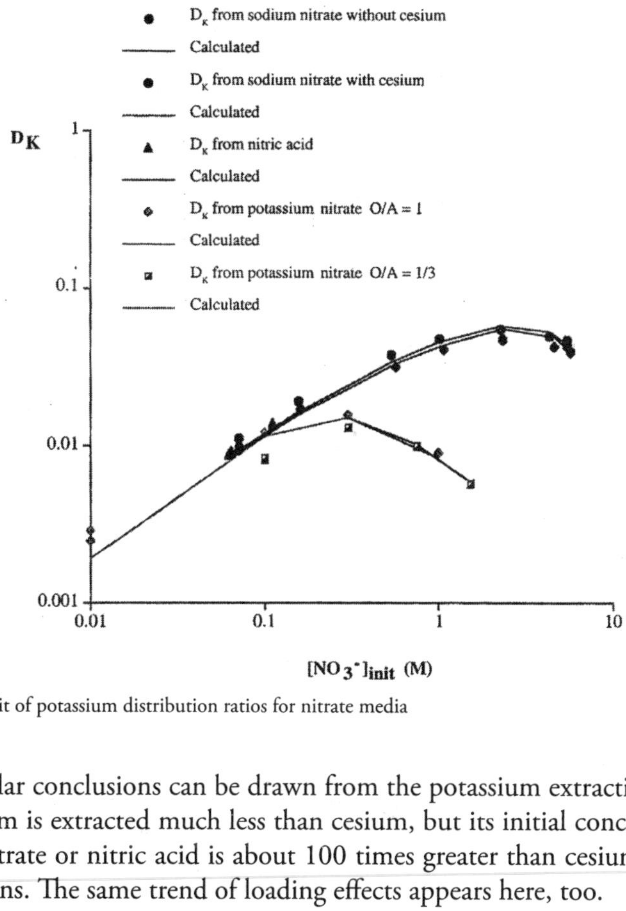

Figure 2. Fit of potassium distribution ratios for nitrate media

Similar conclusions can be drawn from the potassium extraction experiment. Potassium is extracted much less than cesium, but its initial concentration in sodium nitrate or nitric acid is about 100 times greater than cesium under similar conditions. The same trend of loading effects appears here, too.

Extraction Modeling From Hydroxide Media

The second set of data involves the fitting of cation extraction from sodium hydroxide. The results found previously for the nitrate system are required since all the potassium and the cesium were added as spikes of nitrate solutions. Preliminary results on sodium extraction showed that the presence of calixarene was not required to extract this cation, as the amount of sodium present in the organic phase was the same whether or not the calixarene was present in the solvent.

Table 5. Species and formation constants for the model derived for hydroxide data

Species	Formation constant
CsOH(o)	$Log_{10} K = -2.264$
CsOHCalix(o)	$Log_{10} K = 3.332$
KOHCalix(o)	$Log_{10} K = 1.549$
NaOH(o)	$Log_{10} K = -0.565$

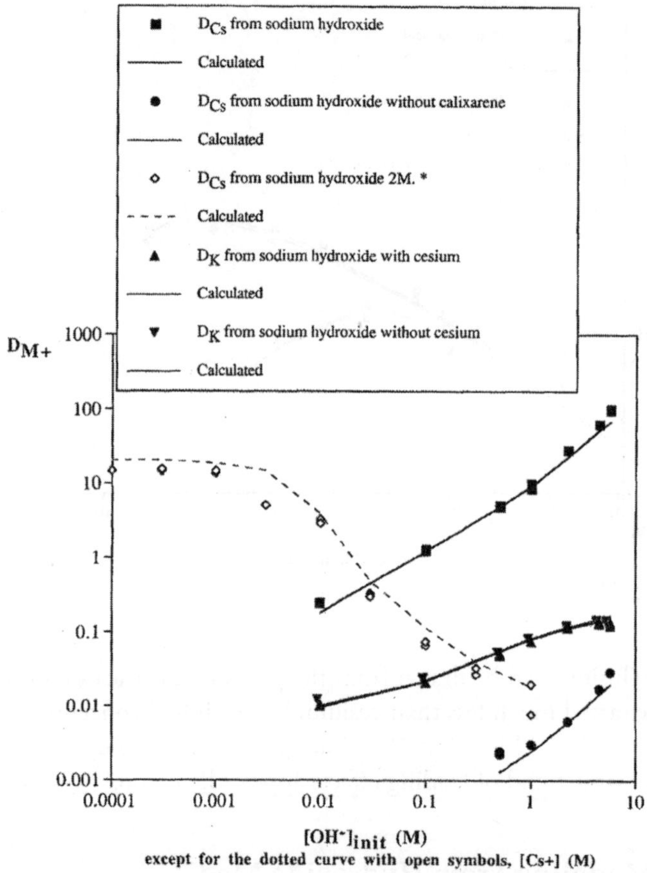

Figure 3. Fit of data points for hydroxide media

The species listed Table 5 were added to the model to achieve a fit of data obtained in hydroxide systems. Figure 3 summarizes the fit as a function of hydroxide concentration. The fit for all the data is rather good with the exception of the curve for which the concentration of sodium hydroxide was held constant and the concentration of cesium nitrate varied (*dotted curve and open symbols =variable cesium nitrate concentration in NaOH 2 M). The reliability of these results was rather low since a third phase was observed for most of them. However, the results and the corresponding fit are presented to show that the overall trend is followed. Consider also, when a third phase is observed, the distribution coefficients are usually lower than they would be without a third phase since some of the activity present in the organic phase that is subsampled for counting is present in the third phase. This is exactly what is observed in this case.

Extraction Modeling From Nitrite or Chloride Media

The inclusion of the data points containing nitrite salts led to the introduction of three more species in the model:

Table 6. Species and formation constants for the model derived for nitrite data

Species	Formation constant
$CsNO_2Calix(o)$	$Log_{10} K = 3.152$
$KNO_2Calix(o)$	$Log_{10} K = 1.098$
$NaNO_2Calix(o)$	$Log_{10} K = -1.313$

Among the anions in the study, chloride has the highest hydration energy, and therefore is not as extractable as the other three. The formation constants of the species involving this anion are expected to be lower than those found earlier. Extraction tests showed that sodium chloride is not extracted detectably when the calixarene is absent. The inclusion of the data points containing chloride salts led to comparable species:

Table 7. Species and formation constants for the model derived for chloride data

Species	Formation constant
CsClCalix(o)	$Log_{10} K = 2.587$
KClCalix(o)	$Log_{10} K = 0.575$
NaClCalix(o)	$Log_{10} K = -1.455$

Figures 4 and 5 present the fit obtained with the model for the systems containing chloride and nitrite, respectively.

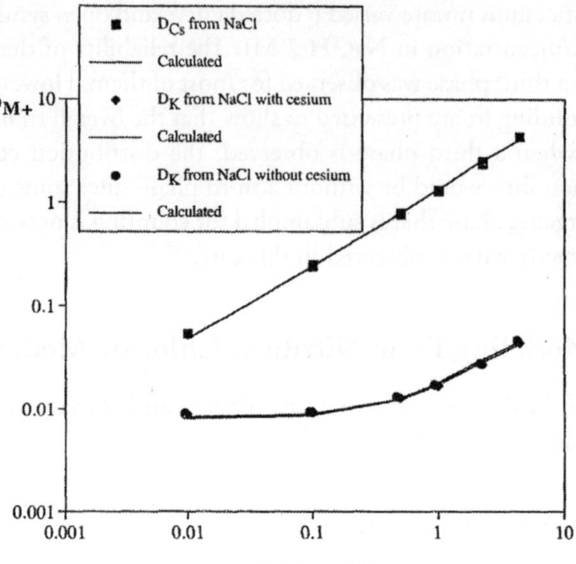

Figure 4. Fit of cesium and potassium distribution ratios for chloride media

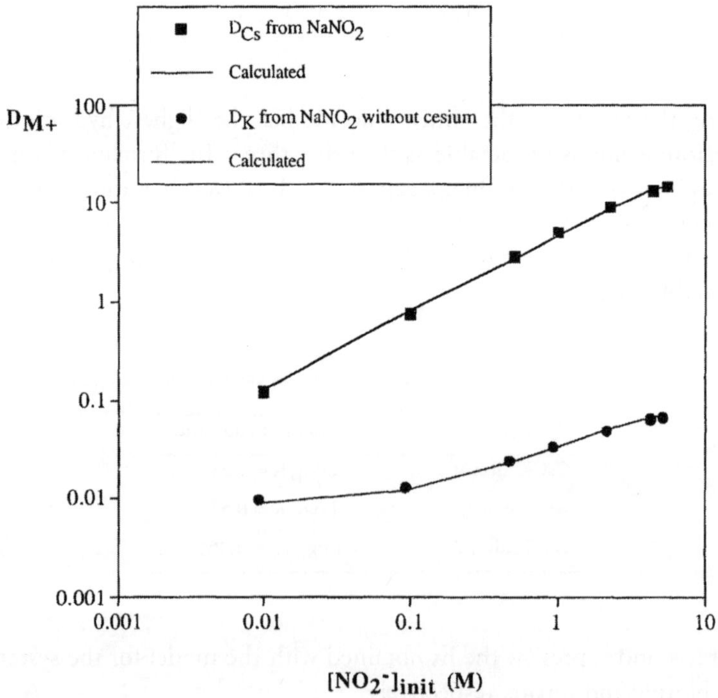

Figure 5. Fit of cesium and potassium distribution ratios for nitrite media

These experiments conclude our tests from simple systems to determine the best model at 25°C. The subsequent experiments attempted to validate the findings with more complicated systems.

Variable Temperature Tests

Two sets of experiments (extraction of cesium from sodium nitrate or sodium hydroxide) were carried out a two temperatures, 20°C and 35°C. It is shown that lower distribution ratios are obtained for higher temperatures, which agrees with the exothermic character of the extraction reaction.

The modeling of these data involved only the determination of the formation constants. Although the activity coefficients are also temperature dependent, the values for 25°C were also used at 20°C and 35°C, since the change is small. Formation constants are mentioned for information purposes only. All experiments carried out at 25°C that allowed the determination of the model to this point would have been needed at other temperatures to determine an accurate model at these temperatures.

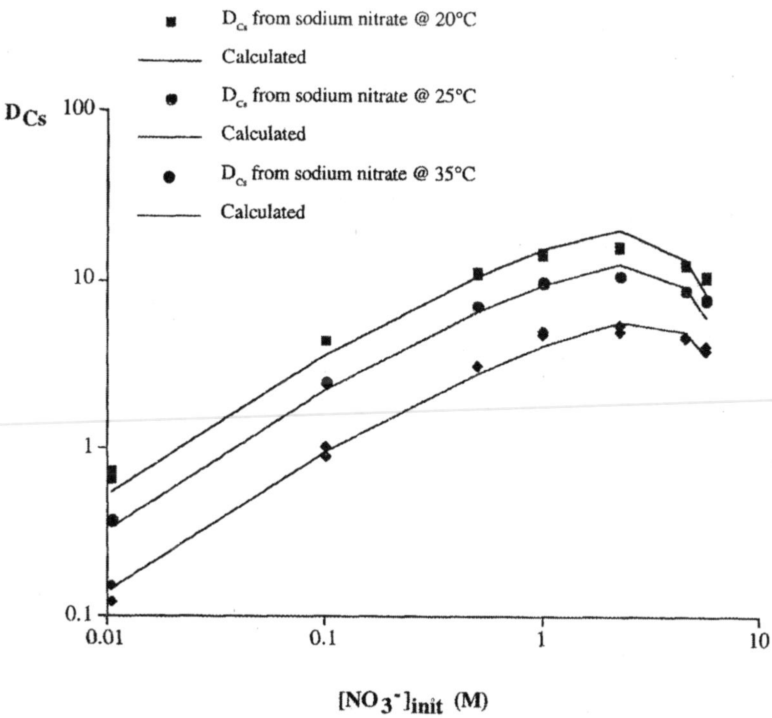

Figure 6. Fit of cesium distribution ratios for nitrate media at different temperature

Table 8. Forniation constant for the model including nitrate data at different temperatures

Species	Formation constant @ 20 °C	Formation constant @ 25 °C	Formation constant @ 35 °C
$CsNO_3$Calix(o)	$Log_{10} K = 3.831$	$Log_{10} K = 3.615$	$Log_{10} K = 3.251$
$(CsNO_3)_2$Calix(o)	kept constant	$Log_{10} K = 4.317$	kept constant
$NaNO_3$Calix(o)	$Log_{10} K = -0.851$	$Log_{10} K = -0.943$	$Log_{10} K = -1.152$

The formation constant of the complex involving two cesium ions is held constant for the three temperatures since there are almost no data points supporting this species in the data sets collected at 20°C or 35°C. Formation constants of complexes containing cesium and sodium nitrate (Table 8) and cesium hydroxide (Table 9) were included in these models for the corresponding temperatures. Figures 6 and 7 present the fit obtained as a function of nitrate and hydroxide concentrations, respectively, at the different temperatures. The expected trend of the formation constant values follow very well the exothermic behavior observed for this system. The fit could be improved, particularly for the data at 35°C, which suggests that some of the assumptions are not valid and the appropriate data sets need to be collected to reduce the uncertainties.

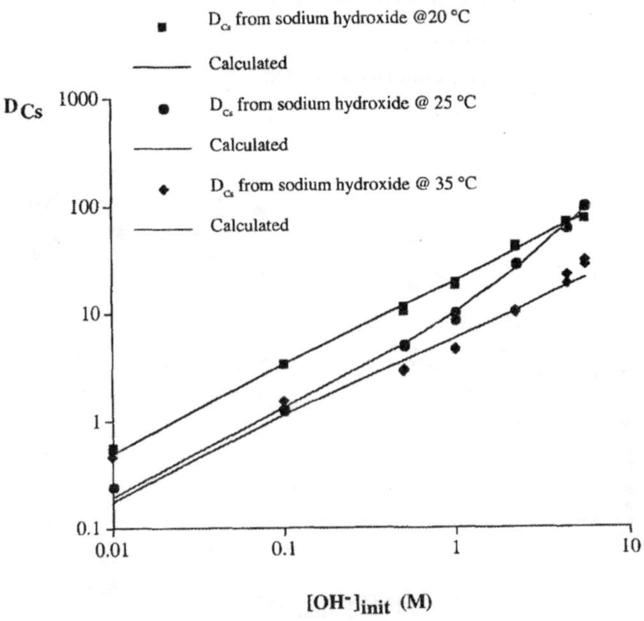

Figure 7. Fit of cesium distribution ratios for hydroxide media at different temperature

Table 9. Formation constant for the model including hydroxide data at different temperatures

Species	Formation constant @ 20 °C	Formation constant @ 25 °C	Formation constant @ 35 °C
CsOHCalix(o)	$Log_{10} K = 3.531$	$Log_{10} K = 3.332$	$Log_{10} K = 3.008$

Tests Involving Tank Simulants

Real waste batch tests conducted at the Savannah River Site provided the concentrations of major components (sodium, potassium, cesium, hydroxide, and nitrate) of the waste. From these results, simulants were prepared using either nitrite or chloride. These initial concentrations were entered into a data file, and the program SXFIT predicted the cesium distribution ratio based on all the species and formation constants presented above. Table 10 presents the comparison between the measured distribution ratios and those that were predicted. The O/A volume ratios were all 113. All experiments were carried out at 25°C.

Table 10. Comparison of measured and predicted distribution ratios for tank simulants

Tank	Composition	D_{Cs} measured	D_{Cs} predicted	error
Tank 13	$[Na+] = 5.6$ M $[K^+] = 0.067$ M $[Cs^+] = 0.512$ mM $[OH^-] = 2.29$ M $[NO_3^-] = 0.767$ M $[NO_2^-] = 2.6$ M	8.38	9.75	16.3%
Tank 13 (with chloride, substituted for nitrite)	$[Na+] = 5.6$ M $[K^+] = 0.067$ M $[Cs^+] = 0.512$ mM $[OH^-] = 2.29$ M $[NO_3^-] = 0.767$ M $[Cl^-] = 2.6$ M	6.68	6.88	3.0%
Tank 26	$[Na+] = 5.6$ M $[K^+] = 0.041$ M $[Cs^+] = 0.219$ mM $[OH^-] = 4.71$ M $[NO_3^-] = 0.956$ M $[NO_2^-] = 0$ M	16.8	15.0	10.7%
Tank 33	$[Na+] = 3.3$ M $[K^+] = 5$ mM $[Cs^+] = 8.03$ µM $[OH^-] = 1.47$ M $[NO_3^-] = 1.44$ M $[NO_2^-] = 0.4$ M	16.3	20.5	25.7%
Tank 35	$[Na+] = 5.6$ M $[K^+] = 0.01$ M $[Cs^+] = 0.188$ mM $[OH^-] = 2.93$ M $[NO_3^-] = 1.4$ M $[NO_2^-] = 1.3$ M	22.8	23.9	4.6%
Tank 46	$[Na+] = 5.6$ M $[K^+] = 0.032$ M $[Cs^+] = 0.378$ mM $[OH^-] = 3.98$ M $[NO_3^-] = 0.606$ M $[NO_2^-] = 1.0$ M	17.6	17.3	1.7%
Tank 46 (with chloride substituted for nitrite)	$[Na+] = 5.6$ M $[K^+] = 0.032$ M $[Cs^+] = 0.378$ mM $[OH^-] = 3.98$ M $[NO_3^-] = 0.606$ M $[Cl^-] = 1.0$ M	15.4	14.8	3.8%

The prediction is very good in all the cases, for any composition of the simulant. The only overestimation occurred for the simulant that contains a lower concentration of sodium. Otherwise, all the predictions are within an average of 7% of error, which is excellent for a model that contains a minimum number of product species. This comparison validates the model described previously.

Prediction of Cesium Extraction from the Full Simluant

Simplified tank simulants were prepared fcontaining only those ions assumed to be playing major roles in the systems. Following this assumption, we decided to go further and test our model by creating a data file containing the initial concentrations representing the full simulant. In order to test our model with the full simulant, a "dummy non-extractable" anion (X^-) was introduced to ensure the electroneutrality of the aqueous phase:

$$[Na^+]+[K^+]+[Cs^+]=[NO_3^-]+[OH^-]+[NO_2^-]+[Cl^-]+[X^-].$$

This "dummy" anion is intended to represent all the anions present in the simulant that are not extracted and assumed to play only minor roles in the extraction process. However, they are important from a global ionic strength standpoint.

Under the aqueous initial conditions:

$$[Na^+]=5.6M,[K^+]=0.015M,[Cs^+]=1.4\times10^{-4}M$$
$$[NO_3^-]=2.03M,[OH-]=2.06M,[NO_2^-]=0.5M,[Cl^-]=0.024M \text{ and}$$

and the species with their respective formation constants listed in Table 11, the predicted cesium distribution ratio is 16.8. The average of all cesium distribution ratios obtained in extraction conditions is 16.9[10] This comparison allows a still greater confidence in the reliability of the model, and confirms that the anions present in the simulant other than hydroxide, nitrate, nitrite, and chloride do not play a major role.

Table 11. Model used in this work

Table 11. Model used in this work

Species	Formation constant
$CsNO_3Calix(o)$	$Log_{10}K = 3.615$
$(CsNO_3)_2Calix(o)$	$Log_{10}K = 4.317$
$CsOH(o)$	$Log_{10}K = -2.263$
$CsOHCalix(o)$	$Log_{10}K = 3.332$
$CsNO_2Calix(o)$	$Log_{10}K = 3.152$
$CsClCalix(o)$	$Log_{10}K = 2.587$
$KNO_3Calix(o)$	$Log_{10}K = 1.387$
$KOHCalix(o)$	$Log_{10}K = 1.549$
$KNO_2Calix(o)$	$Log_{10}K = 1.098$
$KClCalix(o)$	$Log_{10}K = 0.575$
$NaOH(o)$	$Log_{10}K = -0.565$
$NaNO_3Calix(o)$	$Log_{10}K = -0.943$
$NaNO_2Calix(o)$	$Log_{10}K = -1.313$
$NaClCalix(o)$	$Log_{10}K = -1.455$

Conclusion

The model gives a very good overall fit for a large number of data points (almost 300) obtained from simple systems. The overall agreement is adequate for such a large data set and the number of species assumed in the organic phase is very small. In addition, all the formation constants are consistent within themselves. They follow the values of the Gibbs energy of partitioning for the four anions. The definitive test to predict the cesium distribution ratios based on initial concentrations is extremely satisfactory. In addition, the prediction of the distribution coefficient obtained with the full simulant is very close to the value obtained experimentally. We can say that not only does the model fit the data very well, but it also includes the cations and anions that play major roles in more complicated mixtures.

References

1. R. A. Peterson, Savannah River Technology Center, Aiken, SC, private communication, Nov., 2000.
2. C.F. Baes, Jr, W.J. McDotvell, S.A. Bryan, The Interpretation of Equilibriunz Data from Synergistic Solvent Extraction systems, Soh. Extr. Ion Exch., 5, 1–28 (1987).
3. C.F. Baes, Jr, B.A. Moyer, G.N. Case, F.I. Case, SXLSQA, A Computer Program for Including Both Complex Formation and Activity Eflects in the Interpretation of Solvent Extraction Data, Sep. Sci. Technol., 25, 1675–1688 (1990).
4. C.F. Baes, Jr, SXLSQI: A Program for Modeling Solvent Extraction Systems, Oak Ridge National Laboratory report OWM-13604. December 1998.
5. C.F. Baes, Jr, Modeling Solvent Extraction Systems with SXFIT, Solv. Extr. Ion Exch., 19, 193–213 (2001).
6. F. J. Milero, in Water and Aqueaus Solutions, R. A. Horne, Ed., Wiley-Interscience, New York (1972).
7. K.S. Pitzer, Activity Coefficients in Electrolyte Solutim, Yd Ed. K.S. Pitzer Ed., CRC Press, Boca Raton (1991).
8. A.F.M.Barton, Handbook of solubility param-other cohesion pa rameters, 2"d Ed., CRC Press, Boca Raton (1983).
9. R. A. Peterson, Preparation of Simulated Waste Solutions for Solvent Extraction Testing, Report WSRC-RP-2000-00361, Westinghouse Savannah River Company, Aiken, SC, May 1, 2000.
10. Result presented in report by Moyer et al. Caustic-Side Solvent Extraction Chemical and Physical Properties. Progess in FY 2000 and FY 2001, Table 3.3.

Physical Characterization of RX-55-AE-5: A Formulation of 97.5 % 2,6-Diamino-3,5-Dinitropyrazine-1-Oxide (LLM-105) and 2.5% Viton A

Randall K. Weese, Alan K. Burnham, Heidi C. Turner and Tri D. Tran

ABSTRACT

With the use of modern tools such as molecular modeling on increasingly powerful computers, new materials can be evaluated by their structural activity relationships, SAR, and their approximate physical and chemical properties can be calculated in some cases with surprising accuracy. These new capabilities enable streamlined synthetic routes based on safety, performance and processing requirements, to name a few [1]. Current work includes both *understanding properties of old explosives and measuring properties of new ones. The necessity to know and understand the properties of energetic materials is driven by the need to improve performance and enhance stability to various stimuli, such as thermal, friction and impact insult. This review will concentrate on the physical properties of RX-55-AE-5, which is formulated from heterocyclic explosive, 2,6-diamino-3,5dinitropyrazine-1-oxide, LLM-105, and 2.5 % Viton A. Differential scanning calorimetry, DSC, was used to measure a specific heat capacity, C_p, of ≈ 0.950 J/g•°C, and a thermal conductivity, κ, of ≈ 0.160 W/m•°C. The Lawrence Livermore National Laboratory (LLNL) code Kinetics05 and the Advanced Kinetics and Technology Solutions (AKTS) code Thermokinetics were both used to calculate Arrhenius kinetics for decomposition of LLM-105. Both obtained an activation energy barrier E ≈ 180 kJmol^{-1} for mass loss in an open pan. Thermal mechanical analysis, TMA, was used to measure the coefficient of thermal expansion, CTE. The CTE for this formulation was calculated to be ≈ 61 µm/m•°C. Impact, spark, friction and evolved gases are also reported.*

Introduction

A new, less sensitive explosive, RX-55AE-5, has been prepared and subjected to physical characterization tests. RX-55AE-5 is formulated from 97.5% 2,6-diamino-3,5dinitropyrazine-1-oxide (LLM-105) and 2.5% Viton A. Some of its properties are summarized in Table 1 along with a few for LLM-105. LLM-105 has a density of 1.913 g/cm^3, a Dh_{50} of 90-150 cm, and is insensitive to both spark and friction. It has an onset temperature for decomposition of 348°C at a linear heating rate of 10°C/min as measured by Differential Scanning Calorimetry (DSC). It is generally insoluble in common organic solvents. In this work, we measure and report the thermal expansion coefficient, thermal decomposition kinetics, and sensitivity to spark, friction, and impact for the RX-55-AE-5 formulation of LLM-105.

Substantial previous work by Pagoria [2] and associates has attempted to discover new insensitive high explosives (IHE) such as LLM-105 that have higher energy densities than 1,3,5-triamino-2,4,6-trinitrobenzene (TATB). TATB is well noted for its thermal stability (m.p. > 370°C); low insensitivity to external stimuli such as drop hammer, spark, and friction; and its very low solubility in a wide array of solvents. We compare our results for RX-55-AE-5 to PBX-9502, LX-17 and TATB [3]. PBX 9502 is formulated from 95% TATB and 5% Kel-F 800 and LX-17 is formulated from 92.5% TATB and 7.5% Kel-F 800.

Table 1. Properties of RX-55-AE [1, 2, 4]

Molecular weight, LLM-105	216.04 g/mol
Color	Yellow
Crystal structure, LLM-105	monoclinic
Crystal density, LLM-105	1.913 g/cm^3
Coefficient of thermal expansion, CTE	60μm/m•°C
Heat capacity	0.931 J/g/°C
Electrostatic sensitivity	Insensitive

Experimental

RX-55-AE-5 Sample Preparation

We used RX-55-AE-5 sample material (LLNL lot PR#1640) for this work; the designation RX-55-AE-5 refers to a research explosive synthesized and processed at LLNL. Pellets of each explosive were uniaxially-pressed in a conventional compaction die without mold release, using a two pressing cycles of 5 minutes at 200 MPa for the larger pellets. The RX-55-AE-5 pellets were pressed at 105°C and obtained densities 1.74 g/cc to 1.75 g/cc (91.0% to 91.5 % theoretical maximum density), TMD) [5]. Pellet properties are summarized in Table 2.

Table 2. RX-55-AE-5 sample mass, volume, density and dimensions

Sample	length, cm	diameter, cm	mass, g	volume, cm^3	density, g/cm^3
1	0.643	0.635	0.356	0.204	1.75
2	0.642	.0633	0.356	0.205	1.74

Thermal Expansion Measurements

We measured the coefficient of thermal expansion, CTE, of RX-55-AE-5 using a TA Instruments Model 2940 TMA that was controlled by a TA 500 Thermal Analyzer equipped with a TMA Mechanical Cooling Accessory [6,7]. A quartz micro-expansion probe sat on top of all samples with a force of 0.01 Newtons (N). The change in the length of the sample was as it was heated or cooled was measured by a linear transformer that converted the vertical distance of the quartz motion probe and was recorded by the TA Instrument software. Ultra high purity nitrogen carrier gas was used at a constant flow rate of 100 cm^3/min. Samples were heated at a linear heating rate of 3°C /min.

Temperature, force, probe and cell constant calibrations were carried out as prescribed [7], using NIST standards of indium, lead, tin and zinc metals along with aluminum standard reference material. CTE measurements using a NIST certified aluminum standard had less than ± 2 % drift associated over the temperature range of –20 to 100°C.

Decomposition Kinetics by Thermal Analysis [8]

Weight loss measurements were carried out using a TA Instruments Simultaneous Differential Thermogravimetric Analyzer (SDT), model 2960, manufactured by TA Instruments. The SDT instrument is capable of performing both thermogravimetric analysis (TGA) and differential thermal analysis (DTA) at the same time. We also measured thermal decomposition kinetics and specific heat capacity, Cp, using differential scanning calorimetry (DSC). DSC analyses of RX-55-AE-5 were carried out using a TA Instruments Model 2920 and TA Instruments pinhole hermetic aluminum pan that had a small perforation in the pan lid to allow generated gases to escape during decomposition. Linear heating rates of 0.1, 0.34, and 1.0 °C per minute and a purge gas flow of 50 cm3/min of ultra high purity nitrogen were used for decomposition kinetics. A linear heating rate of 3°C per minute was used for Cp measurements. Samples sizes were limited to <0.5 mg to prevent bursting the pan. Data was analyzed using the LLNL Kinetics05 program and the AKTS Thermokinetics program.

Friction Sensitivity [9]

The frictional sensitivity of RX-55-AE-5 was evaluated using a B.A.M. high friction sensitivity tester. The tester employs a fixed porcelain pin and a movable porcelain plate that executes a reciprocating motion. Weight affixed to a torsion arm allows for a variation in applied force between 0.5 and 36.0 kg, and our tests used a contact area of 0.031 cm^2. The relative measure of the frictional sensitivity of a material is based upon the largest pinload at which less than two ignitions (events) occur in ten trials. No reaction is called a "no-go", while an observed reaction is called a "go."

Spark Sensitivity [10]

The sensitivity of RX-55-AE-5 toward electrostatic discharge was measured on a modified Electrical Instrument Services Electrostatic Discharge (ESD) Tester. Samples were loaded into Teflon washers and covered with a 1-mm thick Mylar tape. The density of this packed material was 1.4 cm^3/g. The ESD threshold is defined as the highest energy setting at which a reaction occurs for a 1 in 10 series of attempts. Tests were run on powder and pellets at 68°F and a relative humidity of 56%.

Impact Sensitivity (Drop Hammer) [11]

An Explosives Research Laboratory Type 12-Drop Weight apparatus, more commonly called a "Drop-Hammer Machine" was used to determine the impact sensitivity of CP relative to the primary calibration materials PETN, RDX, and Comp B-3 at 68°F and 56% relative humidity. The apparatus was equipped with a Type 12A tool and a 2.5-kg weight. The 35-mg ± 2-mg powder sample was impacted on a Carborundum "fine" (120-grit) flint paper. A "go" was defined as a microphone response of 1.3 V or more as measured by a model 415B Digital Peakmeter. A sample population of 15 was used. The mean height for "go" events, called the "50% Impact Height" or $Dh5_0$, was determined using the Bruceton up-down method.

Results

CTE Measurements of RX-55-AE-5

Figure 1 shows two independent TMA analyses and their plots of dimensional change versus temperature over the temperature range of -20°C to 100°C. The plotted dimensional change of the two samples is reproducible. CTE values, α, were calculated using equation 1

$$\alpha = \frac{dL}{dT * L_o} \tag{1}$$

where dL is the change in length (µm), dT is the change in temperature (°C) and L_o is the initial sample length (meters). Results are plotted in Figure 2, and average values are listed in Table 3 for six specific temperature intervals.

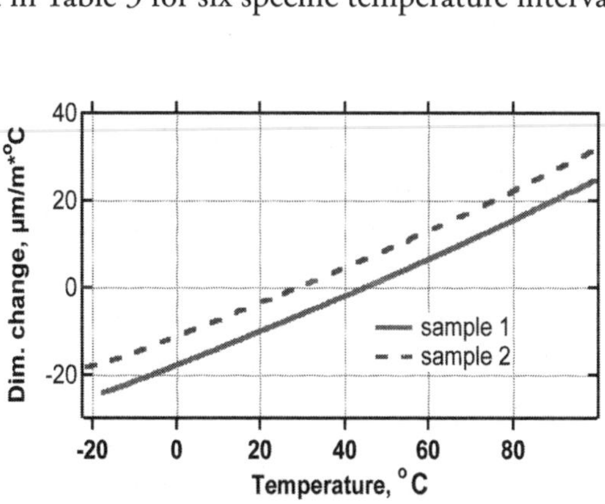

Figure 1. RX-55-AE-5 dimensional change versus temperature.

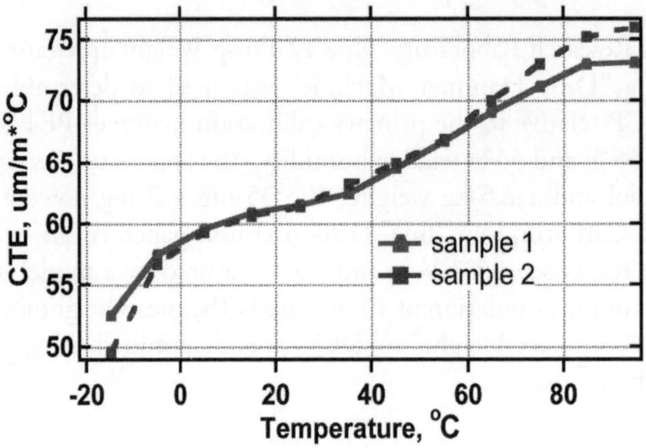

Figure 2. CTE versus temperature for RX-55-AE-5 over the temperature range of -20°C to 100°C.

Table 3. RX-55-AE-5 CTE values, α, μm/m°C

Sample	-20°C to 0°C	0°C to 20°C	20°C to 40°C	40°C to 60°C	60°C to 80°C	80°C to 100°C
1	55	60	62	66	70	73
2	56	60	62	66	72	76

Decomposition kinetics

The basic starting equation gives the rate of reaction in terms of a rate constant times a function of the extent of reaction [12, 13]:

$$\frac{d\alpha}{dt} = k(T)f(\alpha) \qquad (2)$$

where $f(\alpha)$ describes the conversion dependence of the rate and the temperature dependence of k is typically described by an Arrhenius law (k=Aexp(-E/RT)), where A is a frequency factor, E is an activation energy, and R is the gas constant.

The simplest methods of kinetic analysis used in Kinetics05 is Kissinger's method [12], in which the shift of temperature of maximum reaction rate (Tmax) with heating rate (β) is given by

$$\ln\left(\beta / T_{\max^2}\right) = -E / RT_{\max} + \ln(AR / E) \qquad (3)$$

Friedman's isoconversional method [14] involves an Arrhenius analysis at constant levels of conversion, and we determined the apparent first-order frequency factor and activation energy at 1% intervals using both LLNL and AKTS kinetics analysis programs.

Model fitting used the extended Prout-Tompkins model,

$$f(\alpha) = \left(1 - q(1-\alpha)\right)^m (1-\alpha)^n \tag{4}$$

where α is the fraction reacted, n is the reaction order, m is a nucleation-growth parameter, and q is an initiation parameter set equal to 0.99. When m is zero, this model reduces to an nth-order reaction.

For the SDT data, we considered only mass loss for kinetic analysis. Instability of the DTA baseline meant that results were inconclusive as to whether the mass loss corresponds to an endothermic or exothermic reaction or some combination thereof. Kissinger's method yielded A = 2.19×1013 s-1 and E = 173.5 kJ/mol, with a standard error of 8.7 kJ/mol on the activation energy. The Freidman parameters are shown in Figure 3 and are approximately equal to the Kissinger value. The AKTS code with its baseline optimization feature has less noise at low conversion, but the two programs agree very well overall.

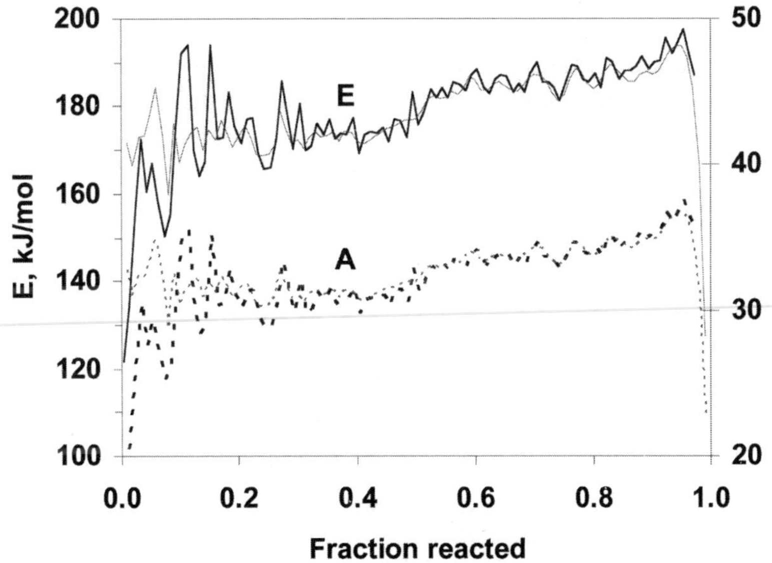

Figure 3. Conversion dependence of A (dotted line) and E (solid line) determined by Friedman's method. The bold lines used Kinetics05 and the thin lines used Thermokinetics.

For the model fitting approach, the shape of the reaction profile—the sharp decline past the maximum reaction rate and the direct approach to baseline—suggests an nth-order reaction with n<1. Simultaneously fitting the three cumulative reaction profiles to such a model gave reaction parameters of A = 6.20×1013 s-1, E = 172.9 kJ/mol, and n = 0.65. A comparison of the measured fractions reacted with those calculated from both the isoconversional and nth-order fits is shown in Figure 3. The reaction profile is not an ideal nth-order reaction, so the nth-order fit shows significant deviation.

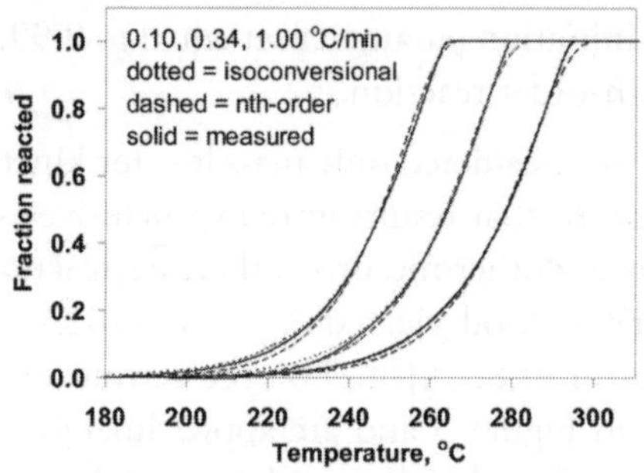

Figure 4. Comparison of integrated experimental data with that calculated from the isoconversional and nth-order fits.

A better fit can be obtained to the latter stages of reaction by fitting the reaction rates instead of the cumulative reacted, but the fit to the early portion of the reaction is worse. Consequently, m was also optimized against reaction rates, and the results are shown in Figure 5. The measured and calculated fraction reacted curves are essentially superimposable. The negative nucleation-growth parameter has no physical meaning; it merely serves as a method of fitting the profile shape and width simultaneously.

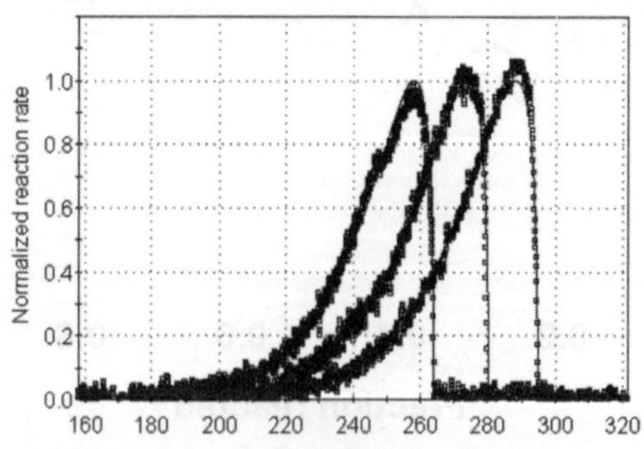

Figure 5. Comparison of measured reaction rates with those calculated from a fit to an extended Prout-Tompkins model, with A = 9.06×10^{13} s^{-1}, E = 182.8 kJ/mol, n = 0.315, and m = -0.32.

The DSC experiments had variability in peak shape, possibly due to confinement conditions of the sample. Two runs at 0.1 and 1.0 oC/min were selected as most representative of pinhole hermetic pan conditions and analyzed using the Kissinger and Friedman methods. Kissinger's method gave A = 8.65×1017 s-1 and E = 244.3 kJ/mol. The Friedman parameters are shown in Figure 6 and a comparison of measured and calculated reaction rates is given in Figure 7. The activation energy varies generally between 200 and 250 kJ/mol, and the value near mid conversion agrees with Kissinger's method. The difference in sharpness of the reaction profiles at the two heating rate causes oscillations in the activation energy near 90% conversion. This also causes some instability in the calculated rates at temperatures above the peak reaction rate, but the results are still generally quite good. A model fitting approach would need at least two reactions.

Comparison of the reaction rates in Figures 5 and 7 indicated that the heat release in a sealed pan occurs at ~50 oC higher than mass loss in an open pan, suggesting that sublimation reactions are reduced and secondary reactions are enhanced. This is also reflected in a ~50 kJ/mol increase in the activation energy deduced from a comparison of Figures 4 and 6.

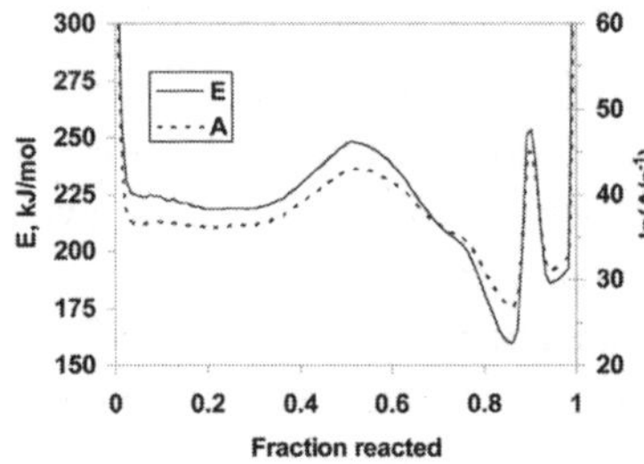

Figure 6. Friedman isoconversional kinetic parameters for DSC heat release from RX-55AE-5 heated at 0.1 and 1.0 °C/min in a pinhole hermetic pan.

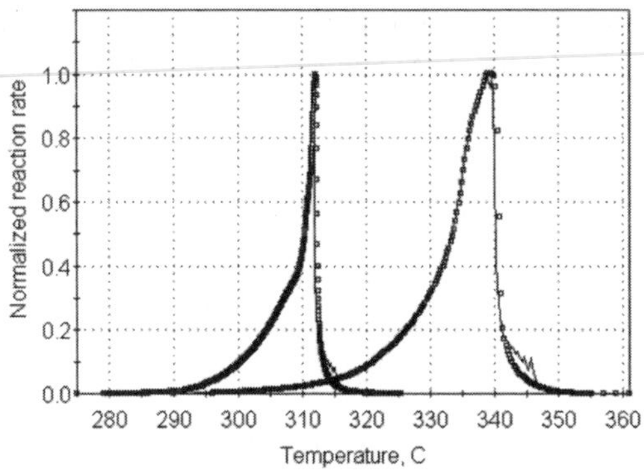

Figure 7. Comparison of measured and calculated reaction rates of RX-55-AE-5 using Friedman's method.

Reaction to Various Stimuli

Small scale testing (SST) of energetic materials and other compounds is done to determine sensitivity to various stimuli, including friction, impact and static spark. These tests establish parameters for the safety in handling and carrying out experiments that will describe the behavior of materials that are commonly stored for long periods of time. In the friction sensitivity test, RX-55-AE-5 was observed to have 1/10 "goes" at 36.0 kg at 22°C and a relative humidity of 64%. RX-55-AE-5 was compared to an RDX calibration sample, which was also found to have 1 event in 10 trials at 12.4 kg. This material is not considered to be friction sensitive. In the spark sensitivity test, no reactions were observed (0/10) at 10 kV (1J). This material is not spark sensitive under these specific conditions. In the impact sensitivity test, the Dh_{50} for RX-55-AE-5 was 170 ± 1.0 cm. For comparison, the Dh_{50} of PETN, RDX, and Comp B-3 were measured at 15.5, 34.5, and 41.4 cm, respectively.

Discussion

Our various determinations of thermal expansion of RX-55-AE-5 as a function of temperature agree well with each other. However, there are no published data on this new formulation, so we cannot compare to literature results.

We can compare our RX-55-AE-5 results to those for TATB formulations, which our new material is intended to replace. Figure 8 shows expansion data for various lots of PBX 9502 (pressed cold and hot) to give an idea of the spread in CTE values depending on preparation conditions. Table 4 gives the measured mass, length, diameter and calculated densities of the PBX 9502 samples shown in Figure 8. Other parameters, such as particle size, wet aminated or dry aminated, should be considered, but they were not varied in the present work. Maienschein and Garcia [15], who show that variations in CTE values can result from factors such as pressing at elevated temperature versus room temperature. Apparent residual strain is incorporated into the sample from the sample pressing process and is released during heating. Irreversible ratchet growth was observed in LX-17 when thermally cycled and continued to expand when held between 250-285°C for 4-5 hours.

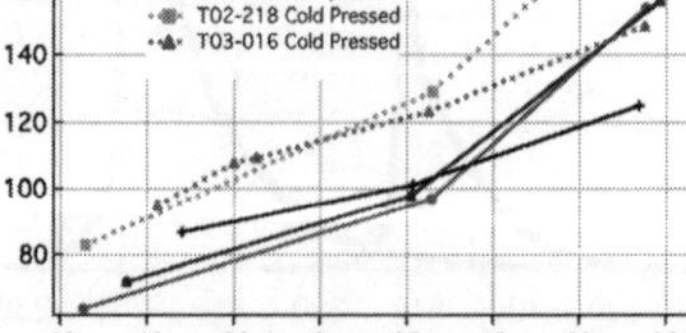

Figure 8. CTE values for various lots of PBX 9502 measured in our laboratory.

Table 4. PBX 9502 sample properties.

Material	Sample I.D.	Lot#	Conditions	mass, g	length, cm	diameter, cm	density, g/cm^3
PBX 9502	T02-158	C-382	Hot pressed	0.374	0.631	0.628	1.91
PBX 9502	T02-219	C-382	Hot Pressed	0.374	0.628	0.631	1.91
PBX 9502	T02-709	C-382	Hot Pressed	0.375	0.630	0.631	1.90
PBX 9502	T02-218	C-382	Cold Pressed	0.375	0.630	0.634	1.88
PBX 9502	T03-016	C-382	Cold Pressed	0.375	0.632	0.634	1.88

Variation in the CTE values for PBX 9502 was not the focus of this work; rather it was how the CTE values for RX-55-AE-5 would compare to a variety of IHE materials. The comparison of RX-55-AE-5, TATB, LX-17 and PBX 9502 in Table 5 shows that RX-55AE-5's CTE values change the least over the temperature range of -25°C to 75°C. Figure 9 makes this comparison graphically. TATB and its formulations have similar slopes that are much greater than that of RX-55--AE-5. While the CTE of RX-55-AE-5 is between the two TATB formulations at -25°C, its weaker temperature dependence makes its CTE substantially smaller than either LX-17 or PBX 9502 at 75°C.

Table 5. CTE values (μm/m•°C) for RX-55-AE-5, TATB, LX-17 and PBX 9502.

Temperature, °C	RX-55-AE-5	TATB	LX-17	PBX 9502
-25	56	74	47	64
25	62	105	76	97
75	71	138	107	154

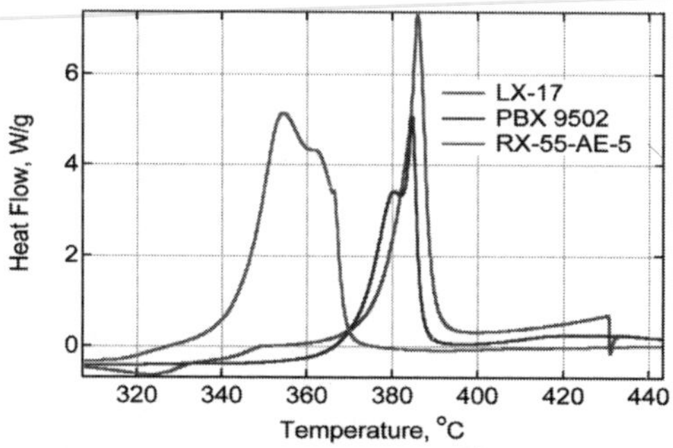

Figure 9. CTE plotted values for RX-55-AE-5, TATB, LX-17 and PBX 9502

Table 6 compares the Small Scale Safety Test (SST) values for RX-55-AE-5, PBX 9502, LX-17, TATB and LLM-105 determined in this work. Thermal decomposition is used to determine the thermal stability of a material with respect to heat [16]. TATB compounds are well known for their stability towards heat and are known to decompose at higher temperatures from 370 to 385°C. In Figure 10 we have overlaid thermal decomposition scans of RX-55-AE-5 with PBX 9502 and LX-17 to be compared for thermal stability. RX-55-AE-5 shows a broad decomposition peak. It is not understood why the decomposition peak is so broad and appears to have at least two peaks present. It is clear that the base of the decomposition peak is approximately 10°C broader than for PBX 9502 and LX-17. The RX-55-AE-5 sample decomposition temperature appears to be approximately 27°C lower than LX-17 and 25°C lower than PBX 9502.

RX-55-AE-5, LLM-105, PBX 9502, LX-17 and PBX 9502 were compared. RX-55-AE-5 showed no sensitivity to friction by the method we used in this experiment. The ESD threshold of 1.0 J of energy applied to or RX-55-AE-5 and the other comparison materials showed no reaction for this experiment. Dh50 for RX-55-AE-5 was slightly higher than the pure LLM-105 material and slightly lower than that of PBX 9502, LX-17 and TATB.

Table 6. Summary of Small Scale Safety test results
a: Samples were approximately 35 mg in mass; no density or sample dimensions are available. #1: LLNL Explosives Handbook. TBD: to be determined later.

Test	**RX-55-AE-5**	**LLM-105**	**PBX 9502**	**LX-17**	**TATB**
DSC (onset of exotherm, °C)	343	350	383	377	385
Friction (# of goes)	0/10	0/10	0/10	0/10	0/10
ESD threshold (1.0 J)	0/10	0/10	0/10	0/10	0/10
Dh_{50} (cm)	170	158	> 177	> 177	> 177
E (kJ/mol)	142	TBD	TBD	TBD	TBD
Cp, J/g°C	0.950	0.931	1.098	1.125	0.991
κ, W/m°C 2 26.85°C	0.160	TBD	0.552	0.799	0.691
CTE, μm/m°C @ 21°C	61	57	93	60	85

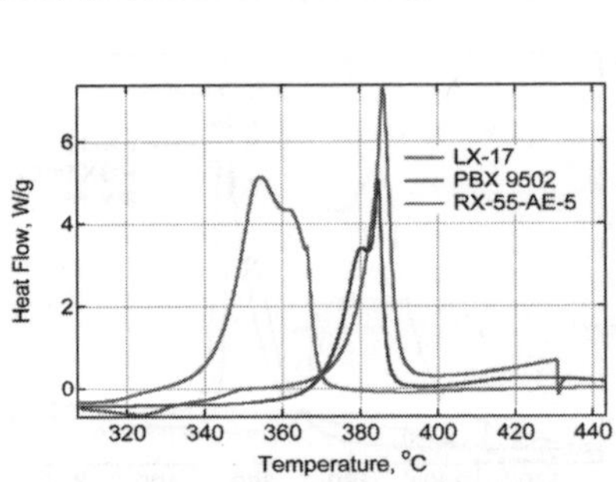

Figure 9a. Overlay of DSC decomposition scans for RX-55-AE-5, PBX 9502 and LX-17.
The heat capacity of RX-55-AE-5, PBX 9502 and LX-17 is shown in Figure 10. The heat capacity of RX-55-AE-5 is less than PBX 9502 or LX-17. Heat capacity for LX-17 is slightly higher than PBX 9502.

The thermal conductivity reported here was measured using a TA Instruments 2920 DSC, ran in the modulated mode, thus is commonly referred to as a MDSC [17, 18, 19]. Future work is planned to measure thermal conductivity, κ, for RX-55-AE-5 by means of a pulse laser system.

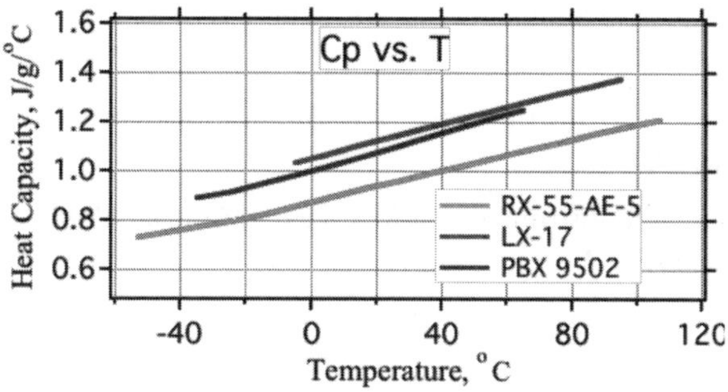

Figure 10. Heat capacity of RX-55-AE-5, PBX 9502 and LX-17 versus temperature.

We have determined that RX-55-AE-5 compares well to PBX 9502 and LX-17. While making this assessment, it became obvious that existing historical data should be compiled in an accessible format, such as the LLNL Explosives Handbook, to facilitate future comparisons.

Acknowledgements

We would like to thank Dr. Craig Tarver and Dr. Phil Pagoria for their technical discussions, help and support on this project. This work performed under the auspices of the U.S. Department of Energy by the University of California, Lawrence Livermore National Laboratory, under Contract number W-7405-Eng-48. UCRL-CONF-212824

References

1. T. Tran, R. Simpson, P. Pagoria, M. Hoffman, Written Communication: Updates on LLM-105 development as an insensitive high explosive booster material, JOWOG 9, Lawrence Livermore National Lab, LLNL report UCRL-PRES-145524 (Oct. 2001).

2. P. F. Pagoria, G. S. Lee, R. D. Schmidt and A. Mitchell, Synthesis and Scale-Up of 2,6diamino-3,5-dinitrpyrazine-1-oxide (LLM 105) and 2,6-diamino-3,5-dinitropyridine-1oxide (ANPyO), Submitted to PEP (Apr. 2005).

3. J. R. Kolb and H. F. Rizzo, Growth of 1,3,5-Triamino-2,4,6-trinitrobenzene, Part I. Anisotropic Thermal Expansion, PEP 4, 10–16 (1979).
4. R. D. Gilardi and R. J. Butcher, 2,6-Diamino-3,5-dinitro-1,4-pyrazine 1-oxide, Acta Cryst. E57, o657-o658 (2001).
5. B. M. Dobratz, "LLNL Explosives Handbook-Properties of Chemical Explosives and Explosive Simulants"; LLNL Report UCRL-52997, March 16, 1982; with Errata of Jan. 28, 1982.
6. Anonymous, TA Instruments, TA Applications Brief, TA 019, 1998, New Castle, DE.
7. Anonymous, TA Instruments, TMA 2940 Manual, 1996, TA Instruments, New Castle, DE.
8. J. W. Dodd and K. H. Tonge, "Thermal Methods," Jon Wiley and Sons, Crown Copyright, 1987.
9. L. R. Simpson and M. F. Foltz, LLNL Small-scale Friction Sensitivity (BAM) Test, LLNL Report UCRL-ID-124563 (Jun. 1996).
10. L. R. Simpson and M. F. Foltz, LLNL Small-scale Static Spark Machine: Static Spark Sensitivity Test, LLNL Report UCRL-ID-135525 (Aug. 1999).
11. L. R. Simpson and M. F. Foltz, LLNL Small-scale Drop-hammer Impact Sensitivity Test, LLNL Report UCRL-ID-119665 (Jan. 1005).
12. H. E. Kissinger, Reaction Kinetics in Differential Thermal Analysis, Anal. Chem. 29, 1702–1706 (1957).
13. T. M. Massis, P.K. Morenus, D. H. Huskisson and R. M. Merrill, Stability and Compatability Studies With The Inorganic Explosive CP, J. of Hazardous Materials 5, 309–323 (1982).
14. H. L. Friedman, "Kinetics of Thermal Degradation of Char-Forming Plastics from Thermogravimetry. Application to a Phenolic Plastic," J. Polymer Sci. C6, 183–195 (1964).
15. J. L. Maienschein, F. Garcia, Thermal Expansion of TATB-based explosives from 300 to 566 K, Thermochim. Acta 384, 71–83, (2002).
16. A. K. Burnham, R. K. Weese and W. J. Andrzejewski, Kinetics of HMX and CP Decomposition and Their Extrapolation for Lifetime Assessment, LLNL Report UCRLTR-208411 (Dec. 2004); also, these proceedings.
17. Modulated DSC Compendium, Basic Theory and Experimental Conditions, TA Applications Brief, TA Instruments, New Castle, Delaware.

18. S. M. Marcus and R. L. Blaine, "Thermal conductivity of polymers, glasses and ceramics by modulated DSC," Thermochim. Acta 243, 231–239 (1994).
19. R. K. Weese, Thermal conductivity of Tetryl by modulated differential scanning calorimetry, Thermochim. Acta 429, 119–123 (2005).

Ballistic Protons and Microwave-Induced Water Solitons in Bioenergetic Transformations

Reuven Tirosh

ABSTRACT

Active streaming (AS) of liquid water is considered to generate and overcome pressure gradients, so as to drive cell motility and muscle contraction by hydraulic compression. This idea had led to reconstitution of cytoplasm streaming and muscle contraction by utilizing the actin-myosin ATPase system in conditions that exclude a continuous protein network. These reconstitution experiments had disproved a contractile protein mechanism and inspired a theoretical investigation of the AS hypothesis, as presented in this article. Here, a molecular quantitative model is constructed for a chemical reaction that might generate the elementary component of such AS within the pure water phase. Being guided by the laws of energy and momentum *conservation and by the physical chemistry of water, a vectorial electro-mechano-chemical conversion is considered, as follows: A ballistic H^+ may be released from H_2O-H^+ at a velocity of 10km/sec, carrying a kinetic energy of 0.5 proton*volt. By coherent exchange of microwave photons during 10-10 sec, the ballistic proton can induce cooperative precession of about 13300 electrically-polarized water molecule dimers, extending along 0.5 μm. The dynamic dimers rearrange along the proton path into a pile of non-radiating rings that compose a persistent rowing-like water soliton. During a lifetime of 20 msec, this soliton can generate and overcome a maximal pressure head of 1 kgwt/cm^2 at a streaming velocity of 25 μm/sec and intrinsic power density of 5 Watt/cm^3. In this view, the actin-myosin ATPase is proposed to catalyze stereo-specific cleavage of H_2O-H^+, so as to generate unidirectional fluxes of ballistic protons and water solitons along each actin filament. Critical requirements and evidential predictions precipitate consistent implications to the physical chemistry of water, enzymatic hydrolysis and synthesis of ATP, trans-membrane signaling, intracellular transport, cell motility, intercellular interaction, and associated electro-physiological function. Sarcomere contraction is described as hydraulic compression, driven by the suction power of centrally-oriented AS. This hydraulic mechanism anticipates structural, biochemical, mechanical and energetic aspects of striated muscle contraction, leading to quantitative formulation of a hydrodynamic power-balance equation yielding a general force-velocity relation.*

Introduction

The Motor-Protein Paradigm and the Active-Streaming Hypothesis

The phenomena of cytoplasm streaming and muscle contraction are generally related to enzymatic catalysis of ATP hydrolysis by the actin-myosin (A-M) system [1]. Therefore, by assuming a common driving mechanism, a simple question is raised: What comes first -streaming or contraction?

Up to 1954, tension generation in striated muscle was related to contraction of the filamentous protein network within each sarcomere. However, during sarcomere shortening, the actin and myosin filaments where observed to slide past each other, while maintaining constant length. This seminal observation had actually revealed no protein contraction. Yet, the contractile-protein paradigm prevailed, being confined to a tensile force or a lever-like action by the A-M cross-bridges [2, 3]. The new hypothesis invited a terminological shift from "contractile proteins" to "motor proteins". This established view have raised minor interest in cytoplasm streaming, being considered as a passive flow caused by contraction of the A-M network.

Since 1972, the present author had advanced the investigation of a cause-effect reversal between streaming and contraction, by first raising the following argument: A mechanical protein action, unlike a fluid action, must rely on a continuous protein network. This argument was tested by reconstitution of protein transport, cytoplasm streaming and muscle contraction, utilizing the A-M ATPase system without forming a continuous protein network [4-13]. All these experiments had clearly revealed that the A-M system is uninvolved mechanically in tension generation. An early theoretical model for the alternative fluid mechanism was suggested [14, 15], as further developed in this article.

In a reciprocal relation to the hydrodynamic concept of passive streaming down pressure gradients, the new concept of active streaming (AS) is defined as being able to generate and overcome pressure gradients. The basic question is -what might drive AS within the pure water phase?

A Proton-Induced Water Soliton

A vectorial electro-mechano-chemical transformation into AS in water is described by the classical laws of momentum and energy conservation, while quantitatively relying and even reflecting on the physical-chemical properties of liquid water. Consider the cleavage of a high-energy complex into two products of relatively small and large masses (m). By energy and momentum conservation, the two products gain an equal and opposite momentum (p), while the low-mass product carries most of the kinetic energy (Ek) released, namely:

$$p_1 = -p_2 \rightarrow Ek_1 / Ek_2 = \left(p_{1^2} / (2 * m_1)\right) / \left(p_{2^2} / (2 * m_2)\right) = m_2 / m_1 . \quad (1)$$

The same laws, however, impose some restrictions that must be overcome for further efficient energy transfer from a ballistic product into massive streaming. First, molecular and electrical scattering must be avoided. Therefore, atoms and electrons are unsuitable products. A unique, electrically charged, product in water is a ballistic H+ released from H2O-H+. As shown below, the H+ kinetic energy is small enough to avoid also quantum-mechanical dissipation by electron excitation of surrounding molecules. Second, by the basic laws, a direct energy transfer into linear translation of massive streaming is not allowed. Therefore, electromagnetic transformation is considered to induce a persistent intermediate state of cooperative molecular circulation, which could drive AS by a rowing-like mechanism. Thus, by exchange of microwave photons, a ballistic proton is proposed to induce cooperative precession of many, electrically-polarized, water molecule dimers.

Dimer precession, rather than single molecular rotation, is considered for several reasons. First, by dimer precession, the proton kinetic energy is coherently shared by circulation of the whole mass of each water molecule. Second, the electrically-polarized dimers can reorganize to form non-radiating axial rings, so as to compose a persistent rowing soliton. Notice that the plane of circular precession is orthogonal to that of the rings, thus avoiding their disruption. Third, persistent cooperative precession can produce a peripheral rowing-like action, whereby the soliton might generate and overcome pressure gradients along the original proton path. Fourth, dimer precession must rely on intrinsic spin angular momentum. This spin is related by the virial theorem to a chain-like electrical interaction of the polarized dimers. Fifth, the frequency of dimer precession accounts for the unique mode of microwave absorption by liquid water. Thus AS might be driven by the inertia of molecular circulation rather than translation.

All these ideas precipitate major structural and energetic aspects of the pwason model. They are quantitatively evaluated in Supplementary Box 1, and further discussed below.

The High-Energy Complex of H_2O-H^+

The free energy of H_2O-H^+ is obtained by the Maxwell-Boltzmann relation:

$$\begin{aligned}&[H_2O-H^+]/[H_2O]=10^{-7}/55.6=EXP\{-\Delta E_1/kT\}\rightarrow\\&\Delta E_1=11.5kcal/mole=0.5proton*volt=8*10^{-13}erg=20kT \text{ at } 20°\text{C.}\end{aligned} \tag{2}$$

H2O-H+ is therefore a minor high-energy component of liquid water at thermal equilibrium. The pH-temperature dependence of liquid water is closely described by Eq.2. The de Broglie quantum-mechanical relation entails a classical view of a protonic molecular orbit enclosing the two lone-pair electrons of the oxygen atom in H2O-H+ (Fig1a). In this molecular view, a bound kinetic H+ is held ready to be released upon cleavage of the parent H2O molecule. Supportively, dissociative photoionization of water reveals a minor fraction of ballistic protons, having a kinetic energy of up to 0.5 proton*volt [16].

The Electrically Polarized Dimers

Water-molecule dimers are proposed to be the main, low-energy, component of liquid water. The molecular distances under interaction of two polarized dimers are presented in Fig.1a and in Supplementary Table1. This molecular structure is consistent with measurements of X-ray and neutron scattering [17], and with recent reports of X-ray absorption and emission spectroscopy, revealing proton delocalization [18], strong electron sharing [19] and double, rather than tetrahedral, hydrogen bonding between water molecules [20]. Theoretical simulation of liquid water has also uncovered a new role for 2-fold hydrogen bonds [21].

An extended electrically-polarized rigid dimer is required by the AS model. This molecular compound is proposed to rely on a central pair of double-hydrogen bonds, formed under covalent electron resonance of orthogonal 2p molecular orbitals. The shorter end extensions of the electrically polarized dimer preserve the corresponding covalent configuration of a single water molecule dipole. The dimer dipole length (0.42nm) is three times that of a free water molecule dipole (0.14nm). Assuming the same electric charges of a water-molecule dipole (1.8 D along 0.14 nm), the triple extension of the dimer configuration gives rise to an electric dipole of 5.4 D, which is effectively 2.7 D per liquid water molecule, as measured. The dimer dipole extension strengthens short-distance electrical attraction between adjacent dimers through electrovalent double hydrogen bonds. This hybrid double hydrogen bond configuration allows both for internal spin and precession within each rigidly bound dimer, along with dynamic flexible orientation due to Coulomb attraction between dimers. All these features are required for cooperative precession of axial rings that may effectively compose the rowing soliton (Fig. 1b).

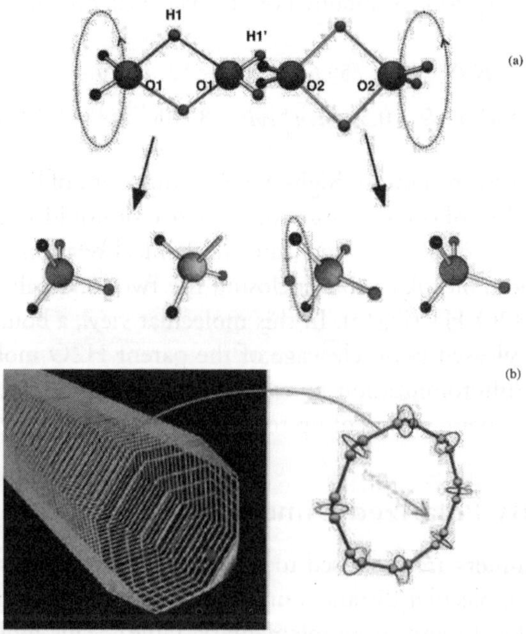

Figure 1. The water dimers and the pwason (a) Internal spinning of two water molecule dimers, and their dissociation into water molecules and ions. Note the proton orbital in $H_2O\text{-}H^+$. $R_{O1\text{-}O1} = R_{H1\text{-}H1} = 2.8*10^{-8}$ cm, $R_{H1'\text{-}H1'} = R_{O1\text{-}O2} = 1.4*10^{-8}$ cm $_{RO1\text{-}H1} = 2*R_{O1\text{-}H1'} = 1.9*10^{-8}$ cm (b) Proton-driven precession of water-molecule dimers forming octal rings that compose a rowing soliton.

The Dimer's Spin and Precession

Dimer precession must rely on an intrinsic angular momentum, which is related to axial spinning of the four bound protons around the O-O axis (Fig.1a). A spin angular momentum of 25$\hbar$ is obtained by the Virial Theorem, related to the electrical cohesive energy of double hydrogen bonding between water molecule dimers. Therefore, 25 states of dimer precession are anticipated, where in each state, the angular momentum of precession tends to counterbalance the projection of the spin angular momentum on the precession axis. Thus, at the highest energy state, this dimer precession has a total angular momentum of 1$\hbar$ and kinetic energy of $6*10^{-17}$ ergs. These values correspond to the angular momentum and energy of microwave photons at a frequency of about 10^{10} sec^{-1}. In this state of dimer precession, each water molecule circulates at the maximal available radius, at an angular momentum of $\hbar/2$, and at a velocity of 14 m/sec. (This dynamic state of dimer spin and precession is a vivid presentation of an intrinsic quantum-mechanical coupling between a boson and a couple of fermions, respectively.) The above values determine the H_+ flight duration of 10-10 sec, the path length of 0.5 µm, and the number of 13300 water molecule dimers that share the H_+ energy.

The Rowing Soliton

The soliton's active propulsion against tangential friction forces can generate a maximal pressure-head of 1kgwt/cm^2, corresponding to the kinetic energy density of water dimer precession. Compared to the gravitational hydrostatic pressure-gradient in water, the localized pressure-gradient within a single pwason is twenty million times greater (10m / 0.5µm). The distant spreading of the dipolar pressure-gradient field of a single pwason is proposed to induce backward buoyant-like body forces and passive, fountain-like, streaming. This cooperative molecular mechanism of the soliton is incorporated into a reverse active version of the hydrodynamic laws of Archimedes, Bernoulli, Newton and Poiseuille, yielding for the pwason a streaming velocity of 25 µm/sec and a life-time of 20 msec.

The Pwason Model

The above arguments are integrated to construct the following model: Under spontaneous cleavage of the high-energy H_2O-H^+ complex, a ballistic H^+ is released at initial velocity of 10 km/sec, corresponding to a kinetic energy $\Delta E1 = 8*10^{-13}$ ergs. By coherent exchange of microwave photons of $6*10^{-17}$ ergs, the H^+ kinetic energy is transformed during 10-10 sec, along 0.5µm, into cooperative precession of about 13300 electrically-polarized water molecule dimers. The dynamic polarized dimers rearrange into non-radiating octal rings, forming a persistent rowing soliton having a length of 500nm and a diameter of 1.4 nm. During a lifetime of 20 msec, this rowing soliton develops a streaming velocity of 25 μm/sec, while being able to generate and overcome a maximal pressure-head of 1kgwt/cm^2, with an intrinsic power density of 5 Watt/cm^3. These quantitative features of the pwason model will be examined throughout this article.

Further Physical-Chemical Implications for Liquid Water

In Scheme #1, endothermic water dimer dissociation is related to the molecular ionization and evaporation energies (ΔE_2, ΔE_3), which are revealed, respectively, by the heat released in the reverse processes of acid-base neutralization and water condensation.

Scheme 1

$$\begin{array}{c} \qquad\qquad +2^*\Delta E_2 \qquad\qquad +2*\Delta E_3 \\ H_2O-H^+ + OH^- \Leftarrow H_2O = H_2O \Rightarrow H_2O + H_20 \end{array}$$

ΔE2 and ΔE3 are approximately equal. Notably, the same molar value is also found for the free energy of H2O-H+ ions (ΔE1 in Eq.2, above), for ATP hydrolysis (ΔE4 in Eq.4, below), and for the heat + work per mole of ATP hydrolysis (ΔE5) as measured during closed cycles of muscle contraction [22]. Namely:

$$\Delta E_1 = \Delta E_2 = \Delta E_3 = \Delta E_4 = \Delta E_5 = \Delta E \qquad (3)$$

where ΔE=8*10^{-13} ergs = 0.5 proton*volt = 48kJoules/mole=11.5kcal/mole.

The relatively high proton mobility in water is related to spontaneous release of ballistic protons from H2O-H+. The life-time of a ballistic proton in water is related to the period of microwave photons in the range of 100psec, as recently verified [23]. A longer range and life-time of proton mobility is anticipated within the ice phase, lacking microwave absorption by dimer precession.

A flow of water towards the cathode is anticipated under electric field application. This flow may be related to electrical orientation of spontaneous ballistic protons that induce AS.

The reversible function of a pH glass electrode in water is elucidated by a free passage of ballistic protons through a thin glass membrane, when protons are released from and recaptured on both sides as H2O-H+ ions, yielding the pH-indicative electric potential at thermal equilibrium. A similar electrochemical performance is expected across biological membranes, where passage of ballistic protons might induce transient gating of fluid, molecular, and ionic currents [24].

The first water anomaly, i.e. volume contraction upon ice melting, may be due to a transformation from tetrahedral hydrogen bonding to the more compact packing of the electrically polarized dimers. This phase transition elucidates the anomalous phenomenon of high and low density states of water and of ice, at lower temperatures and under higher pressures [25, 26].

The 25 energy states of dimer precession predict a testable thermal ladder for water between 0 and 100 °C. If evenly spaced every 4°C, this ladder might explain the second anomaly of water. This thermal ladder might also explain the unexpected thermal anomalies seen in the temperature responses of living systems at about every 16 °C increments [27].

There are two easy ways to grasp the soliton's pressure-head of ΔP = 1kgwt/cm2 = 1kgwt*cm/cm3, by stemming from the water molecule precession velocity, V = 14m/sec. First, this ΔP value is derived as the kinetic energy density, namely ΔP=½ ρ*V2, where ρ = 1gr/cm3. Second, the same pressure-head is produced under gravitation by a 10 m water column, where such a column can be created by upward water flow with an initial velocity of about 14 m/sec.

By the same reasoning, if surface winds drive long-distance sea waves due to cooperative energy absorption by water dimer precession, then such waves may reach maximal amplitude of 10m, or peakto-peak height of 20m [28]. (Standing waves in closed straits might reach twice this amplitude). Cooperative molecular inertia of water dimer precession at Vw = 14m/sec = 50km/hour, may also account for the velocity of large-scale ocean waves, as well as for their vertical circulation profile. The same mechanism is related to generation and propagation of a water solitary wave following abrupt stopping of a ship in a narrow canal [29]. The same mechanism is proposed for microscopic Bekesy waves, propagating at 14m/sec along the cochlear canal of the ear [30]. Similarly, microscopic turbulence can cause abrupt changes in micro-particle velocities up to 14m/sec [31]. Thus, the liquid water phase mediates diverse macroscopic and microscopic effects that are quantitatively related to the quantum of water dimer precession.

In this view, it is tempting to speculate that even the atmospheric pressure of 1kgwt/cm2 at sea level may manifest a balanced steady state with surface generation of pwasons.

Furthermore, the spontaneous surface ejection of kinetic protons might leave the oceans at a negative electric potential. Trapped in the atmospheric vapor, the protons can be involved in the cohesive dispersion of clouds. Various lightening effects might therefore be due to massive H+ currents.

Similarly proton-driven micro-atmospheric lightening effects are presumably manifested in single-bubble sono-luminescence in water [32]. Note that the ambient 20 kT ballistic proton is equivalent to a thermal plasma nucleon at 20*300°K = 6000°K.

Biochemical Implications for the Actin-Myosin (A-M) System

Unidirectional vectorial fluxes of protons and solitons may be generated along a single actin filament by stereospecific cleavage of $H_2O\text{-}H^+$. Enzymatic catalysis of this reaction is proposed for ATP hydrolysis by myosin heads, while being attached to the actin filament. The reaction may take place at the P-loop of the apical side, on the molecular clefts of myosin heads attached to actin [33] (Fig.2a). The mechano-chemical proton-burst reaction occurs concomitantly with hydration of the terminal P-O-P bond, as given by:

$$ADP-O-P^4+H_2O-H^+ \rightarrow ADP-O-H---OH-P+H^+ +\Delta E_4. \quad (4)$$

According to Eq.1, the kinetic proton carries the free energy ΔE4, which is equal to ΔE in Eq.3. The ATP anion is proposed to interact with H2O-H+, rather than H2O as generally accepted. Indeed, the rate of spontaneous ATP hydrolysis increases in acidic solutions.

The enzymatic cycle for the A-M system of striated muscle is described in Scheme#2. It is consistent with structural, kinetic, and isotopic measurements [34-36]. The enzymatic cycle is described in three steps between three states:

Scheme 2

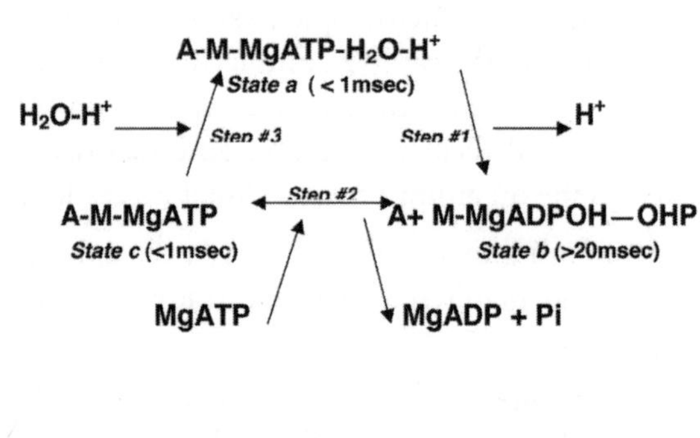

State a is a short-lived high-energy stereo-specific rigid complex. The stereospecific configuration might rely on the quaternary ionic complex of Mg+2 with ATP-4 that is proposed to maintain a ring-like rigid framework of the adenine-ribose-triphosphate complex. This complex is attached to the P-loop of the catalytic site, which is exposed to the water phase, opposite to the A-M binding site. At Step #1, the H+ ejection occurs concomitantly with an equal recoil momentum and with an abrupt increase of negative charge at the myosin head. Both effects enhance the detachment of a stable product-loaded myosin head. In State b, the pwason remains active during 20msec, while preventing effective reattachment of the myosin head to actin. This dominant state of protein dissociation during the enzymatic cycle ensures minimal resistance to filaments sliding. In Step #2, an effective A-M binding enhances product-substrate exchange under thermal equilibrium. Thus, the release of ADP and the inorganic phosphate Pi is delayed after completion of the hydraulic power stroke [37]. (Step #2 is regulated by the troponin-tropomyosin system at micromolar concentrations of calcium ions.) The duration of State c, at physiological millimolar concentrations of MgATP, is relatively short. Indeed, the duration of the A-M complex is estimated to hold less than 5% of the enzymatic cycle period.

This cyclic interaction profile is considered for other types of myosin that determine the pwasons direction along the actin filament [38-40]. A similar mechanism based on proton-induced water solitons is suggested for ATP hydrolysis by dynein or kinesin along microtubules, or by G-actin polymerization [41], as well as for other substrates of various hydrolytic enzymes. Thus, a hydraulic drive due to AS is proposed to translocate the ATPase machinery of DNA replication, RNA transcription, or peptide translation. These processes can be effectively associated with active flow of substrates and products across the catalytic regions.

The microwaves proton-soliton coupling is irreversible. For a reversible reaction of ATP hydrolysis, the ballistic protons must be separated from the water phase, as discussed below.

Biochemical Implications for the Reversible Function of Membrane-Bound Atpases

The proton-burst equation (Eq. 4) suggests a reversible pathway for ATP synthesis. In the direction of synthesis, a ballistic H^+ could strike a catalytic site in-between the juxtaposed anions of ADP and Pi. This H^+ impact might extract the high-energy complex of H_2O-H^+, while leaving attached the anhydride ADP-O-P complex. However, this dehydration pathway requires a hydrophobic compartment in order to uncouple the proton movement from AS induction. In the absence of solitons interference at the hydrophobic catalytic region, a much faster reversible catalysis is anticipated, compared to the soluble A-M system. Such a reversible ballistic proton mechanism may be catalyzed by the membrane-bound

F_O-F_1 ATP-synthase of bacteria, mitochondria and chloroplasts, as discussed by the following arguments.

The proposed mechanism involves two main aspects. First, the trans-membrane concentration difference of H2O-H+ determines the direction and rate of the ballistic proton flux and the corresponding amount of ATP synthesis or hydrolysis. Second, in the direction of ATP synthesis, a threshold electric potential difference is obligatory for compensation of a certain dissipation of the proton's kinetic energy. (e.g. with a total energy of 0.5 proton*volt, 0.1 volt compensates for 20% loss). Thus, the ballistic proton mechanism elucidates the thermodynamic concept of a "proton-motiveforce" in the chemiosmotic hypothesis [42, 43]. It accounts quantitatively for the elementary energetic event; it bypasses the problem of trans-membrane proton transportation; and it differentiates between independent roles of chemical potential and electrical potential of H2O-H+ across the membrane [44].

Furthermore, rotational catalysis was proposed for the FO-γε-(αβ)3-F1 complex [45]. This rotation might be electrically driven by the reversible ballistic proton mechanism, as follows. In ATP synthesis, each ADP-Pi loaded αβ-site of the water exposed F1 head is bound in turn to the hydrophobic, topically bent, γε axis. This internal axis is inserted into the FO membrane component so as to form an effective channel for ballistic protons. The αβ catalytic unit is comparable to a myosin head, while the biochemical role of the γε axis is quite similar to that of the actin filament in the enzymatic cycle (Scheme #2). Thus, in the direction of synthesis, the impact of a trans-membrane ballistic H+ within the hydrophobic catalytic region is proposed to drive three concomitant effects: First, dehydration of the terminal phosphate bond results in ATP synthesis. Second, molecular recoil upon the H+ impact dissociates the γε-αβ complex. Third, the abrupt increase of electric charge at the hydrophobic site drives fast relative rotation at 120° towards hydrophobic γε interaction with the next, ADP-Pi loaded, αβ-site. Simultaneous exchange of products and substrates, carried out at the other two, water exposed, αβ sites, might electrically dictate ongoing rotation in the appropriate direction.

Inversely, opposite rotation might be similarly driven under ATP hydrolysis, while generating transmembrane efflux of ballistic protons, which are trapped as H2O-H+ ions. Thus, the reversible ballistic proton mechanism can consume or generate trans-membrane electro-chemical potential of H2O-H+ ions. In this view, the proton-induced electric drive of rotation is considered to share a minor part in energy consumption, compared to the lion share of P-O-P dehydration in ATP synthesis, or of building up the trans membrane electrochemical gradient of H2O-H+ under ATP hydrolysis.

Stepwise rotation was first demonstrated by attaching a relatively long actin filament to an exposed γ axis of a fixed F1 head [46]. For each hydrolytic event, the long flexible filament exhibited a fast 1msec step of unbent rotation through 120°, followed by a pause during about 20msec. Such a fast rotation step may be easily driven by a single pwason, acting along 500nm during a lifetime of 20msec. Therefore, the observation of fast 120° steps of rotation, while preserving radial extension of the long flexible filament, should be considered as another compelling evidence for the AS hypothesis.

Molecular demonstration of ATP synthesis was carried under artificially-forced opposite rotation of the γε axis [47]. This rotation is considered to re-occupy the hydrophobic compartment with a MgADP-Pi loaded αβ-site. Thereby, the hydrophobic catalytic site is being prepared toward spontaneous impact of a ballistic proton for ATP synthesis.

The reversible ballistic proton mechanism is directly acting at the intact catalytic site. In this view, there is no need for intermediate power transmission between two motor proteins, as currently assumed for the FO and F1 components. Like the FO-F1 system, the reversible rotation of bacterial flagella under trans-membrane pH difference [48] could be driven by molecular electrostatic forces, as induced by the ballistic proton mechanism. Thus, intermittent switching between opposite rotations could be controlled by switching between opposite pH gradients.

The biochemical similarity between the A-M and the FO-F1 systems, as well as other soluble and membrane-bound ATPases, is revealed in their utilization of the quaternary MgATP complex within the P-loop catalytic site [49]. This evolutionary conserved framework is probably required for a stereospecific interaction that ensures an appropriate channeling of the ballistic protons. Thus, in reconstitution experiments in the absence of Mg ions, ATP hydrolysis by A-M or FO-F1 is carried out as intensively as by M or F1 alone, but producing no myofibril contraction or outward trans-membrane flux of protons, respectively. In such reconstitution experiments, the γ-axis rotation on the F1 head is still present under hydrolysis of CaATP [50]. This demonstration is consistent with the ballistic proton mechanism, where the electrostatic rotational drive is not the main energy consumer.

Trans-membrane fluxes of ballistic protons through specific channels could induce a flow of water, together with specific, loosely bound, molecules and ions, moving up or down gradients of hydrostatic pressure, chemical concentration, and electric potential. Like the FO-F1 system, these proton-driven pumps may be coupled in closed vesicles to membrane-embedded redox and photonic reactions that produce, or consume, trans-membrane changes in pH and electric potential. This mechanism is related to the photosynthetic pathway of bacterio-rhodopsin, to visual signaling by rhodopsin, and to photosynthetic water oxidation in thylakoids [51]. Various signaling pathways are associated with the hydration or dehydration of phosphate bonds that may be related to a release or absorption of ballistic H+ from or onto H2O-H+. Through changes in pH and/or electric potential, transmembrane signaling by ballistic protons might trigger various physiological effects, such as calcium ion mobilization from internal or external sources.

Molecular and Cellular Implications for the A-M System

Molecular Motility Assays and Intracellular Transport

The development of AS along a free actin filament is predicted to drag the filament along with attached myosin heads. Such a random motion of individual actin filament "rockets" in dilute solution was verified by Doppler broadening of laser-light scattering [13]. A similar drag of a free actin filament in one or the other direction is demonstrated also when the myosin molecules are anchored. By immobilizing the actin filament, the myosin heads might be propelled along with AS. However, a microscopic cargo attached to a myosin head might rather be pushed down the pressure gradient (gradP) due to AS. (According to the relation F = V*gradP, a buoyancy-like hydraulic body force -F is proportional to the body volume – V.) In a single hydrolytic event, this hydraulic force could pull the myosin head along the actin filament in a stepwise sliding movement due to occasional A-M interactions that remain catalytically ineffective, as observed [52]. The integral duration and extent of this stepwise drag due to a single hydrolytic event further verify the theoretical values for a single pwason, namely -its 20msec life-time and its possible extension up to 500nm [53]. In the case of double-head myosin, a hand-over-hand drag along actin is anticipated. Such "walking" movement of the double-head Myosin VI reveals a single head stepping of 60 nm, which is much larger than its 10 nm lever arm. This demonstrates sufficient power of AS to stretch the stalk's dimeric twist [54, 55].

Further evidence for the pwason model is gained by several observations that confirm the prediction of unidirectional drag movement by AS and dynamic electrical polarization of actin filaments due to the ballistic proton flux (Fig.2a), as discussed in the following text.

A free electrically-polarized actin filament, moving under AS, might tend to reorient in the direction of a relatively small electric field, as observed [56]. Indeed, due to the electric force acting on their net negative charge, the filaments moving towards the anode go faster than without an external field, while those moving towards the cathode go slower, as expected.

Attractive hydrodynamic and electrical polarization forces among soluble actin filaments, are expected to orient the filaments so as to drive massive cytoplasm streaming. An imitation of such massive streaming, as observed in dumbbell-shaped micro-plasmodia of Physarum, was reconstituted in stretched micro-capillaries [4].

Lateral hydraulic compression, due to Bernoulli's effect of AS, can form and maintain a narrow channel in the dumbbell-shaped micro-plasmodia of Physarum. Notably, the velocity of AS across the elongated channel might reach up to 1mm/sec, which is much greater than the pwason's theoretical value of 25 μm/sec. This vital increase in streaming velocity in proportion to the length of the streaming pathway is simply explained as follows: serial pwasons generate the same pressure head (ΔP) along a greater distance (L), thus reducing the pressure gradients ($\Delta P/L$) that oppose the AS. Similarly, a relatively high drag velocity is predicted for elongated actin filaments, as observed in motility assays [57].

Shuttle cytoplasm streaming along a narrow channel of Physarum, with a period of about two minutes, was recorded by simultaneous kinetic measurement of two oscillating variables: the opposing pressure head, which is required to stop it, and the autonomic buildup of a phase-lagged electric potential difference, referred to as the "electro-dynamo-plasmo-gram" [58]. Each streaming reversal occurred at the peak of the electric field, and it was associated with a concomitant record of a nerve-like action potential. This complex profile is simply explained by unidirectional association of proton-driven AS along oriented actin filaments, and the autonomic development of electric field, until it induces a cooperative flipping of the electrically polarized actin filaments.

A similar cytoplasmic mechanism was therefore considered for bulk generation and propagation of a nerve action potential and for the hydraulic release of neurotransmitters from synaptic vesicles [59]. Thus, in a neuron at rest, a threshold flux of protons and solitons towards the cell body may be generated along oriented actin filaments in the dendrites and the axon. A negative electric potential and a lower hydrostatic pressure are thereby maintained at the nerve terminals. Stimulating signals at the dendrites will enhance proton fluxes that will integrate to increase and even positively reverse the electric potential at the cell body. A local strong electric field is thereby created at the axon entrance, where it might locally flip over the electrically polarized actin filaments. Concomitantly, a localized increase in electric potential and in hydrostatic pressure might develop by opposing fluxes of protons and solitons. The localized increase in electric potential is further regenerated by gating the sodium ion influx, thus safely propagating the action potential down the axon under gradual flipping of the actin filaments. Upon reaching the nerve terminal, a momentary increase of hydrostatic pressure might induce a hydraulic release of neurotransmitters from the terminal vesicles.

Excess reuptake is driven by sub-pressure recovery due to reorientation of the actin filaments. According to this mechanism, the action potential is predicted to involve bulge propagation and longitudinal compressive tension due to AS (like in a virtual sarcomere), as was indeed observed [60]. Furthermore, associated impulses of birefringence reduction, pH decrease, increase in Ca2+ ion concentration, and a rise of heat production, are anticipated to propagate along with the action potential, as observed [61]. In sensory organelles, similar mechanisms might induce electro-mechanical modulation of bulk proton currents and AS along oriented actin filaments [62].

Fish electric organ is composed of 100 μm units, containing unidirectional actin filaments [63]. Coherent stimulation of a bulk proton flux, that can develop 0.5 volt per each serial unit, might generate electric shock impulses of up to 5000 volt/m.

Similarly, the localized stimulation of centrally innervated myofibrils might autonomously propagate down the non-innervated regions through differential proton-induced depolarization along the sarcomeres.

Cell Motility and Intercellular Interaction

Pwason fluxes integrate along oriented actin filaments into bulk AS, pressure gradients, and electrical polarization, all in the same direction, as verified also at the cellular level in the following examples.

Amoeboid-like movement is electrically polarized in the direction of streaming [64]. A higher pressure is developed at the anterior area and a sub-pressure at the posterior region [65] (Fig.2b). The forward pressure-gradient can be responsible for three observable effects: frontal protrusion, peripheral fountain-like backward streaming, and either flattening or permanent endocytosis at the posterior region [66]. In case of ground adhesion, the suction power of forward AS can "weigh anchor" by hydraulic thinning and tearing of rear cytoplasmic fibers.

Adjacent electrically-polarized motile cells are predicted to move in a head-to-tail orientation. Upon cells contact, unidirectional AS can generate hydraulic conjugation through dynamic intercellular invaginations (Fig.2c). A forward proton flux through cognitive receptor channels can preferentially stimulate the leading cell, where intensive AS is exhibited. In cytotoxic T-lymphocyte interaction, stimulation of such AS in the target cell is observed as aggressive terminal explosion, termed Zeiosis [67].

Linear and branched polymerization of actin filaments attached to the plasma membrane, or to internal components, is associated with ATP hydrolysis that can locally generate AS and pressure gradients [68]. Similarly, antigenic receptor-mediated cell stimulation is related to internal clustering of membrane-anchored actin filaments, which induce localized orthogonal pwasons. Hydraulic endocytosis can down-regulate this cellular stimulation, which is subsequently communicated by AS to internal organelles in various intracellular pathways. Antibody + complement fixation of the stimulatory antigenic complexes can inhibit this hydraulic endocytosis, thus leading to lethal metabolic exhaustion [69]. Such hydraulic action is indicated in reconstitution experiments of nano-tube extraction from membrane vesicles by kinesin ATPase activity along microtubules [70].

Similarly, in erythrocytes, trans-membrane ventilation of CO2 and O2 may be enhanced by sub-pressure fluctuations, generated by inwardly oriented pwasons due to ATPase activity of branched actin polymerization on membrane-attached filaments [71]. According to the pwason model, the rate and amplitude of these active membrane fluctuations are anticipated also for other membrane-attached hydrophilic ATPases [72]. Such hydraulic compression of the cell membrane might explain the peculiar observation of a higher internal hydrostatic pressure in an intact cellular state of partial inflation. Intracellular vesicles, bacteria or designed objects can utilize this hydraulic machinery for the actin-based propulsion inside the cell or in a cell-like medium [73].

The A-M gel state of amoeboid cells at rest is observed to break down under the power of AS. A similar gel-sol transition is observed in-vitro in A-M preparation from striated muscle at mM concentration of MgATP. At diminishing levels of MgATP, A-M clusters are formed, demonstrating vigorous dehydration, termed super-precipitation [1]. This process of water squeezing is related to a localized hydraulic compression of each protein cluster under the suction power of AS.

Centrosomes' and chromosomes' separation during mitosis is related to viscous drag by opposite outwardly oriented AS due to cooperative action of various ATPase systems [74]. Consequently, central sub-pressure might cause cellular cleavage by equatorial hydraulic compression.

Motility assays of free actin filaments verify that AS in striated muscle is generated toward the sarcomere's center. The suction power of centrally-oriented AS in a smooth muscle cell, or in each sarcomere of striated muscle, might generate sub-pressure at the opposite end regions of each unit, leading to its hydraulic compression, as elaborated below.

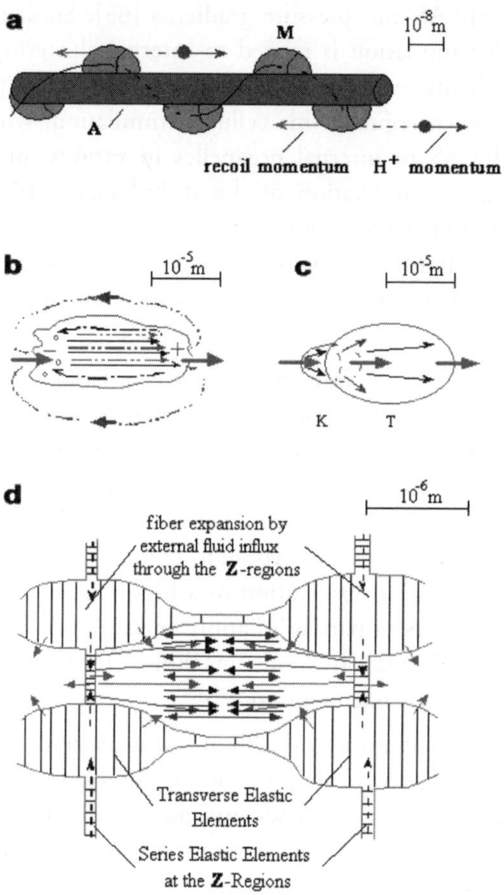

Figure 2. Ballistic protons and water solitons in biology (a) Vectorial flux of protons from myosin (M) heads along actin (A) filament. (b) Unidirectional electrical polarization and active streaming against pressure gradients in amoeboid movement. (c) Hydraulic conjugation and trans-membrane protonic stimulation in killer-target (K-T) interaction. (d) Hydraulic compression and lateral expansion of a muscle sarcomere due to centrally oriented active streaming.

Hydraulic Compression of the Sarcomere

Essential structural features of striated muscle fibers and significant physiological functions are further anticipated by the proposed mechanism of hydraulic compression due to active streaming:

The primordial pwason dimensions -1.4nm diameter and 0.5 μm length -explain the spatial evolution of optimal packing of sliding filaments in the sarcomere. The persistent hydrodynamic action throughout the pwasons' lifetime ensures minimal friction and efficient mechano-chemical performance by preventing cross-bridges interaction during most of the enzymatic cycle, while preserving the individual straight extension of the sliding filaments. This simply means that only a very small fraction of myosin heads are attached to actin filaments at any single time during contraction. This dynamic state under detachment of the A-M filaments is recognized in X-ray interference reports, where it is considered to be highly problematic in terms of the cross-bridge mechanism [75].

The hydraulic compression of each sarcomere requires an efficient peripheral buffering of hydrostatic pressure inside the Z-regions. This pressure buffering is related to intensive plasma membrane invaginations, which penetrate at the level of the Z-regions and extend around the I-regions of each sarcomere [76]. The cellular development and maintenance of this micro-fluidic network might itself be due to hydraulic AS processing. The serial dense elastic elements at the Z-regions hold the hydraulic compressive tension. Similarly, in order to resist lateral hydraulic compression, transverse elastic elements must be formed alongside between adjacent sarcomeres. These transverse elements also preserve the orthogonal invagination pattern which is essential for the hydraulic compression mechanism.

According to the hydraulic model, the isometric tension is held by a series of elastic elements at the Z-regions. Therefore, no forces, comparable to the isometric tension, are expected to stretch the actin and myosin filaments. This prediction is verified by X-ray interference measurements during fast force transients [77]. The absence of filament stretching under tension is also revealed in the sarcomeres' wave-like pattern that is developed under isometric contraction. (see Ref.3, p.332, Fig.20,I). This non-stretched profile under tension is a paradox in conventional terms, but it is an important prediction directly derived from the hydraulic mechanism. Its implications will be further discussed below.

Hydraulic suction of external water influx through the transverse network of plasma membrane invaginations is expected during isometric contraction. This influx allows for the ongoing overall expansion under lateral compression of the I-regions, while stretching the transverse elastic elements. Concomitantly, the longitudinal pumping action by AS causes a central inflation of each sarcomere (Fig.2d). This swelling mechanism explains the observed wave-like (barrel-like) pattern mentioned above, whereby tubular sarcomeres become constricted at the I-regions. This active swelling process is proposed to have four most significant physiological functions:

First, during isometric tetanus, an early development of an approximately maximal pressure-head tension of Po = 1kgwt/cm2, is followed by an ongoing slower expansion of the cross-section area, from Ao to Ae. This expansion process entails a proportional increase of the hydraulic force up to Fe = Po·Ae. Usually, the effective tension, Teff = Fe/Ao = Po·Ae/Ao, is reported. Thus, the Teff value may reach more than twice the actual Po tension [15, 78]. A similar increase in isometric force is obtained under osmotic expansion [79].

Second, the pressure-head development and relaxation precede the slower effects of initial fiber swelling and terminal elastic deflation, respectively. Thereby, three effects are anticipated for isometric force development, known as the creep phase, post-tetanic potentiation, and stretch potentiation. When these effects are taken into account, a nearly flat force-length relation is obtained, which is quite independent of the degree of cross-bridges' overlap [15, 80]. This creep-phase effect is another fundamental problem for the established model,

Third, the swelling process can provide hydraulic protection against series stretching of weaker sarcomeres, especially at a decreasing cross-bridges overlap. Evidently, and quite luckily, a homogenous steady state is maintained [81]. (Notice that by a cross-bridge mechanism, a "tearing catastrophe" is rather anticipated.) This protective mechanism is demonstrated by a whole muscle response during isotonic contraction, when the load is suddenly increased somewhat above the isometric tension; then, an initial "give" effect of rapid elongation slows down, stops, and even overcomes the initial overload [82]. Such a protective hydraulic response is obviously important also in cardiac and smooth muscle contraction.

Fourth, repetitive contraction is associated with pumping of external fluid fluxes around each sarcomere. These external fluid pulses, together with internal circulation of AS, provide for an effective metabolic exchange and heat dissipation, over and above the slow rates of diffusion and convection. Evidently, a relatively quick fatigue is felt under isometric tetanus.

Similarly, the internal circulation of AS in each sarcomere might enhance the regulatory function of Ca2+ ions in inverse relation to the muscle power. Thus, a suitable response rate is autonomously secured upon contraction and relaxation.

By initial coherent stimulation of AS, a transient lateral compression, due to the Bernoulli Effect, entails the well-known early events of latency elongation and relaxation of myofibrils. Thus, by Scheme#2, under photo-release of caged ATP at sub-threshold levels of Ca2+ ions, a single pulse of coherent pwasons can generate an early isometric relaxation, followed by a 20 msec phase of hydraulic compression throughout the ongoing state of protein dissociation, as observed [83].

Coherent interaction of cross-bridges might occasionally interfere with filaments sliding, as revealed in histograms of stepwise shortening and elongation [84]. In skeletal muscle, this resistive interaction is smoothed out by a differential Vernier scaling of monomers with different lengths between the sliding filaments.

In contrast, monomers of equal lengths are found in the sliding filaments of asynchronous insect flight muscles, which exhibit a slow rate of Ca2+ reuptake by the sarcoplasm reticulum. Therefore, a relatively low frequency of stimulation can maintain persistent Ca2+ activation, where simultaneous cross-bridges attachment might catalyze bulk coherent pulses of pwasons. Such repetitive coherent pulses of AS might generate fast and efficient cycles of hydraulic compression and relaxation. This pulsating contraction has a relatively large amplitude of 70 nm per half sarcomere, with no interference of cross-bridges' interaction all along the period of each cycle [85].

The established sliding filament model relies on two effects, which are measured in relation to an increasing sarcomere length (L) at a region of decreasing cross-bridges overlap: The first effect is the linear decrease in isometric force (Fo), when neglecting the creep-phase. The second effect is the constant value of unloaded shortening velocity (Vo) [2]. The AS hypothesis anticipates both effects by taking the hydraulic tension (Po) and the sarcomere volume (N = L*A) to be independent of L, and by reasonably assuming an inverse relation between the flow-rate (FR = Vo*A) and L at a decreasing cross-bridges overlap. Then:

$$Fo = Po * A = Po * N / L = \text{constant} * 1 / L$$
$$Vo = FR / A = FR * L / N = \text{constant}.$$

Notice that at a constant overlap, FR is constant, so that Vo is predicted to increase with L, as observed.

The sub-pressure development of 1kgwt/cm2 in about half a sarcomere volume also explains the observed initial effect of bulk expansion according to the coefficient of water compression.

The opposite circulatory propulsion of AS along actin and myosin filaments keeps them well extended for optimal sliding.

Mechanical energy absorption and unbalanced heat release are related, respectively, to stretching and relaxation of serial and transverse elastic elements. Thus, during isotonic shortening, the heat released per unit length due to ongoing relaxation of transverse elastic elements depends on the load. This heat is quantitatively related to the so-called "heat of shortening" [86].

Another unbalanced energy component is quantitatively related to entropy changes of water associated with either sub-pressure development or relaxation within each sarcomere throughout an isometric cycle of hydraulic compression. This bulk baro-entropic reversible heat production cancels out during closed cycles, and it accounts to the "thermo-elastic heat" [87]. The baro-entropic component is quantitatively related to the coefficient of thermal water expansion. It is therefore predicted to change sign above and below 4°C, as observed.

These unbalanced heat components, and the tension creep phase mentioned above, are not taken into account in the following treatment.

Quantitative Formulation of Striated Muscle Contraction

Five energetic parameters of striated muscle contraction are directly related to the quantitative aspects of the pwason model in the pure water phase:

1. The electro-mechano-chemical power of ATP hydrolysis is related to the free energy of H_2O-H^+, namely: ΔE = 0.5 proton*volt per ATP molecule = 11.5 kcal per mole ATP (See Eq.3).
2. The general isometric tension, Po = 1kgwt/cm^2, is related to the kinetic energy density of water dimer precession. The same isometric tension is produced also by smooth muscle [88].
3. The maximal possible value of unloaded shortening velocity per unit muscle length is V1o (max) = 25 unit length/sec.
4. The maximal possible value of power density in unloaded or isometric contraction is Hh1o (max) = 2.5 Watt/cm^3.
5. The friction coefficient per unit volume is given by the following expression: η1 = Po/V1o (max) = 40 grwt·cm·sec/cm^3 = 4 mJoule·sec/cm^3.

The process of hydraulic compression of a single sarcomere is further formulated in Supplementary Box 3. The normalized power-balance equation describes the mechanical (Hm) plus heat (Hq) components, and the input hydrolytic (Hh) component (Eq.6i), where Hq is composed of translation and circulation contributions of AS. The corresponding normalization factors are: the isometric tension, Po = 1kgwt/cm2; the unloaded shortening velocity per unit length, V1o; and the isometric power density for Hm = 0, Hh1o. By the following argument we prove that Hh1o has the same value in isometric and unloaded contractions. The soliton's life-time of 20 msec may be reduced upon external work production, thus increasing the hydrolytic rate, as observed in the Fenn-Effect. This effect is introduced in its differential and integral forms in Eq.6ii. A general expression is thereby obtained for the normalized force-velocity (P-V) relation in Eq.6iii. The single parameter, a, has several meaningful expressions in Eq. 6iv, and its value is proportional to the P-V curvature:

$$Hm + Hq = Hh$$

$$a * P * V + V^2 + P^2 = Hh$$

$$dHh / dHm = -dHq / dHm \rightarrow Hh = 1 + Hm/2; Hq = 1 - Hm/2$$

$$a * P * V + V^2 + P^2 = 1 + 0.5a * P * V$$

$$a = Po * Vlo / Hclo = Po / (\eta 1 * Vlo) = Vlo(\max) / Vlo = 25 / Vlo = (Hhlo(\max) / Hhlo)^{1/2}$$

where Hhlo = $\eta 1 * Vlo^2 = 4 * Vlo^2 mWatt / cm^3$.

Analytical solution of Eq.6iii, for various values of a, yields the whole spectrum of power balance and force-velocity relations for isotonic contraction (Fig.3a, b). Kinetic profiles of isometric contraction against elastic elements of relative compliance C are obtained by numerical integration of the P-V relation (Fig.3c). (Thus, an effective internal compliance and power input of a given muscle can be evaluated.) The variables of closed cycles of twitch contraction are displayed against the relative tension attained by various elastic loads during the twitch. The twitch duration is related to an effective calcium regulation due to a whole-sarcomere-volume circulation of AS. The optimum total energy values of maximum Eh and Em, and minimum Eq, are obtained at about 0.5Po. Equal but oppositely directed changes are obtained for Eh and Eq with respect to Eho. The Eho value is found to be equal for the unloaded twitch and for the fully developed isometric twitch (Fig.3d). This theoretical profile is closely verified by measurements of isotonic twitch contractions [89]. The high efficiency values are further verified when taking into account the surplus of mechanical energy consumed and the heat released by internal elastic elements (longitudinal as well as transversal), while the baro-entropic component is canceled throughout the closed cycle. A simple analysis of Eq.6iii, compared to the force-velocity relation of A.V. Hill [87, 90] is included in the Supplementary Interactive Workbook, where all the computational procedures are demonstrated. The present model may become a new useful tool for experts in muscle contraction.

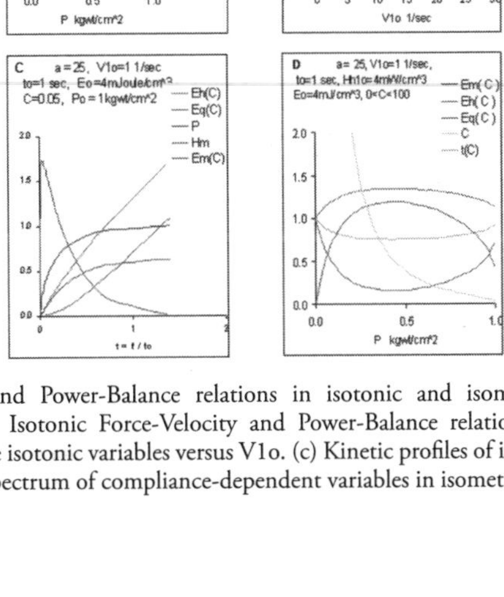

Figure 3. Force-Velocity and Power-Balance relations in isotonic and isometric contractions (see also Supplementary Box 3). (a) Isotonic Force-Velocity and Power-Balance relations for two values of a. (b) Quantitative spectrum of the isotonic variables versus V1o. (c) Kinetic profiles of isometric contraction against a load of compliance C. (d) Spectrum of compliance-dependent variables in isometric twitch contractions, versus the final tension.

Summary

Electrical, chemical, and hydraulic aspects of the AS model were consistently examined, covering a wide spectrum of novel and testable implications to the physics, chemistry and biology of water. The detailed presentation calls for critical experimental and theoretical investigations, as well as for engineering applications.

Acknowledgement

I would like to thank Ilia Stambler for manuscript editing, and Avshalom Tirosh for the video-animations. A special appreciation is raised for the critical inspiration of the reviewers.

References

1. Szent-Gyorgyi, A.G. The early history of the biochemistry of muscle contraction. J. Gen. Physiol. 2004, 123, 631–641.
2. Huxley, A.F. Mechanics and models of the myosin motor. Philos. Trans. R. Soc. Lond. B Biol Sci. 2000, 355, 433–440.
3. Huxley, H.E. Molecular basis of contraction in cross-striated muscles. In The Structure and Function of Muscle; Bourne, G.H., Ed.; Academic Press: New York, 1973; Vol.1, pp. 301–387.
4. Tirosh, R.; Oplatka, A.; Chet, I. Motility in a "cell sap" of the slime mold physarum polycephalum. FEBS Lett. 1973, 34, 40–42.
5. Oplatka, A.; Tirosh, R. Active streaming in actomyosin solutions. Biochim. Biophys. Acta. 1973, 305, 684–688.
6. Cohen, I.; Tirosh, R.; Oplatka, A. Active streaming in human thrombosthenin solutions. Pflugers Arch. 1974, 352, 81–85.
7. Oplatka, A.; Borejdo, J.; Gadasi, H.; Tirosh, R.; Liron, N.; Reisler, E. Contraction of Glycerinated Muscle Fibers, Myofibrils and Actin Threads Induced by Water-Soluble Myosin Fragments. Proc. 9th FEBS Meeting. 1974, pp. 41–46.
8. Oplatka, A.; Gadasi, H.; Tirosh, R.; Lamed, Y.; Muhlrad, A.; Liron, N. Demonstration of mechanochemical coupling in systems containing actin, ATP and non-aggregating active myosin derivatives. J. Mechanochem. Cell Motil. 1974, 2, 295–306.

9. Oplatka, A.; Borejdo, J.; Gadasi, H. Tension development in stretched glycerinated muscle fibers and contraction of 'ghost' myofibrils induced by irrigation with heavy meromyosin. FEBS Lett. 1974, 45, 55–58.

10. Oplatka, A.; Gadasi, H.; Borejdo, J. The contraction of "ghost" myofibrils and glycerinated muscle fibers irrigated with heavy meromyosin subfragment-1. Biochem. Biophys. Res. Commun. 1974, 58, 905–912.

11. Borejdo, J.; Oplatka, A. Tension development in skinned glycerinated rabbit psoas fiber segments irrigated with soluble myosin fragments. Biochim. Biophys. Acta. 1976, 440, 241–258.

12. Tirosh, R.; Oplatka, A. Active streaming against gravity in glass microcapillaries of solutions containing acto-heavy meromyosin and native tropomyosin. J. Biochem. 1982, 91, 1435–1440.

13. Tirosh, R.; Low, W.Z.; Oplatka, A. Translational motion of actin filaments in the presence of heavy meromyosin and MgATP as measured by Doppler broadening of laser light scattering. Biochim. Biophys. Acta. 1990, 1037, 274–280.

14. Tirosh, R. Ph.D. Thesis: Elementary Aspects of the Mechano-Chemical Coupling in the Actomyosin System. The Weizmann Institute of Science: Rehovot, 1978 (in Hebrew).

15. Tirosh, R.; Liron, N.; Oplatka, A. A hydrodynamic mechanism for muscular contraction. In Cross-Bridge Mechanism in Muscle Contraction, Proceedings of the International Symposium on the Current Problems of Sliding Filament Model and Muscle Mechanics, Tokyo, Japan, 1978. University of Tokyo Press: Tokyo, 1979; pp. 593–609.

16. Cairns, R.B.; Harrison, H.; Schoen, R.I. Dissociative Photoionization of H2O. J. Chem. Phys. 1971, 55, 4886–4889.

17. Head-Gordon, T.; Hura, G. Water structure from scattering experiments and simulation. Chem Rev. 2002, 102, 2651–2670.

18. Bakker, H.J.; Nienhuys, H.K. Delocalization of protons in liquid water. Science 2002, 297, 587–590.

19. Kashtanov, S.; Augustsson, A.; Luo, Y.; Guo, J.H.; Sathe, C.; Rubensson, J.E.; Siegbahn, H.; Nordgren, J; Agren, H. Local structures of liquid water studied by x-ray emission spectroscopy. Phys. Rev. B. 2004, 69, 024201.

20. Wernet, Ph.; Nordlund, D.; Bergmann, U.; Cavalleri, M.; Odelius, M.; Ogasawara, H.; Naslund, L.A.; Hirsch, T.K.; Ojamae, L.; Glatzel, P.; Pettersson, L.G.M.; Nilsson, A. The Structure of the First Coordination Shell in Liquid Water. Science 2004, 304, 995–999.

21. Bushuev, Y.G.; Davletbaeva, S.V.; Muguet, F. The 3-Attractor Water Model: Monte-Carlo simulations with a new, effective 2-Body potential (BMW). Molecules 2003, 8, 226–242.

22. Wilkie, D.R;. Heat work and phosphorylcreatine break-down in muscle. J. Physiol. 1968, 195, 157–183.

23. Mohammed OF, Pines D, Dreyer J, Pines E, Nibbering ET; Sequential proton transfer through water bridges in acid-base reactions. Science 2005, 310, 83–86.

24. Kaufmann, K.; Silman, I. The induction by protons of ion channels through lipid bilayer membranes. Biophys. Chem. 1983, 18, 89–99.

25. Soper, A.K. Thermodynamics. Water and ice. Science 2002, 297, 1288–1289.

26. Giovambattista, N.; Stanley, H.E.; Sciortino, F. Relation between the high density phase and the very-high density phase of amorphous solid water. Phys. Rev. Lett. 2005, 94, 107803.

27. Drost-Hansen, W. Temperature effects on cell-functioning--a critical role for vicinal water. Cell Mol. Biol. (Noisy-le-grand). 2001, 47, 865–883.

28. "Waves of the Sea." Encyclopedia Britannica 1970, 23, 316–317.

29. Allen, J E. The Early History of Solitons (Solitary Waves). Phys. Scr. 1998, 57, 436–441.

30. Olson, E.S. Direct measurement of intra-cochlear pressure waves. Nature 1999, 402, 526–529.

31. La Porta, A.; Voth, G.A.; Crowford, A.M.; Alexander, J.; Bodenschatz, E. Fluid particles accelerations in fully developed turbulence. Nature 2001, 409, 1017–1019.

32. Flannigan, D.J.; Suslick, K.S. Plasma formation and temperature measurement during single-bubble cavitation. Nature 2005, 434, 52–55.

33. Holmes, K.C.; Schroder, R.R.; Sweeney, H.L.; Houdusse, A. The structure of the rigor complex and its implications for the power stroke. Philos. Trans. R. Soc. Lond. B. Biol Sci. 2004, 359, 18191828.

34. Lymn, R.W.; Taylor, E.W. Mechanism of adenosine triphosphate hydrolysis by actomyosin. Biochemistry 1971, 10, 4617–4624.

35. Sleep, J.A.; Hackney, D.D.; Boyer, P.D. The equivalence of phosphate oxygens for exchange and the hydrolysis characteristics revealed by the distribution of [18O]Pi species formed by myosin and actomyosin ATPase. J. Biol. Chem. 1980, 255, 4094–4099.

36. White, H.D.; Belknap, B.; Webb, M.R. Kinetics of nucleoside triphosphate cleavage and phosphate release steps by associated rabbit skeletal actomyosin, measured using a novel fluorescent probe for phosphate. Biochemistry 1997, 36, 11828–11836.

37. Takagi, Y.; Shuman, H.; Goldman, Y.E. Coupling between phosphate release and force generation in muscle actomyosin. Philos. Trans. R. Soc. Lond. B. Biol. Sci. 2004, 359, 1913–1920.

38. Wells, A.L.; Lin, A.W.; Chen, L.Q.; Safer, D.; Cain, S.M.; Hasson, T.; Carragher, B.O.; Milligan, R.A.; Sweeney, H.L. Myosin VI is an actin-based motor that moves backwards. Nature 1999, 401, 505–508.

39. De La Cruz, E.M.; Ostap, E.M. Relating biochemistry and function in the myosin superfamily. Curr. Opin. Cell Biol. 2004, 16, 61–67.

40. Forkey, J.N.; Quinlan, M.E.; Shaw, M.A.; Corrie, J.E.; Goldman, Y.E. Three-dimensional structural dynamics of myosin V by single-molecule fluorescence polarization. Nature 2003, 422, 399–404.

41. Fletcher, D.A.; Theriot, J.A. An introduction to cell motility for the physical scientist. Phys. Biol. 2004, 1, T1–10.

42. Mitchell, P. A commentary on alternative hypotheses of protonic coupling in the membrane systems catalysing oxidative and photosynthetic phosphorylation. FEBS Lett. 1977, 78, 1–20.

43. Harold, F.M. Gleanings of a chemiosmotic eye. BioEssays 2001, 23, 848–855.

44. Kaim, G.; Dimroth, P. ATP synthesis by F-type ATP synthase is obligatorily dependent on the transmembrane voltage. EMBO J. 1999, 18, 4118–4127.

45. Boyer, P.D. A research journey with ATP synthase. J. Biol. Chem. 2002, 277, 39045–39061.

46. Noji, H., Yasuda, R., Yoshida, M. & Kinosita, K. Jr Direct observation of the rotation of F1ATPase. Nature 1997, 386, 299–302.

47. Rondelez, Y.; Tresset, G.; Nakashima, T.; Kato-Yamada, Y.; Fujita, H.; Takeuchi, S.; Noji, H. Highly coupled ATP synthesis by F1-ATPase single molecules. Nature 2005, 433, 773–777.

48. Berg, H.C. The rotary motor of bacterial flagella. Annu. Rev. Biochem. 2003, 72, 19–54,

49. Smith, C.A.; Rayment, I. Active site comparisons highlight structural similarities between myosin and other P-loop proteins. Biophys. J. 1996, 70, 1590–602.

50. Tucker, W.C.; Schwarz, A.; Levine, T.; Du, Z.; Gromet-Elhanan, Z.; Richter, M.L.; Haran, G. Observation of Calcium-dependent unidirectional rotational

motion in recombinant photosynthetic F1-ATPase molecules. J. Biol. Chem. 2004, 279, 47415–47418.

51. Haumann, M.: Liebisch, P.; Muller, C.; Barra, M.; Grabolle, M.; Dau, H. Photosynthetic O2 formation tracked by time-resolved x-ray experiments. Science 2005, 310, 1019–1021.

52. Kitamura, K.; Tokunaga, M.; Iwane, A.H.; Yanagida, T. A single myosin head moves along an actin filament with regular steps of 5.3 nanometres. Nature 1999, 397, 129 –134.

53. Sakamoto, T.; Amitani, I.; Yokota, E.; Ando, T. Direct observation of processive movement by individual myosin V molecules. Biochem. Biophys. Res. Commun. 2000, 272, 586–590.

54. Ali, M.Y.; Homma, K.; Iwane, A.H.; Adachi, K.; Itoh, H.; Kinosita, K.Jr.; Yanagida, T.; Ikebe, M. Unconstrained steps of myosin VI appear longest among known molecular motors. Biophys. J. 2004, 86, 3804-3810.

55. Rock, R.S.; Ramamurthy, B.; Dunn, A.R.; Beccafico, S.; Rami, B.R.; Morris, C.; Spink, B.J.; Franzini-Armstrong, C.; Spudich, J.A.; Sweeney, H.L. A flexible domain is essential for the large step size and processivity of myosin VI. Mol. Cell. 2005, 17, 603–609.

56. Riveline, D.; Ott, A.; Julicher, F.; Winkelmann, D.A.; Cardoso, O.; Lacapere, J.J.; Magnusdottir, S.; Viovy, J.L.; Gorre-Talini, L.; Prost, J. Acting on actin: the electric motility assay. Eur. Biophys. J. 1998, 27, 403–408.

57. Landesberg, A.; Landesberg, Y.; Sideman, S.; Ter Keurs, H.E. Molecular motion and cardiac muscle motor dynamics. Ann. NY Acad. Sci. 2002, 972, 119–126.

58. Kamiya, N.; Abe, S. Bioelectric phenomena in the myxomycete plasmodium and their relation to protoplasmic flow. J. Colloid. Sci. 1950, 5, 149.

59. Tirosh, R. Axoplasmic transport and the nerve impulse explained by cytoplasmic axial flux of energetic protons along actin filaments. In Biological Structures and Coupled Flows, Proceedings of the Symposium Dedicated to Aharon Katzir Katchalsky; Oplatka A., Balaban, M., Eds.; Intern. Sci. Serv: Rehovot, 1983; pp. 397–400.

60. Tasaki, I.; Iwasa, K. Further studies of rapid mechanical changes in squid giant axon associated with action potential production. Jpn. J. Physiol. 1982, 32, 505–518.

61. Tasaki, I. Rapid structural changes in nerve fibers and cells associated with their excitation processes. Jpn. J. Physiol. 1999, 49, 125–138.

62. Mermall, V.; Post, P.L.; Mooseker, M.S. Unconventional myosins in cell movement, membrane traffic, and signal transduction. Science 1998, 279, 527–533.

63. Zakon, H.H.; Unguez, G.A. Development and regeneration of the electric organ. J. Exp. Biol. 1999, 202, 1427–1434.

64. Nuccitelli, R.; Poo, M.M.; Jaffe, L.F. Relations between ameboid movement and membrane-controlled electrical currents. J. Gen. Physiol. 1977, 69, 743–763.

65. Charras, G.T.; Yarrow, J.C.; Horton, M.A.; Mahadevan, L.; Mitchison, T.J. Non-equilibration of hydrostatic pressure in blebbing cells. Nature 2005, 435, 365–369.

66. Chapman-Andresen, C. Endocytosis in freshwater amebas. Physiol. Rev. 1977, 57, 371–385.

67. Tirosh, R.; Berke, G. Immune cytolysis viewed as a stimulatory process of the target. Adv. Exp. Med. Biol. 1985, 184, 473–492.

68. Giardini, P.A.; Fletcher, D.A.; Theriot, J.A. Compression forces generated by actin comet tails on lipid vesicles. Proc. Natl. Acad. Sci. USA. 2003, 100, 6493–6498.

69. Tirosh, R.; Degani, H.; Berke, G. Prelytic reduction of high-energy phosphates induced by antibody and complement in nucleated cells. 31P-NMR study. Complement 1984, 1, 207–212.

70. Leduc, C.; Campas, O.; Zeldovich, K.B.; Roux, A.; Jolimaitre, P.; Bourel-Bonnet, L.; Goud, B.; Joanny, J.F.; Bassereau, P.; Prost, J. Cooperative extraction of membrane nanotubes by molecular motors. Proc. Natl. Acad. Sci. USA. 2004, 101, 17096–17101.

71. Tuvia, S.; Levin, S.; Bitler, A.; Korenstein, R. Mechanical fluctuations of the membrane–skeleton are dependent on F-Actin ATPase in human erythrocytes. J. Cell Biol. 1998, 141, 1551–1561.

72. Girard, P.; Prost, J.; Bassereau, P. Passive or active fluctuations in membranes containing proteins. Phys. Rev. Lett. 2005, 94, 088102.

73. Bernheim-Groswasser, A.; Prost, J.; Sykes, C. Mechanism of actin-based motility: a dynamic state diagram. Biophys. J. 2005, 89, 1411–1419.

74. Uyeda, T.Q.; Nagasaki, A. Variations on a theme: the many modes of cytokinesis. Curr. Opin. Cell. Biol. 2004, 16, 55–60.

75. Huxley, H.E. A personal view of muscle and motility mechanisms. Annu. Rev. Physiol. 1996, 58, 1–19.

76. Franzini-Armstrong, C. Membraneous systems in muscle fibers. In The Structure and Function of Muscle; Bourne, G.H., Ed.; Academic Press: New York, 1973; Vol. 2, pp. 531–619.

77. Piazzesi, G.; Reconditi, M.; Linari, M.; Lucii, L.; Sun, Y.B.; Narayanan, T.; Boesecke, P.; Lombardi, V.; Irving, M. Mechanism of force generation by myosin heads in skeletal muscle. Nature 2002, 415, 659–661.

78. Tirosh, R. 1 kgf/cm2 -The isometric tension of muscle contraction: implications to cross-bridge and hydraulic mechanisms. Adv. Exp. Med. Biol. 1984, 170, 531–539.

79. Edman, K.A.P. The force bearing capacity of frog muscle fibres during stretch: its relation to sarcomere length and fibre width. J. Physiol. 1999, 519, 515—526.

80. ter Keurs, H.E.; Iwazumi, T.; Pollack, G.H. The length-tension relation in skeletal muscle. J. Gen. Physiol. 1978, 4, 565–592.

81. Rassier, D.E.; Herzog, W.; Pollack, G.H. Dynamics of individual sarcomeres during and after stretch in activated single myofibrils. Proc. R. Soc. Lond. B. Biol. Sci. 2003, 270, 1735–1740.

82. Katz, B. The relation between force and speed in muscular contraction, J. Physiol. 1939, 96, 45–64.

83. Goldman, Y.E.; Hibberd, M.G.; McCray, J.A.; Trentham, D.R. Relaxation of muscle fibres by photolysis of caged ATP. Nature 1982, 300, 701–705.

84. Delay, M.J.; Ishide, N.; Jacobson, R.C.; Pollack, G.H.; Tirosh, R. Stepwise shortening: analysis by high-speed cinemicrography. Science 1981, 213, 1523–1525.

85. Dickinson, M.; Farman, G.; Frye, M.; Bekyarova, T.; Gore, D.; Maughan, D.; Irving, T. Molecular dynamics of cyclically contracting insect flight muscle in vivo. Nature 2005, 433, 330–334.

86. Hill, A.V. The effect of load on the heat of shortening of muscle. Proc. R. Soc. Lond.B. Biol. Sci. 1964, 159, 297–318.

87. Woledge, R.C.; Curtin, N.A.; Homsher, E. Energetic Aspects of Muscle Contraction. Monogr. Physiol. Soc. 1985, 41, 1–357.

88. Csapo, A.I. Smooth muscle. In The Structure and Function of Muscle; Bourne, G.H., Ed.; Academic Press: New York, 1973; Vol.2, p. 16.

89. Hill, AV. The variation of total heat production in a twitch with velocity of shortening. Proc. R. Soc. Lond. B. Biol. Sci. 1964, 159, 596–605.

90. Hill, A.V. First and Last Experiments in Muscle Mechanics. Cambridge University Press: London, 1970.

Surface Photochemistry: Benzophenone as a Probe for the Study of Modified Cellulose Fibres

L. F. Vieira Ferreira, A. I. Costa, Ferreira Machado,
T. J. F. Branco, S. Boufi, M. Rei-Vilar and A. M. Botelho do Rego

ABSTRACT

This work reports the use of benzophenone, a very well characterized probe, to study new hosts (i.e., modified celluloses grafted with alkyl chains bearing 12 carbon atoms) by surface esterification. Laser-induced room temperature luminescence of air-equilibrated or argon-purged solid powdered samples of benzophenone adsorbed onto the two modified celluloses, which will be named C12-1500 and C12-1700, revealed the existence of a vibrationally structured phosphorescence emission of benzophenone in the case where ethanol was used for sample preparation, while a nonstructured emission of benzophenone exists when water was used instead of ethanol. The decay times of the benzophenone emission vary greatly with the solvent used for sample preparation and do not change with the alkylation degree in the range of 1500–1700 micromoles of alkyl chains per gram of cellulose. When water was used as a solvent for sample preparation, the shortest lifetime for the benzophenone emission was observed; this result is similar to the case of benzophenone adsorbed onto the "normal" microcrystalline cellulose surface, with this latter case previously reported by Vieira Ferreira et al. in 1995. This is due to the more efficient hydrogen abstraction reaction from the glycoside rings of cellulose when compared with hydrogen abstraction from the alkyl chains of the modified celluloses. Triplet-triplet transient absorption of benzophenone was obtained in both cases and is the predominant absorption immediately after laser pulse, while benzophenone ketyl radical formation occurs in a microsecond time scale both for normal and modified celluloses.

Introduction

Diffuse reflectance laser flash photolysis and laser-induced luminescence, both in time-resolved mode or ground-state absorption spectroscopy in the diffuse reflectance mode, are important techniques that have been used by several research groups to study opaque and crystalline systems [1–8]. These solid-state photochemical methods have been applied by us to study several organic compounds adsorbed onto different hosts such as microcrystalline cellulose [7, 8], p-tertbutylcalix[n]arenes (n = 4, 6, and 8) and their derivatives [10–12], silicalite, cyclodextrins [7, 12, 13], and silica [14].

Benzophenone (BZP) is an extremely useful molecule for probing new hosts. The n $\rightarrow \pi^*$ absorption transition was found to be very sensitive to the environment characteristics and also exhibits a photochemistry which depends on the host properties [10, 12, 14]. In a recent paper [14], we reported a comparative study of the luminescent properties of BZP adsorbed onto reversed phase silicas, "normal" silica, and silicalite (a de-aluminated zeolite). Apart from the triplet-state luminescence observed in all cases, in the case of "normal" silica the emission of an excited form of hydrogen-bonded benzophenone was also detected [14].

With this work we intend to contribute to the study of the mechanism by which modified cellulose fibres are able to trap dissolved organic pollutants from water. Indeed, as we have shown previously [15, 16], grafting of linear alkylchains on the fibre's surface boosts its capacity to uptake organic solutes from aqueous solution. By ensuring grafted alkyl chains, one gives rise to hydrophobic domains on which organic solutes can be accumulated. The adsorption process occurs by transfer of the sparingly soluble organic molecules from water to organic zones of the modified cellulose where the most significative adsorbent-adsorbate interactions occur.

This work reports the use of BZP, a very well characterized probe, to study new hosts (i.e., modified celluloses). The modification consists in grafting with alkyl chains, bearing 12 carbon atoms, by surface esterification with a high density of alkyl chains [15], therefore, transforming the polar surface of the normal cellulose into surfaces with a certain degree of nonpolar character. A comparison of the photochemical behaviour of BZP in modified and nonmodified celluloses was made.

Experimental

Materials

Microcrystalline cellulose (Fluka DS0) was used as powdered solid support, as received. Benzophenone (Koch-Light, Scintillation grade) and ethanol (Merck, LiChrosolv grade) were also used as received. The preparation of the modified cellulose fibres started with the use of microcrystalline cellulose and involved an acylation reaction based on a solvent exchange procedure, as described in detail in [15]. The aliphatic anhydrides have 12 carbon atoms (C12) per alkyl chain and the final modified cellulose has 1500 (C12-1500) or 1700 (C12-1700) micromoles of alkyl chains per gram of cellulose [15].

Sample Preparation

Benzophenone adsorption on samples was performed using two methods: the solvent evaporation method for the case of ethanol [14], and, also, adsorption from water [15]. The former method consists in the addition of an ethanolic solution containing the probe to the previously dried powdered solid substrate, followed by solvent evaporation from the stirred slurry in a fume cupboard. In the case of water, the fibers were first swollen for at least two hours, and the addition of BZP was done by adding 500 micromoles of this probe dissolved in ethanol (saturated solution so that the added amount of ethanol was minimized). The water suspensions were kept under agitation for 24 hours and the modified cellulose (with the adsorbed BZP) was removed by filtration. From the initial 500 μmole g^{-1} of BZP, about 300 μmole g^{-1} remained into the powdered substrate (i.e., an increase in the retention capacity of the powdered substrate of about twelve times when compared to previous reported results) was obtained with nonmodified cellulose [15].

In both cases, the final solvent removal was performed for about two hours in an acrylic chamber with an electrically heated shelf (Heto, Model FD 1.0–110) with temperature control (30 ± 1°C) and under moderate vacuum at a pressure of ca. 10–3 Torr.

Methods

Ground-State Diffuse Reflectance Absorption Spectra (GSDR)

Ground-state absorption spectra for the solid samples were recorded using an OLIS 14 spectrophotometer with a diffuse reflectance attachment. Further details are given in [1, 7].

Laser-Induced Luminescence (LIL) and Diffuse Reflectance Laser Flash Photolysis (DRLFP) Systems

Schematic diagrams of the LIL and of the DRLFP systems are presented in [1]. Laser flash photolysis experiments were carried out with the third or the fourth harmonic of an Nd : YAG laser (355 and 266 nm, ca. 6 ns FWHM, ~10–30 mJ/pulse) from B. M. Industries (Thomson-CSF, model Saga 12-10) in the diffuse reflectance mode. The light arising from the irradiation of solid samples by the laser pulse is collected by a collimating beam probe coupled to an optical fiber (fused silica), and is detected by a gated intensified charge-coupled device Oriel model Instaspec V (Andor ICCD, based on the Hamamatsu S57 69-0907). The ICCD is coupled to a compact fixed imaging spectrograph (Oriel, model FICS 77441). The system can be used either by capturing all light emitted by the sample or in a time-resolved mode by using a delay box (Stanford Research Systems, model D6535). The ICCD has high speed gating electronics (2.2 nanoseconds) and intensifier, and covers the 200–900 nm wavelength range. Time-resolved absorption and emission spectra are available in a time scale ranging from nanosecond to second. Transient absorption data are reported as percentage of absorption (%Abs.), defined as $100\Delta J_t/J_0 = (1 - J_t/J_0)100$, where J_0 and J_t are diffuse reflected light from sample before exposure to the exciting laser pulse and at time t after excitation, respectively. Laser-induced luminescence experiments were performed with an N_2 laser (PTI model 2000, ca. 600 ps FWHM, ~1.1mJ per pulse). In this case, the excitation wavelength is 337 nm. With these setups, both fluorescence and phosphorescence spectra are easily available (by the use of the variable time gate width and start delay facilities of the ICCD).

Results and Discussion

In both cases (benzophenone adsorption from water or from ethanol), similar results were obtained for the two modified celluloses C12-1500 or C12-1700 (within experimental error) (i.e., with degree of modification ranging in 1500–1700 μmol of alkyl chains per gram of cellulose). This is valid for all the experimental techniques used in this work. This paper reports data obtained for this specific degree of alkylation and compares samples obtained with two different solvents.

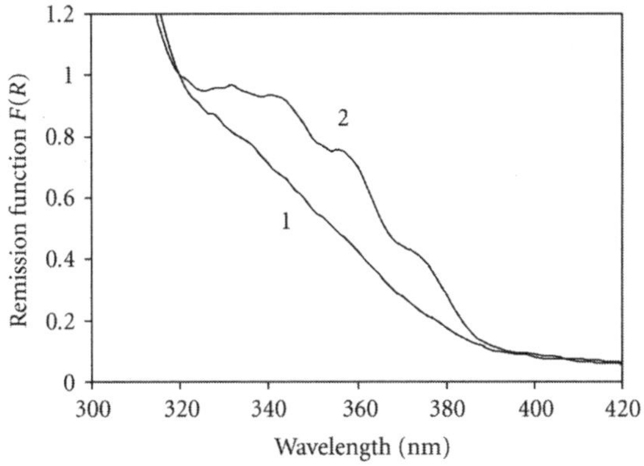

Figure 1: Ground-state diffuse reflectance absorption spectra for samples of benzophenone adsorbed onto C12-1500 (curve 1) and C12-1700 (curve 2) modified celluloses. Curve 1 refers to benzophenone adsorbed from water and curve 2 refers to benzophenone adsorbed from ethanol.

Ground-State Diffuse Reflectance Absorption Spectra

Ground-state diffuse reflectance absorption spectra for BZP adsorbed onto C12-1500, C12-1700, and microcrystalline cellulose [7] were obtained with the use of an integrating sphere [1, 7].

For BZP/C12-1700 samples prepared by solvent evaporation method with ethanol, the ketone's ground-state absorption S0 → S1 transition (n → π*) has a maximum at about 347nm and exhibits a clear vibronic structure characteristic of the excited carbonyl group of BZP in a hydrophobic environment [17].

For BZP adsorbed from water onto C12-1500 modified cellulose, the ground-state absorption curves appear now as broad bands, shifted hypsochromically when compared with BZP/C12-1700 ethanol case, where the vibronic absorption bands of the carbonyl group can be seen (Figure 1). These hypsochromic shifts are quite characteristic of the n → π* transition with increasing polarity of the surface

[7, 11, 14, 17]. The broadening of the spectra is probably also related both with heterogeneity of the adsorbent and a much smaller rigidity of the adsorbed probe. Solution absorption spectra of BZP, for instance in cyclohexane and ethanol, also exhibit this type of influence of polarity, characteristic of the n $\rightarrow \pi*$ transition [17].

Clearly, going from C12-1700/EtOH to C12-1500/H2O an increase in the surface polarity is observed, quite consistent with the surface characteristics and different adsorption sites to BZP: long alkyl chains with 12 carbons in the C12-1700/EtOH sample, and contact with the hydroxyl groups of the cellulose polymer chains in the C12-1500/H2O case.

Room Temperature Laser-Induced Phosphorescence

Figure 2 presents the room temperature phosphorescence spectra of BZP onto the surfaces of the C12-1500/H_2Omodified cellulose while Figure 3 refers to BZP within C12- 1700/EtOH.

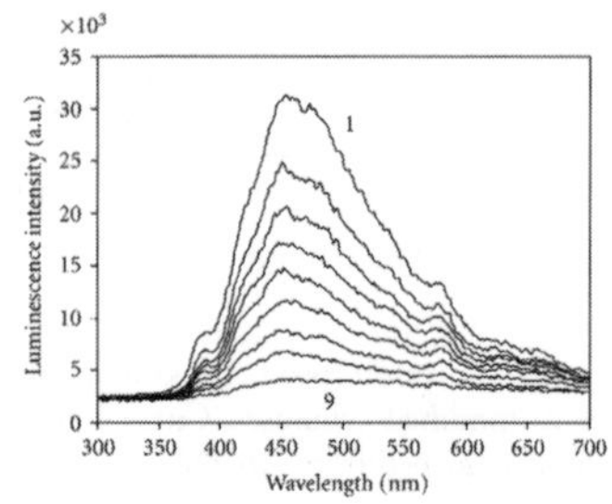

Figure 2: Laser-induced phosphorescence spectra for argon-purged samples of benzophenone adsorbed onto C12-1500 modified cellulose from water. Curves were recorded 0.5 microsecond, 2.0 microseconds, 4.5 microseconds, 8 microseconds, 16 microseconds, 30 microseconds, 70 microseconds, 120 microseconds, and 380 microseconds after the laser pulse. The excitation wavelength was 337 nm.

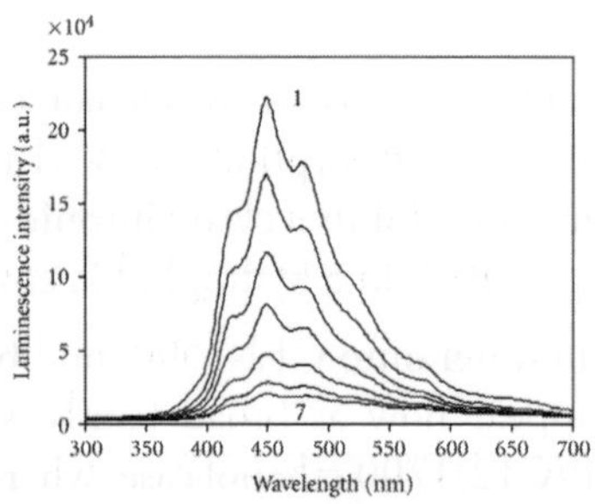

Figure 3: Laser induced phosphorescence spectra for argon-purged samples of benzophenone adsorbed onto C12-1700 modified cellulose from ethanol. Curves were recorded 1 microsecond, 5 microseconds, 20 microseconds, 45 microseconds, 75 microseconds, 150 microseconds, and 225 microseconds after the laser pulse. The excitation wavelength was 337 nm.

Those time-resolved spectra were obtained with air-equilibrated conditions and were identical to the ones obtained with argon-purged samples within experimental error. Half-lives f about 20 μs can be obtained from time resolved spectra shown in Figures 2 and 3.

For comparison purposes, lifetimes of about 80 microseconds were determined for the calixarene inclusion [10, 11] and 3.1 milliseconds for inclusion into the narrower channels of silicalite [12, 18], as compared to about 40 microseconds for benzophenone microcrystals, all determined at the maximum emission wavelength (about 448 nm) [14].

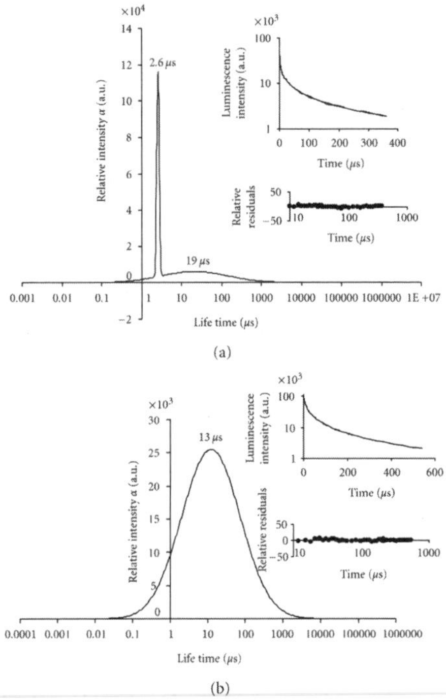

Figure 4: Lifetime distributions for benzophenone within C12- 1500/H_2O (a) and for benzophenone within C12-1700/EtOH (b) recovered from luminescence decays observed at 450 nm. The insets show the fitting of the recovered decay to the experimental data and residuals.

A new tool for a lifetime distributions analysis of emissions of probes adsorbed onto heterogeneous surfaces was recently developed by our research group [18]. This new methodology allows for asymmetric distributions and uses pseudo-Voigt profiles (Gaussian-Lorentzian product) instead of pure Gaussian or Lorentzian distributions. A very simple and widely available tool for fitting has been used, the Microsoft Excel Solver. This is a convenient way to treat the emission decay data because it reflects the multiplicity of adsorption sites available for the probe onto each specific surface. The use of a sum of a few exponentials to analyse the decay of probes onto heterogeneous surfaces is a description without physical meaning [18].

A detailed study of the luminescence decay curves of pyrene included within p-tert-butylcalix[4]arene cavities and benzophenone into silicalite channels has been reported recently [18, 19].

When applied to the C12-1700/EtOH and C12-1500/H2O cases (see Figures 4(a) and 4(b)), and after recording the decay traces in various instrumental time scales (experimental information was obtained starting in the 0.02 microsecond or longer timescales), the lifetimes distribution analysis evidences a single broad band centred at ca. 13 microseconds, which we assigned to BZP adsorbed onto the long alkyl chains with 12 carbons in the C12-1700/EtOH case; but for the case of C12-1500/H2O, another band (in this case a sharp band) centred at 2.6 microseconds predominates, which we assigned to BZP in close contact with the hydroxyl groups of the cellulose polymer chains. The emission from BZP in contact with the alkyl chain still exits (peaking now at 19 microseconds) but it is of much less importance than the other component.

This kinetic information is consistent with previous data from ground-state absorption spectra as well as the spectroscopic information from laser-induced luminescence experiments.

The laser-induced emission experiments for BZP/C12- 1500/H2O also showed a special emission of BZP in the nanosecond time scale, very similar to the one reported in [20, Figure 3(b)] for BZP on MCM-41. This emission peaks at ca. 430 nm, and therefore it originates from hydrogenbonded BZP (data not shown). In the case of the BZP/C12- 1700/EtOH sample, no hydrogen-bonded BZP emission in the nanosecond time scale could be detected.

Previous work indicates that calixarene [10–12], cellulose [7, 8], or reversed phase silicas [14] are good hydrogen atom donors towards BZP (in solid powdered samples), so it sounds reasonable to assume that ketyl radical formation of benzophenone may also occur here. The emission maxima at 575nm presented both in spectra of Figures 2 and 3 are a reasonable indication that this is also the case.

In order to perform the lifetime distribution analysis, several decay curves in different instrumental time scales were recorded. A superposition of those decay traces was made by normalization of each decay curve at a time range where they overlap, in order to produce a composite decay with closely spaced data at short times and larger spaced values at long times. This procedure was adopted before [14, 19], and is necessary because the abscissa is ln t, therefore a very large time range had to be used.

Diffuse Reflectance Laser Flash Photolysis

Time-resolved absorption spectra of samples of BZP/C12- 1700/EtOH and C12-1500/H_2O samples were obtained by the use of diffuse reflectance laser flash photolysis technique, developed by Wilkinson et al. [2–4]. In this study, the use of an intensified charge-coupled device as a detector allowed us to obtain time-resolved absorption spectra with nanometer spectral spacing (i.e., where the 200–900 scale is defined by the 512 pixels used for recording spectra in the array of the ICCD) [1, 8–14].

Both transient absorption spectra of the BZP/C12- 1700/EtOH and C12-1500/H2O samples have shown the simultaneous formation of triplet benzophenone and also of hydroxylbenzophenone radical (BZP•OH) (data not shown). The triplet-triplet absorption spectra of benzophenone (max. at 530 nm) was easily identified from comparison with the one published by Wilkinson et al. [4]. The transient absorption which peaks at 390nm can be assigned to BZP•OH radical by comparison with previously reported spectra in solution or on the MCM-41 surface [20]. Ketyl radical formation peaking at about 550nm was observed for both BZP/C12-1700/EtOH and C12-1500/H2O samples a few microsecond after laser pulse, showing that hydrogen atom abstraction can occur either when adsorbed onto the long alkyl chains or when BZP is close to the main polymer chain of cellulose where the hydrogen atoms, bound to the secondary carbons, are the most easily abstractable ones [7].

Conclusions

The photochemistry of BZP onto the modified celluloses is determined by the nature of the adsorption site (long alkyl chains or the hydroxyl groups of the polymer chains). The adsorption site of the probe depends on the solvent used for sample preparation: ethanol privileges the first case while water leads to the second situation. As a consequence of the detected BZP phosphorescence, BZP ketyl radical fluorescence and hydrogen-bonded BZP luminescence reflect the different sites for adsorption.

A lifetime distributions analysis provided important information and revealed an important quenching effect in the case of adsorption from water, comparable to the case of adsorption onto microcrystalline cellulose [7], due to the more efficient hydrogen abstraction reaction from the glycoside rings of cellulose when compared with hydrogen abstraction from the alkyl chains of the modified celluloses. Diffuse reflectance transient absorption spectra revealed the presence of the triplet state of BZP in all supports under study, and also of the diphenylketyl radical and BZP•OH radicals.

Acknowledgements

The authors thank Fundacao para a Ciencia e Tecnologia (FCT) for financial support (Project POCI/QUI/57491/2004), and A. I. Costa thanks Instituto Superior de Engenharia de Lisboa, (ISEL) for a Doctoral fellowship.

References

1. A. M. Botelho do Rego and L. F. Vieira Ferreira, "Photonic and electronic spectroscopies for the characterization of organic surfaces and organic molecules adsorbed on surfaces," in Handbook of Surfaces and Interfaces of Materials, H. S. Nalwa, Ed., vol. 2, chapter 7, pp. 275–313, Academic Press, Boston, Mass, USA, 2001.
2. F. Wilkinson and G. P. Kelly, "Laser Flash Photolysis on Solid Surfaces" in Photochemistry on Solid Surfaces, M. Anpo and T. Matsuara, Eds., pp. 31–47, Elsevier, Amsterdam, The Netherlands, 1989.
3. F. Wilkinson and G. P. Kelly, "Diffuse reflectance flash photolysis," in Handbook of Organic Photochemistry, J. C. Scaiano, Ed., vol. 1, chapter 12, pp. 293–314, CRC Press, Boca Raton, Fla, USA, 1989.
4. F. Wilkinson and C. J. Willsher, "Detection of triplet-triplet absorption of microcrystalline benzophenone by diffusereflectance laser flash photolysis," Chemical Physics letters, vol. 104, no. 2–3, pp. 272–276, 1984.
5. G. Cosa and J. C. Scaiano, "Laser techniques in the study of drug photochemistry," Photochemistry and Photobiology, vol. 80, no. 2, pp. 159–174, 2004.
6. J. K. Thomas, "Effect of SiO2 and zeolite surfaces on the excited triplet state of benzophenone, BT; a spectroscopic and kinetic study," Photochemical & Photobiological Sciences, vol. 3, no. 5, pp. 483–488, 2004.
7. L. F. Vieira Ferreira, J. C. Netto-Ferreira, I. V. Khmelinskii, A. R. Garcia, and S. M. B. Costa, "Photochemistry on surfaces: matrix isolation mechanisms study of interactions of benzophenone adsorbed on microcrystalline cellulose investigated by diffuse reflectance and luminescence techniques," Langmuir, vol. 11, no. 1, pp. 231–236, 1995.
8. L. F. Vieira Ferreira, I. Ferreira Machado, J. P. Da Silva, and A. S. Oliveira, "A diffuse reflectance comparative study of benzil inclusion within microcrystalline cellulose and β- cyclodextrin," Photochemical & Photobiological Sciences, vol. 3, no. 2, pp. 174–181, 2004.
9. L. Ilharco, A. R. Garcia, J. Lopes da Silva, M. J. Lemos, and L. F. Vieira Ferreira, "Ultraviolet-visible and Fourier transform infrared diffuse reflectance studies of

benzophenone and fluorenone adsorbed onto microcrystalline cellulose," Langmuir, vol. 13, no. 14, pp. 3787–3793, 1997.

10. L. F. Vieira Ferreira, M. R. Vieira Ferreira, J. P. Da Silva, I. Ferreira Machado, A. S. Oliveira, and J. V. Prata, "Novel laser-induced luminescence resulting from benzophenone/Opropylated p-tert-butylcalix[4]arene complexes. A diffuse reflectance study," Photochemical & Photobiological Sciences, vol. 2, no. 10, pp. 1002–1010, 2003.
11. L. F. Vieira Ferreira,M. R. Vieira Ferreira, A. S. Oliveira, T. J. F. Branco, J. V. Prata, and J.C.Moreira, "Diffuse reflectance studies of β-phenylpropiophenone and benzophenone inclusion complexes with calix[4], [6] and [8]arenes," Physical Chemistry Chemical Physics, vol. 4, no. 2, pp. 204–210, 2002.
12. L. F. Vieira Ferreira,M. R. Vieira Ferreira, A. S. Oliveira, and J. C. Moreira, "Potentialities of diffuse reflectance laser-induced techniques in solid phase: a comparative study of benzophenone inclusion within p-tert-butylcalixarenes, silicalite and microcrystalline cellulose," Journal of Photochemistry and Photobiology A, vol. 153, no. 1-3, pp. 11–18, 2002.
13. J. P. Da Silva, L. F. Vieira Ferreira, A. M. Da Silva, and A. S. Oliveira, "A comparative study of the photophysics and photochemistry of 4-chlorophenol adsorbed on silicalite and β- cyclodextrin," Journal of Photochemistry and Photobiology A, vol. 151, no. 1–3, pp. 157–164, 2002.
14. L. F. Vieira Ferreira, I. Ferreira Machado, J. P. Da Silva, and T. J. F. Branco, "Surface photochemistry: benzophenone as a probe for the study of silica and reversed-phase silica surfaces," Photochemical & Photobiological Sciences, vol. 5, pp. 665–673, 2006.
15. S. Boufi and M.N. Belgacem, "Modified cellulose fibers for adsorption of dissolved organic solutes," Cellulose, vol. 13, no. 1, pp. 81–94, 2006.
16. F. Aloulou, S. Boufi, and J. Labidi, "Modified cellulose fibres for adsorption of organic compound in aqueous solution," Separation and Purification Technology, vol. 52, no. 2, pp. 332–342, 2006.
17. N. J. Turro, Modern Molecular Photochemistry, Benjamin Cummings, Menlo Park, Calif, USA, 1978.
18. T. J. F. Branco, A. M. Botelho do Rego, I. Ferreira Machado, and L. F. Vieira Ferreira, "Luminescence lifetime distributions analysis in heterogeneous systems by the use of Excel's Solver," Journal of Physical Chemistry B, vol. 109, no. 33, pp. 15958–15967, 2005.
19. T. J. F. Branco, L. F. Vieira Ferreira, A. M. Botelho do Rego, A. S. Oliveira, and J. P. Da Silva, "Pyrene-p-tert-butylcalixarenes inclusion complexes formation:

a surface photochemistry study," Photochemical & Photobiological Sciences, vol. 5, no. 11, pp. 1068–1077, 2006.

20. J. P. Da Silva, I. Ferreira Machado, J. P. Lourenço, and L. F. Vieira Ferreira, "Photochemistry of benzophenone adsorbed on MCM-41 surface," Microporous and Mesoporous Materials, vol. 84, no. 1–3, pp. 1–10, 2005.

Study of Polymer Material Aging by Laser Mass Spectrometry, UV-Visible Spectroscopy, and Environmental Scanning Electron Microscopy

Junien Exposito, Claude Becker, David Ruch and Frédéric Aubriet

ABSTRACT

Dyed natural rubber (NR) and styrene butadiene rubber (SBR), designed for outdoor applications, were exposed to an accelerated artificial aging in xenon light. The aging results in the deterioration of the exposed surface material properties. The ability of dyed polymers to withstand prolonged sunlight *exposure without fading or undergoing any physical deterioration is largely determined not only by the photochemical characteristics of the absorbing dyestuff itself but also by the polymer structure and fillers. Results obtained by laser mass spectrometry, UV-visible spectroscopy, and environmental scanning electron microscopy indicate that dyed filled NR and SBR samples behave differently during the photo-oxidation. The fading of the dyed polymers was found to be promoted in the NR sample. This can be correlated with LDI-FTICRMS results, which show the absence of [M-H]$^-$ orange pigment pseudomolecular ion and also its fragment ions after aging. This is confirmed by both EDX and UV/Vis spectroscopy. EDX analysis indicates a concentration of chlorine atoms, which can be considered as a marker of orange pigment or its degradation products, only at the surface of SBR flooring after aging. Reactivity of radicals formed during flooring aging has been studied and seems to greatly affect the behavior of such organic pigments.*

Introduction

Under mechanical stress, temperature, moisture, radiations, and corrosive and aggressive environments, material performances decrease over the time. The alteration of use properties is named "aging" [1]. Understanding and predicting these degradation processes are really a great challenge. Consequently, the evaluation of material lifetime is very complex; this difficulty increases when compositematerials are considered. In this paper, our attention is focused on a specific composite material: orange polymer-based flooring. This flooring is amixture of different organic and inorganic fillers with an organic orange azoic pigment ($C_{32}H_{24}N_8O_2Cl_2$) in a polymer matrix. The polymer matrix can correspond to different homopolymers or copolymers: natural rubber (NR), styrene butadiene rubber (SBR), or polybutadiene (PB).

Three different analytical techniques have been used to achieve a precise description of the material properties before and after aging. The laser desorption ionization coupled to Fourier transform ion cyclotron resonance mass spectrometry (LDI-FTICRMS) allows molecular information to be obtained [2]. Scanning electron microscopy (SEM) is used to investigate, after aging, the surface topology in secondary electron (SE) mode and to determine the distribution of atoms at the interface of aged/nonaged compositematerial in energy dispersive X-ray spectrometry (EDX). Moreover, UV visible spectroscopy is used as a qualitative and quantitative approach to investigate the degradation of the material.

In this study, two composite floorings have been investigated before and after accelerated aging. The first one is associated to SBR polymer and darkened under

UV/Vis irradiation; the other one is based on NR polymer matrix and faded during aging.

Scheme 1: Developed formula of orange pigment, natural rubber (NR), and styrene–butadiene rubber (SBR).

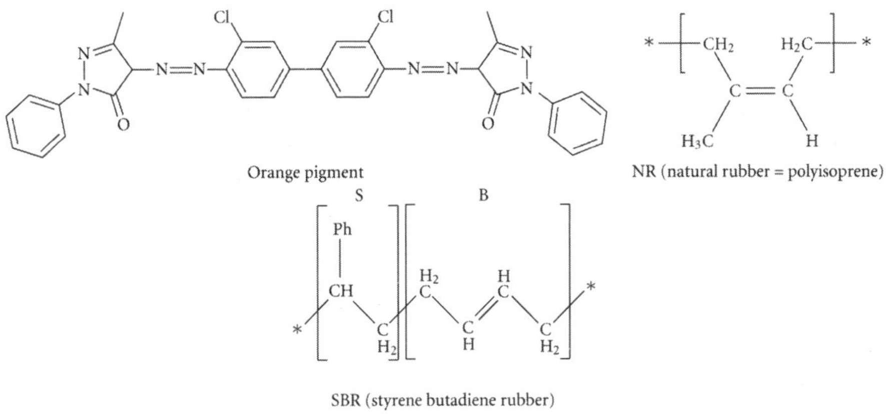

Materials

The investigated floorings are a mixture of different organic and inorganic fillers with an organic orange azoic pigment ($C_{32}H_{24}N_8O_2Cl_2$) in a polymer matrix. Both compounds are constituted by 28.6% of polymer matrix, 8.7% of chemical products (stearic acid, coumarone resin, zinc and calcium oxides, polyethylene and microcristaline wax), 53.4% of inorganic fillers (kaolin kick, kaolin GTY, silicium VN3, Argirec B24), 2.6% of titanium dioxide, 1.2% of orange pigment, and 1.4% of polymerization accelerators (CBS, sulfur, ZBEC).

The two investigated polymer matrices are natural rubber (NR) and styrene butadiene rubber (SBR). The structure of organic pigment is described in Scheme 1.

Methods

Artificial Aging

To accelerate degradation processes artificial aging with an Xe lamp is used [3]. The artificial simulated weathering was carried out in an Atlas Weatherometer Ci 5000 chamber equipped with a 6500W borosilicate glass-filtered xenon arc source. The artificial aging program consists of a continuous irradiation of 168 hours. The intensity of radiation is auto adjusted to assure a global irradiation of 240 MJ/m^2 for wavelengths in the 290 to 800nm range.

LDI-FTICR MS Experiments

All mass spectra are acquired in negative detectionmode using an LDI-FTICR mass spectrometer (IonSpec, Lake Forest, Calif, USA) equipped with an actively shielded 9.4-tesla superconducting magnet (Cryomagnetics, Oak Ridge, Tenn, USA) [4–6]. The external ion source ProMaldi card is used. Small pieces of samples are put on a specific sample holder prior to be introduced into the source region of the mass spectrometer. Ions were generated by LDI of the sample with an ORION air-cooled Nd:YAG laser system (New Wave Research Inc, Fremont, Calif, USA) working at the 355nm wavelength (laser pulse duration 5 nanoseconds, outpout energy 4mJ). The ions resulting from 8 successive laser-sample interactions are stored in an RF-only hexapole before being transferred to the FTICR cell. After transfer, ions are trapped in the FTICR cell with a 0.2V trapping potential. m/z 100 to 1500 ions are then excited by the application of an arbitrary excitation wave function on the excitation plates. The resulting image current is detected, amplified, digitized, apodized (Blackman), and Fourier-transformed to produce amass spectrum. The signal is sampled during 2.097 seconds with 4096 Ko data points. The obtainedmass accuracy is typically better than 0.75 ppm, and the mass resolution at m/z 500 close to 400 000.

SEM Measurements

SEM observations are carried out by using Quanta field emission gun environmental scanning electron microscope (SEM) from FEI (FEG-200) operating at 20.00 kV and 0.70 mbar and 1.16 mbar. The origin of environmental SEM is directly linked to the high vacuum needed in electron microscopes, that introduce restrictions on the way that insulating specimens as polymer may be imaged without performing a preliminary conductive coating.

UV/Vis Spectroscopy Measurements

Experiments on solid state samples are performed using a Perklin Elmer UV/Vis spectrometer Lambda 14. Absorbance is recorded for the 200–900nm wavelength range.

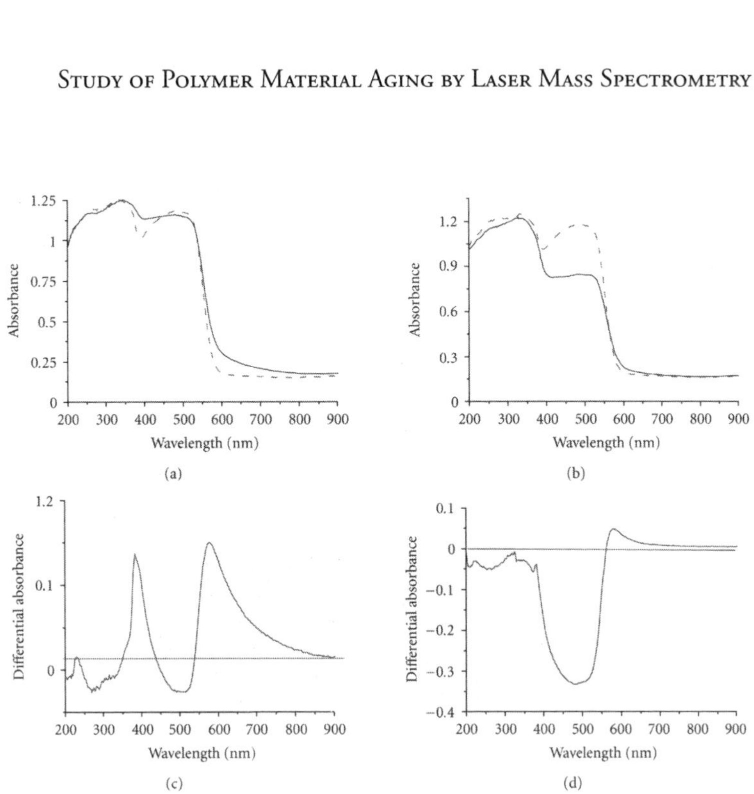

Figure 1: UV/Vis spectra of (a) SBR flooring and (b) NR flooring before (dashed line) and after aging (full line). Differential absorbance spectra for SBR (c) and NR (d) floorings (nonaged flooring is taken as the reference).

Results and Discussion

UV-Visible Spectroscopy

The absorbance of NR, SBR, and orange pigment is first investigated. Both polymers present a significant absorption in the UV, whereas orange pigment strongly absorbs visible wavelength. The maximum of absorption is in the 200– 350nm range for both polymers and in the 350–550nm range for the orange pigment. Before aging, the absorption behavior of the orange-colored NR and SBR-based flooring could be considered as the additive behavior of polymer on the one hand and orange pigment on the other hand as it is reported in Figure 1. After aging, the behavior of both samples is modified. Absorption bands broaden for both samples. This is confirmed by differential UV/Vis spectroscopy reported in Figures 1(c) and 1(d). Small differential absorption peaks are systematically observed and could be considered as small pseudo- "red shift" after aging. This could be interpreted as an increasing number of conjugated unsaturations and/or incorporation of oxygen atoms in the hydrocarbon chains.

The UV/Vis spectrum of NR flooring reported in Figure 1(b) also demonstrates a significant decrease of the absorption associated to the pigment after aging. This behavior is not observed for SBR-based flooring. This is confirmed by the visual aspect of both floorings: the color of NR-based flooring fades, whereas SBR-based flooring darkens.

Figure 2: SEM analysis in secondary electron mode of aged (a) NRand (b) SBR-based floorings. Kα chlorine mapping of aged (c) NR and (d) SBR-based floorings by EDX measurements.

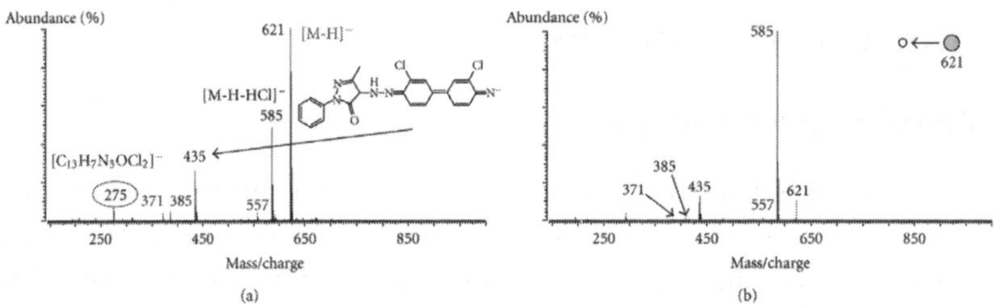

Figure 3: (a) LDI-FTICRMS analysis of standard orange pigment in negative-ion detection mode at the wavelength of 355nm and (b) tandem mass spectrum in SORI-CID mode of [M-H]⁻ pseudomolecular ion.

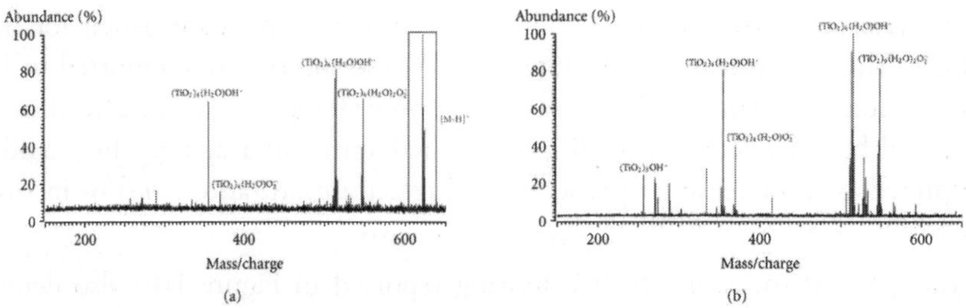

Figure 4: LDI-FTICRMS spectrum of NR-based orange flooring in negative-ion detection mode at the wavelength of 355nm (a) before and (b) after aging.

Surface Analysis and EDX Measurements

The topology of the surface of both floorings has been investigated by SEM in secondary electron mode (Figures 2(a) and 2(b)). Cracks can be observed for the first 75 and 300 μm of depth with SBR and NR floorings, respectively. Consequently, it can be argued that aging processes lead only to surface or near surface degradation and that depth degradation is dependant of the polymer (NR or SBR). The distribution of different atoms in the degraded portion of flooring as well as at the interface of altered/nonaltered compositematerial is investigated by EDX mapping and is reported in Figures 2(c) and 2(d). The more important information is obtained when chlorine atom mapping is considered. Significant differences appear when NR- and SBR-based aged flooring behavior is compared. Aged NR flooring Cl mapping shows a homogeneous distribution of chlorine atoms whereas aged SBR denotes a higher concentration of Cl at the surface compared to the bulk. Reader has to keep in mind that two chlorine atoms are present in the orange pigment molecule (see Scheme 1). Due to the fact that chlorine is not present in the structure of other organic and inorganic components of the composite, chlorine is thought to be relevant of pigment or at least of its degradation products. Consequently, it is assumed that aging of SBR induced an enrichment of the surface by the orange pigment.

LDI-FTICRMS Measurements

Prior to perform analysis of SBR and NR floorings, LDIFTICRMS experiments has been conducted on the standard orange pigment to obtain an MS fingerprint of this compound. Its LDI-FTICRMS negative mass spectrum is presented in Figure 3. Different negative ions are detected. More specifically, m/z 621, 585, 435, and 275 ions are observed on the mass spectrum. m/z 621 ion is associated to the $[M\text{-}H]^-$ pseudomolecular ion of orange pigment. m/z 585 and 435 correspond to fragments of pseudomolecular ion. The first one is thought to be $[M\text{-}H\text{-}HCl]^-$ whereas the second one corresponds to the release of a more important moieties of the pigment molecule (see assignment in Figure 3).

Table 1: S/N ratio of detected ions in the LDI-FTICRMS study of SBR and NR flooring samples before and after aging (nd: not detected).

		Fragment m/z 435	$[M\text{-}H\text{-}HCl]^-$ m/z 585	$[M\text{-}H]^-$ m/z 621
SBR flooring	Before aging	125	46	108
	After aging	98	60	250
NR flooring	Before aging	nd	nd	25
	After aging	nd	nd	nd

Figure 4 gathers LDI-FTICRMS spectra of NR flooring before and after aging. The main detected species on the mass spectrum before aging are associated to orange pigment and TiO2 filler. [M-H]– pseudomolecularion is detected at m/z 621. Cluster ion associated to titanium dioxide mainly consists of association of TiO2 and H2O building blocks with OH– or O2– charge carriers. After aging, these latter species are the solely detected ions. The [M-H]– pseudomolecular ion is not more detected. Similar experiments have been conducted with SBR-based flooring. Results for NR and SBR before and after aging are summarized in Table 1. In contrast to what it is observed with NR, the aging of SBR flooring does not lead to a decrease of the [M-H]– signal on the mass spectra but to an increase of its S/N (signal/noise) ratio by a factor of 2. Consequently, the huge difference in LDI-FTICRMS behavior for both floorings is the evolution of the [M-H]– with the aging treatment, which clearly indicates the influence of the polymermatrix on the degradation of the compositematerial, especially the degradation of the orange pigment.

Conclusion

The LDI-FTICRMS analysis of NR- and SBR-based flooring before aging clearly demonstrates the ability of this technique to detect the orange pigment in compositematerial bymeans of the observation of $[M-H]^-$ pseudomolecular ion and also its fragment ions. The LDI-FTICRMS behavior of floorings after aging is highly dependant of the polymer matrice. Indeed, after UV irradiation, the abundance of $[M-H]^-$ ion increases for SBR whereas it dramatically decreases for NR, which means that orange pigment is significantly destroyed at the surface of NR flooring whereas it concentrates at the surface for SBR flooring. This is confirmed by both EDX and UV/Vis spectroscopy. EDX analysis indicates a concentration of chlorine atoms, which is a marker of orange pigment or of its degradation products at the surface of SBR flooring after aging, which is not observed with NR. Moreover, the UV/Vis absorption behavior is only slightly modified after aging for SBR flooring whereas NR absorption decreased significantly between 350–600nm (pigment absorption wavelengths). As a consequence, it is assumed that both polymer matrices undergo degradation processes, which is confirmed by the broadening of polymer absorption band and the observed pseudo- "red shift" formation of unsaturation and incorporation of oxygen atoms. Degradation products and especially peroxides and hydroperoxides radicals greatly affect the behavior of the organic pigment. For NR, these species are thought to interact with the orange pigment, which leads to its destruction. In contrast, the radicals formed during SBR flooring aging appear to be inefficient to degrade the pigment. Nevertheless, the partial destruction of SBR polymer matrix allows pigment to migrate and concentrate at the surface, leading to a darker material. The modification of the absorption behavior of the polymer (see in Figure 1) seems to be indicative of unsaturations formation, which only could be obtained by the release of hydrogen radical for SBR and hydrogen and methyl ones for NR. The understanding of the interaction of these radicals with the orange appears consequently to be the key point to better understand the aging processes.

References

1. K. A. M. dos Santos, P. A. Z. Suarez, and J. C. Rubim, "Photodegradation of synthetic and natural polyisoprenes at specific UV radiations," Polymer Degradation and Stability, vol. 90, no. 1, pp. 34–43, 2005.
2. D. Ruch, C. Boes, R. Zimmer, J. F. Muller, and H.-N. Migeon, "Quantitative analysis of styrene butadiene copolymers using SSIMS and LA-FTICRMS," Applied Surface Science, vol. 203-204, pp. 566–570, 2003.
3. L. P. Real, J.-L. Gardette, and A. Pereira Rocha, "Artificial simulated and natural weathering of poly(vinyl chloride) for outdoor applications: the influence of water in the changes of properties," Polymer Degradation and Stability, vol. 88, no. 3, pp. 357–362, 2005.
4. T. Dienes, S. J. Pastor, S. Schürch, et al., "Fourier transform mass spectrometry-advancing years (1992-mid. 1996)," Mass Spectrometry Reviews, vol. 15, no. 3, pp. 163–211, 1996.
5. A. G. Marshall, C. L. Hendrickson, and G. S. Jackson, "Fourier transform ion cyclotron resonance mass spectrometry: a primer," Mass Spectrometry Reviews, vol. 17, no. 1, pp. 1–35, 1998.
6. A. G. Marshall and C. L. Hendrickson, "Fourier transform ion cyclotron resonance detection: principles and experimental configurations," International Journal of Mass Spectrometry, vol. 215, no. 1–3, pp. 59–75, 2002.

Organic Analysis of Peridotite Rocks from the Ashadze and Logatchev Hydrothermal Sites

Marie-Paule Bassez, Yoshinori Takano and Naohiko Ohkouchi

ABSTRACT

This article presents an experimental analysis of the organic content of two serpentinized peridotite rocks of the terrestrial upper mantle. The samples have been dredged on the floor of the Ashadze and Logatchev hydrothermal sites on the Mid-Atlantic Ridge. In this preliminary analysis, amino acids and long chain n-alkanes are identified. They are most probably of biological/microbial origin. Some peaks remain unidentified.

Introduction

The origin of terrestrial life is not yet understood. An accepted hypothesis is that a transition occurred between a molecular prebiotic evolution and a biological evolution and that prebiotic organic matter could have been delivered to Earth within carbonaceous chondrite meteorites, such as the CM2 Murchison meteorite.

At the bottom of the terrestrial oceans, where tectonic forces separate the lithospheric plates along mid-ocean ridges, the ultramafic rocks of the upper-mantle, the peridotites, are exposed to circulating seawater [1]. They encounter various physico-chemical conditions and the hydrolysis of their silicate constituents, the olivine and pyroxenes minerals, into serpentine, occur at different degrees of serpentinization depending on the characteristics of the medium: temperature, pressure, oxygen fugacity, nature and composition of the fluid phase, fluid flux, pH, rock composition, water:rock ratio [2]. The Mid-Atlantic-Ridge, MAR, is covered with several hydrothermal sites and presents black smoker activity. The active Logatchev site, 14° 45'N-43'N, at a water depth of 2,970 m and the active Ashadze site (12° 58'N, 4,080 m) are located on an ultramafic geological environment of serpentinized peridotite rocks, while the Krasnov site (16° 38'N), discovered with the Ashadze site during the 2007 French-Russian Serpentine cruise [3] is inactive and located on a basaltic environment. Ultramafic environments seem enriched in Cu and Zn content compared to the basaltic ones [3].

The Logatchev hydrothermal vent fluids originate from the interaction between the underlying peridotite rocks and seawater. They have been previously analyzed [4]. The H2 concentration is 12 mmol/wkg (data from 1996) and 19 mmol/wkg (data from 2005) and the analyses made in 1996, 2004 and 2005 show a stable composition of the fluids. The analyses of the Ashadze vent fluids [5] also show a great amount of H2. Both these vent fluids, as those of the Rainbow site (36° 14'N on the MAR, 2,300 m) also contain significant amounts of CO2, CH4, N2, CO. Their pH is acidic ~3-4, the temperature of their fluids is ~310-370 °C and the detected saturated hydrocarbons, carboxylic acids and methyl esters in the fluids have been proposed of either abiogenic origin or not [5,6].

An accepted hypothesis to explain the occurence of the carbon-based organic compounds in the fluids is the synthesis of these molecules in the context of catalytic Fischer-Tropsch Type (FTT) reactions involving hydrothermal CO2. The dihydrogen, formed during the hydrolysis of the peridotite terrestrial rocks, which contain ferrous iron-rich minerals, olivine and pyroxenes, could react with hydrothermal CO2, to form methane and saturated hydrocarbons. Hydrocarbons have been synthesized during experimental serpentinization of olivine at 300 °C and 500 bar [7] and methane, ethane and propane were synthesized at 390 °C and 400 bar in an experiment catalyzed with Cr2O3 in combination with FeO [8]. A more recent experiment, at 200 °C and 500 bar, simulating subseafloor serpentinization produced significant amounts of dissolved H2 when artificial seawater reacted with a peridotite rock composed of 62% olivine, 26% orthopyroxene and

10% clinopyroxene. Even during the early stages of the reaction, ~25 mmol/kg of water are produced after 2,000 h of experiment and 77 mmol/wkg after 8,000 h [9]. Experiments conducted at 250 °C and 325 bar on an aqueous solution of formic acid (HCOOH) in the presence of Fe produced a series of n-alkanes with typical FTT distribution. Volatile hydrocarbons (C1-C6), magnetite (Fe3O4) and siderite FeCO3 were also detected [10]. FTT mechanism can be invoked since hydrothermal Fe reacts with water to form magnetite and H2 and formic acid decomposes into CO2 and H2.

The exact factors that control the hydrolysis of peridotite remain unknown. Calculations considering the thermodynamics of fluid mixing between hydrothermal fluids containing dissolved CO2 and H2 at 350 °C, and seawater containing bicarbonate at 2 °C, led to the organic synthesis of carboxylic acids, alcohols, ketones [11]. These calculations depend on the fugacity of O2. They show that the oxidation state of ultramafic rocks, driven by the equilibrium of the FMQ, fayalite-magnetitequartz mineral assemblage, lead to a lower oxygen fugacity and a greater potential for organic synthesis than for the PPM, pyrrhotite-pyrite-magnetite assemblage. Numerical models, considering a rock composed of 80 wt% olivine, 15 wt% orthopyroxene and 5 wt% clinopyroxene predict that, at 35 MPa, a peak production of H2 (a few hundred mmol/kg) occurs approximately at temperatures of 200-315 °C. These models also predict a decrease in pH from ~11 to ~6, when the temperature increases from 50 °C to 400 °C, with pH values of ~9 around 150 °C and ~8 around 200 °C [12].

Analyses of hydrothermally altered peridotites drilled between 14°N and 16°N on the Mid-Atlantic Ridge (MAR) between 1,800 and 4,000 m depth have been reported [13]. They suggest that extensive serpentinization processes occur at all sites and that the transformation of the mineral olivine into serpentine, magnetite and brucite with release of H2 is favored at temperatures below 250 °C, while pyroxene is replaced by talc and tremolite above 350-400 °C [2,13], where olivine is stable. The latitude of these drillings corresponds to the area of the hydrothermal sites Logatchev, Ashadze and Krasnov.

Several experiments have demonstrated the production of hydrothermal organic matter including nitrogen atoms at various temperatures, 100-400 °C, and pressures and with various starting compounds [14-20 and Ref. therein]. A gas mixture of methane and dinitrogen above simulated seawater under ~8 MPa at room temperature was heated to 325 °C. Amino acids were extracted after acid hydrolysis of the products [15]. In experiments conducted at 150 °C and 1 MPa with HCN, CH2O, NH3 in the presence of the PPM redox buffer, amino acids were also detected [16]. Their yields were higher than in previous gaseous spark discharge experiments [14]. Di- and triglycine were synthesized in a flow reactor under 24.0 MPa at 200 °C-250 °C with consecutive quenching at 0 °C.

The presence of copper ions seemed to help synthesize tetraglycine [17]. Using a supercritical water flow reactor with temperature control inside the fluids, it is suggested that condensates of glycine, which yielded amino acids after hydrolysis, formed even in supercritical water at 400 °C, under 25 MPa pressure [18]. When an aqueous mixture of ten amino acids was heated at 200-400 °C, the acid hydrolysis of the products led to a higher content in glutamic acid and α-amino acids, such as α-aminobutyric acid, 5-aminovaleric acid and 6-aminohexanoic acid than in α-amino acids even over supercritical conditions of water suggesting that α-amino acids could be chemical markers of abiotic hydrothermal systems [18]. Reviews report the various conditions of amino acid syntheses [19-21 and Ref. therein].

Recent calculations using measured data of the Rainbow hydrothermal site, show that an abiotic synthesis of the five nucleobases and of the two sugars from formaldehyde and hydrogen cyanide is thermodynamically favored between 0 °C and 150-250 °C [22 and Ref. therein].

Some similarities with the Murchison meteorite can be noticed. The Murchison mineral structure is dominated with a phyllosilicate (serpentine) matrix which contains minerals such as olivine, pyroxenes, calcium carbonates, iron oxides (magnetite), iron-nickel sulfides and sulfates [23-25]. It has been altered by water, by heat, by pressure shock waves, by short-lived radionuclides [26,27]. The transformation of olivine and pyroxene chondrules seems to grow with the extent of mineral hydrolysis and the formation of water-soluble organic compounds is described at temperatures below ~125 °C [28,29]. Aside from any terrestrial contamination, all the classes of organic molecules considered of biological relevance are identified [30-32 and Ref. therein] and also non-terrestrial amino acids and enantiomeric excesses [33-35].

Several hypotheses are proposed for the production of meteoritic organic matter, either solar-nebula processes or secondary processes which occurred after the accretion, on asteroidal parent bodies [3539, 23, 40-46 and Ref. therein]. Among these are FTT reactions; ion-molecule and radical-radical reactions; γ-, proton- and UV-irradiation; Strecker's type reactions involving aqueous processing of simple molecules such as H2O, HCN, H2CO and NH3; internal heating of the parent body produced by the radioactive decay of short-lived nuclides. Although quite significant amounts of glycine are detected in the Murchison meteorite, no ascertained interstellar glycine has yet been identified since its first observational report in 1979 [47] suggesting that molecules formed in the interstellar medium, ISM, underwent further processing. It has been suggested that the primary products from proton irradiation of a mixture of CO, N2/NH3, H2O are amino acid precursors, molecules that provide amino acids after acid hydrolysis [36,38]. Amino acids recovered after acid hydrolysis of products obtained in vacuum

UV-photolysis of H2O, CO, CO2, CH3OH, CH4, NH3, simulating the ISM, do not match the Murchison meteorite distribution, suggesting that the organic molecules found in the meteorite parent bodies experienced contact with water [46]. Indeed, it has been demonstrated that bound amino acids in aqueous solution exposed to γ- and UV- rays are much more photostable than the corresponding free amino acids [39].

Thus, it seems consequently plausible to imagine that the H2, released during the serpentinization processes of the peridotite terrestrial rocks, could react with the CO2 embedded inside the rock, to form methane and saturated hydrocarbons, in the context of catalytic reactions involving hydrothermal CO2. The simple molecules H2O, H2, CO2, CH4, would be present as a consequence of mineral reactions of the terrestrial peridotites with seawater and, with the N2 of the environment and with an activation source such as gamma rays, they could form the simple organic molecules of biological relevance [4850,22,51 and Ref. therein].

These reactions could occur at temperatures ~150-200 °C, where olivine transforms into serpentine, magnetite and brucite with the release of H2. At these temperatures, combined with the pressures encountered at the hydrothermal sites, many compounds are in their supercritical state and peculiar chemistry can occur. In this IJMS issue on the Origin of Life, syntheses of amino acids in a mixture of supercritical CO2-liquid water (10:1) starting with hydroxylamine hydrochloride and pyruvic or glyoxylic acid are reported [52]. A hypothesis for the origin of the living systems could consequently be found at the bottom of the oceans, in ultramafic hosted hydrothermal systems, where tectonic plates separate to leave the upper mantle rock reacts with seawater to form hydrothermally altered peridotites and lead to the necessary molecules for life to emerge.

In this hypothesis, serpentinized peridotite rocks located on hydrothermal sites could contain organic molecules. Here we report organic analyses made on two peridotite rocks of Ashadze (12° 58'N, 4,080 m) and Logatchev (14° 43'N, 2,970 m) hydrothermal sites in the Mid-Atlantic Ridge. The samples have been dredged on the seafloor in March 2007, during the French-Russian Ifremer Serpentine cruise [53]. These organic analyses provide the first observations of organic compounds in the serpentinized peridotite rocks of Ashadze and Logatchev hydrothermal sites. They are reported here for the first time.

Experimental Methods

The analyses have been carried out in the Institute of Biogeosciences of the Japan Agency for Marine-Earth Science and Technology, in Yokosuka. The rock sample was pre-washed by ultra-pure methanol to eliminate possible exogenous compounds from the external surfaces. An aliquot of dried and grounded sample powder (ca 0.5 g) was dispensed into 16 x 100 mm reaction vials with PTFE-lined caps and acid hydrolyzed with 6 M HCl at 110 °C for 12 h. Non-polar fraction was extracted by liquid/liquid separation in HCl solution and 2.0 mL of a hexane/dichloromethane (6:5, v/v) mixture in two portions. The hexane/dichloromethane fraction was recovered and dried under a gentle nitrogen flow, and then 200 □L of dichloromethane was added to the final non-polar fraction.

Another procedure was used for the polar fraction, especially for amino acids. After drying the hydrolysis residue under N2 flow, the samples were adjusted to pH 1 .0with 0.1 M HCl, and the amino acid fraction was isolated with cation-exchange column chromatography. The purification of amino acid fractions via application to an AG-50W-X8 (200-400 mesh; Bio-Rad Laboratories) cation exchange resin column was performed by the procedure described earlier [46]. Briefly, a slurry of resin in deionized water was poured into a disposable glass pipette column plugged with quartz wool. Before the injection of the sample to the column, the resin was cleaned by passing three bed volumes (resin/carrier, 1:3, v/v) of 1 M HCl, H_2O, 1 M NaOH, and H_2O through the column in succession (i.e., 2 mL of AG50 resin requires 6 mL of 1 M HCl for the first prewash). Immediately before the injection of the sample, the resin was reactivated to the H^+ form with three bed volumes of 1 M HCl and then rinsed with three bed volumes of H_2O. The sample solution was loaded and then eluted with three bed volumes of H_2O to retain only the amino acid fraction. Finally, the amino acid fraction was eluted with three bed volumes of 10% NH_3 aqueous solution, and then dried by nitrogen flow for the next derivatization procedure.

The esterification reaction was performed with 500 □L of a thionyl chloride/(S)-(+)-2-butanol mixture (1:4, v/v) at 110 °C for 2 h. After the solution had been cooled to ambient temperature, it was evaporated to dryness under a gentle nitrogen flow at ~80 °C. The acylation reaction was then performed with 500 □L of a pivaloyl chloride/dichloromethane mixture (1:1, v/v) at 110 °C for 2 h. After cooling, the solution was again evaporated to dryness with a gentle nitrogen flow at ~80 °C. The N-pivaloyl-(S)-2-butyl esters (NP/S2Bu) of the amino acid diastereomers [46] were extracted by liquid/liquid separation in 0.5 mL of distilled water and 1.0 mL of a hexane/dichloromethane (6:5, v/v) mixture for two times. The hexane/dichloromethane mixture fraction containing the NP/S2Bu esters was recovered and dried under a gentle nitrogen flow. Then, 200 □L of dichloromethane was added to the final fraction. The NP/S2Bu esters of the amino acid diastereomers (Figure 1) were identified by a gas chromatograph/mass spectrometry (GC/MS; Agilent Technologies 6890N/5973MSD). The capillary column used for GC was an HP-5 (30 m □ 0.32 mm i.d., 0.52 □m film thickness; Agilent Technologies). The GC oven temperature was programmed as follows: initial temperature 40 °C for 4 min, ramped up at 10 °C min–1 to 90 °C, and ramped up at 5 °C min–1 to 220 °C, where it was maintained for 10 min. The MS was scanned over m/z of 50–550 with the electron-impact mode set at 70 eV. Optically active (S)-(+)-2-butanol (purity 99%; boiling point 99-100 °C) was obtained from Sigma-Aldrich Co. All glassware was heated at 450 °C for 4 h before use to eliminate any possible contaminants.

Results and Discussion

As seen in Figure 2, we identify a wide variety of amino acids including protein and non-protein amino acids. Among these, glycine and glutamic acid are more predominant than the others. Although non-proteinous amino acids such as sarcosine, beta-alanine (BALA) and gamma-aminobutyric acid (GABA) have been found as products in laboratory experiments simulating hydrothermal systems [15], in our experiment sarcosine is under detection limit and BALA and GABA are present as minor constituents. The peak at 17.9 min could not be identified. Figure 1 illustrates the mass spectrum of the N-pivaloyl-(S)-2-butyl esters obtained for the identification of the D- and L-alanine of the gas chromatogram (Figure 2). It corresponds to the retention times of the alanine peaks in the chromatogram.

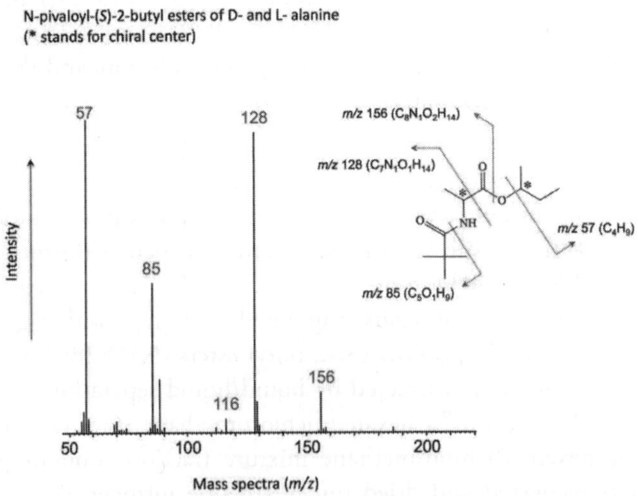

Figure 1. Mass fragment pattern of the N-pivaloyl-(S)-2-butyl esters of the D- and L-alanine diastereoisomers.

For amino acids formed abiotically [15], the D/L ratio of amino acids converges to around 1. On the other hand, large enantiomeric excess of L-form amino acids may indicate that the amino acids are derived from sub-seafloor biogenic processes [54] or abiogenic racemization reaction during the pathway of stereochemical conversion via alpha-hydrogen elimination. The racemization of amino acid standards during 22 hours hydrolysis treatment ranged 0.5-1.3% for D-alanine generated from L-alanine [55]. Here, as seen in Figure 3a for the Ashadze peridotite rock, the molar fraction (%D- and %L-) of D-alanine: L-alanine in the serpentine sample is 15:85, hence D/L ratio is 0.18 and other amino acids are also L-form predominant. On Figure 3b the D/L ratios of the sedimentary amino acids, Ala, Asx (asparagine and aspartate) and Glx (glutamine and glutamate) shows the racemization process during early diagenesis as a function of depth over 10,000 years [56]. The similarities in the values on D/L ratios provide a plausible conclusion of a biological origin for the amino acids identified in the Ashadze peridotite sample and also for the Logatchev sample. Although the prokaryotic community in hydrothermal sediments of the Alvin zone location, ~3,500 m, near the TAG mound, ~26°N on the MAR, seems present with a low total cell count [57], we conclude in a biological origin for the identified amino acid peaks.

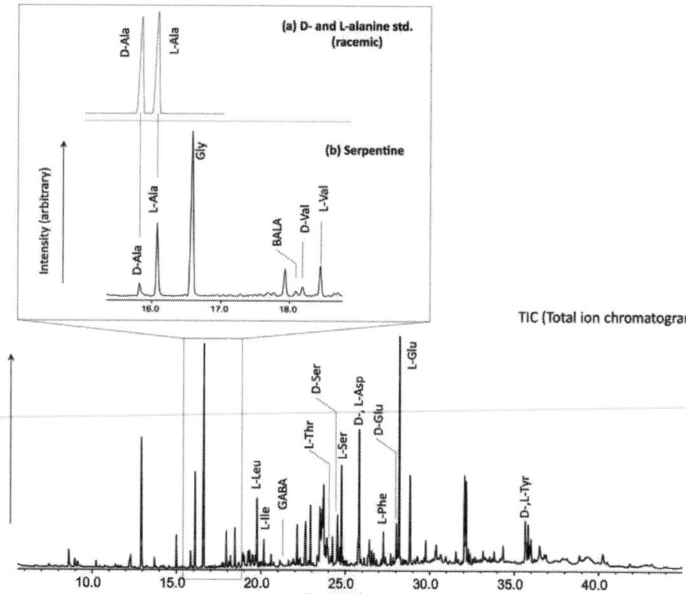

Figure 2. Representative chromatogram of chiral separation for D- and L-amino acids in polar fraction extracted from the Ashadze serpentinized peridotite rock sample by GC/MS analysis. Abbreviations: D-Ala, D-Alanine; L-Ala, L-Alanine; Gly, Glycine; BALA, beta-Alanine; D-Val, D-Valine; L-Val, L-Valine; L-Leu, L-Leucine; L-Ile, L-Isoleucine; GABA, gamma-aminobutyric acid; L-Thr, L-Threonine; D-Thr, D-Threonine; D-Ser, D-Serine; L-Ser, L-Serine; D-, L-Asp, D-, L-Aspartic acid; L-Phe, L-Phenylalanine; D-Glu, D-Glutamic acid; L-Glu, L-Glutamic acid; D-,L-Tyr, D-,L-Tyrosine.

We also detect a long-chain n-alkane compound (< n-C28H58) in the non-polar fraction (Figure 4) under GC conditions up to 220 °C. Although we do not identify lipid compounds in this non-polar fraction, long-chain n-alkanes may have two origins. One can be fossilized past biota and/or present microbes which migrated within hydrothermal fluids and the other can be hydrothermally synthesized and/or altered organic molecules.

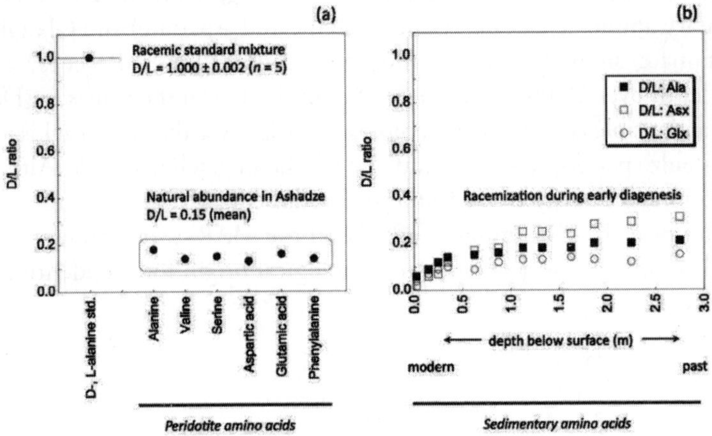

Figure 3. D/L amino acid ratios in the analysed Ashadze peridotite rock and in sedimentary rocks.: Ala (alanine), Asx (asparagine and aspartate) and Glx (glutamine and glutamate).

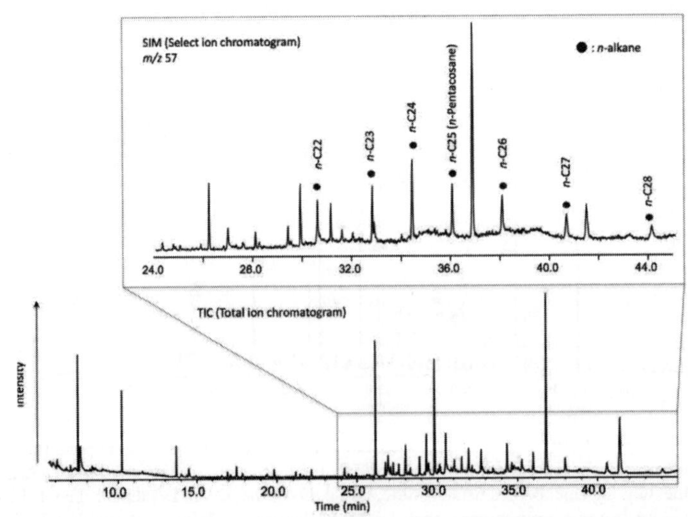

Figure 4. Representative chromatogram of hydrocarbons including n-alkanes in non-polar fraction of the Logatchev rock sample. Select ion monitoring (SIM) was also performed to identify n-alkane chain analogs.

The GC/MS of the n-alkanes shows a decrease in intensity with increasing carbon number, which seems to be a characteristic of abiotic synthesis [10]. Recently, an abiogenic hydrocarbon production by FTT at Lost City hydrothermal field has been proposed wherever warm ultramafic rocks are in contact with water [58]. However, as discussed for the Suiyo Seamount, Izu-Bonin Arc, Pacific Ocean [54] and for the Lost City, Mid-Atlantic Ridge [59] hydrothermal systems, it is difficult to differentiate biotic/abiotic sources. An experimental analysis of the isotopic fractionation of the stable carbon-13 and carbon-12 elements in the organic compounds detected in the Ashadze and Logatchev samples would, as it is widely thought, indicate if these organic compounds derive from microbial decomposition or from an abiotic synthesis. However, it has been demonstrated in laboratory experiments conducted at 250 °C and 350 bar, that organic products, synthesized abiotically in FTT reactions, are depleted in 13C to a degree typically ascribed to biological processes [10]. These experiments indicate that the analysis of the carbon isotopic fractionation is an ineffective diagnostic to distinguish between abiotic and biotic origin of organic compounds. Consequently, we will not proceed to the carbon isotopic analysis of the rocks and we do not conclude yet in a biotic or abiotic origin for the identified n-alkanes.

Conclusions

This preliminary analysis of the organic composition of two peridotite rock samples dredged on the ocean floor of the Logatchev and Ashadze hydrothermal sites on the Mid-Atlantic Ridge allows the identification of amino acids and long-chain n-alkanes. Many peaks of the amino acid gas chromatograms remain unidentified. Further analyses need to be made with non terrestrial amino acids as references. Signals of abiotically formed organic compounds may be present with negligible intensity compared to the intensities of the identified biotical signals. Consequently, we conclude in a biotic origin for the identified amino acids but we do not exclude an abiotic origin for some amino acids which correspond to the not yet identified peaks. Especially because it is difficult to conclude anything about a biotic/abiotic origin for the n-alkanes, since carbon isotopic fractionation is inefficient in distinguishing these sources. It would be more appropriate to analyze samples which are drilled far beneath the ocean floor and which would be less exposed to biological contamination. That could be one goal of a next IODP (Integrated Ocean Drilling Program) cruise.

Acknowledgements

The authors thank Kensei Kobayashi (Yokohama National University) for discussions. They are sincerly grateful to Adélie Delacour and Mathilde Cannat from the Institut de Physique du Globe de Paris and to the Ifremer Serpentine 2007 cruise for giving them the two Logatchev and Ashadze peridotite rock samples. This research was partly supported by the Japan Society for the Promotion of Science (Y.T), and a Grant-in-Aid for Creative Scientific Research (19GS0211). Marie-Paule Bassez deeply acknowldges André Brack for his interest in the project and the french Exobiologie group for some financial support.

References

1. Altérations hydrothermales de la lithosphère océanique In Comptes Rendus Geosciences; Honnorez, J., Ed.; Elsevier: Paris, France, 2003; Volume 335, pp. 777–864.
2. Mével, C. Serpentinization of abyssal peridotites at mid-ocean ridges. C. R. Geoscience 2003, 335, 825–852.
3. Fouquet, Y.; Cherkashov, G.; Charlou, J.; Ondreas, H.; Cannat, M.; Bortnikov, N.; Silantiev, S.; Etoubleau, J. Diversity of ultramafic hosted hydrothermal deposits on the Mid Atlantic Ridge: First submersible studies on Ashadze, Logatchev 2 and Krasnov vent Fields during the Serpentine cruise. AGU Fall Meeting, San Francisco, CA, USA, 10–14 December 2007; Abstract T51F-03.
4. Schmidt, K.; Koschinsky, A.; Garbe-Schönberg, D.; de Carvalho, L.M.; Seifert, R. Geochemistry of hydrothermal fluids from the ultramafic-hosted Logatchev hydrothermal field, 15°N on the Mid-Atlantic Ridge: Temporal and spatial investigation. Chem. Geolog. 2007, 242, 1–21.
5. Charlou, J.; Donval, J.; Konn, C.; Birot, D.; Sudaikov, S.; Jean-Baptiste, P.; Fouquet, Y. High hydrogen and abiotic hydrocarbons from new ultramafic hydrothermal sites between 12°N and 15°N on the Mid Atlantic Ridge-results of the Serpentine cruise (march 2007). AGU-Fall Meeting, San Francisco, CA, USA, 10–14 December 2007; Abstract T51F-04.
6. Konn, C.; Charlou, J.L.; Donval, J.P.; Holm, N.G.; Dehairs, F.; Bouillon, S. Hydrocarbons and oxydized organic compounds in hydrothermal fluids from Rainbow and Lost City ultramafic-hosted vents. Chem. Geolog. 2009, 258, 299–314.
7. Berndt, M.E.; Allen, D.E.; Seyfried, W.E., Jr. Reduction of CO2 during serpentinization of olivine at 300 °C and 500 bar. Geology 1996, 24, 351–354.

8. Foustoukos, D., I.; Seyfried, W., E., Jr. Hydrocarbons in hydrothermal vent fluids: the role of chromium-bearing catalysts. Science 2004, 304, 1002–1005.
9. Seyfried, W.E., Jr.; Foustoukos, D.I.; Fu, Q. Redox evolution and mass transfer during serpentinization: An experimental and theoretical study at 200 °C, 500 bar with implications for ultramafic-hosted hydrothermal systems at Mid-Ocean Ridges. Geochim. Cosmochim. Acta 2007, 71, 3872–3886.
10. McCollom, T.M.; Seewald, J.S. Carbon isotope composition of organic compounds produced by abiotic synthesis under hydrothermal conditions. Earth Planet. Sci. Lett. 2006, 243, 74–84.
11. Shock, L.E.; Schulte, M.D. Organic synthesis during fluid mixing in hydrothermal systems. J. Geophys. Res. 1998, 103, 28513–28527.
12. McCollom, T.; Bach, W. Thermodynamic constraints of hydrogen generation during serpentinization of ultramafic rocks. Geochim. Cosmochim. Acta 2009, 73, 856–875.
13. Bach, W.; Garrido, C.J.; Paulick, H.; Harvey, J.; Rosner, M. Seawater-peridotite interactions: First insights from ODP Leg 209, MAR 15°N. Geochem. Geoph. Geosyst. 2004, 5, 1–22.
14. Miller, S.L.; Orgel, L.E. The Origin of Life on the Earth; Prentice-Hall: Englewoods Cliffs, NJ, USA, 1974.
15. Yanagawa, H.; Kobayashi, K. An experimental approach to chemical evolution in submarine hydrothermal systems. Orig. Life Evol. Biosphere 1992, 22, 147–159.
16. Holm, N.G.; Andersson, E.M. Abiotic synthesis of organic compounds under the conditions of submarine hydrothermal systems: a perspective. Planet. Space Sci. 1995, 43, 153–159.
17. Imai, E.; Honda, H.; Hatori, K.; Brack, A.; Matsuno, K. Elongation of oligoeptides in a simulated submarine hydrothermal system. Science, 1999, 283, 831–833.
18. Islam, M.N.; Kaneko, T.; Kobayashi, K. Reaction of amino acids in a supercritical water-flow reactor simulating submarine hydrothermal systems. Bull. Chem. Soc. Jpn. 2003, 76, 1171–1178.
19. Simoneit, B.R.T. Prebiotic organic synthesis under hydrothermal conditions: an overview. Adv. Space Res. 2004, 33, 88–94.
20. Pascal, R.; Boiteau, L.; Commeyras, A. From the prebiotic synthesis of α-amino acids towards a primitive translation apparatus for the synthesis of peptides. Top. Curr. Chem. 2005, 259, 69–122.

21. Zaia, D.A.M.; Zaia, T.B.V.; De Santana, H. Which amino acids should be used in prebiotic chemistry studies? Orig. Life Evol. Biosphere 2008, 38, 469–488.

22. LaRowe, D.E.; Regnier, P. Thermodynamic potential for the abiotic synthesis of adenine, cytosine, guanine, thymine, uracil, ribose and desoxyribose in hydrothermal systems. Orig. Life Evol. Biosphere 2008, 38, 383–397.

23. Brearley, A.J. Nebular versus parent-body processing. In Meteorites, Comets and Planets. Treatise on Geochemistry; Davis, A.M., Ed.; Elsevier: Oxford, U.K. 2005; Volume 1, p. 180, pp. 247–268.

24. Zolensky, M.; Barrett, R.; Browning, L. Mineralogy and composition of matrix and chondrule rims in carbonaceous chondrites. Geochim. Cosmochim. Acta 1993, 57, 3123–3148.

25. El Amri, C.; Maurel, M.C.; Sagon, G.; Baron, M.H. The micro-distribution of carbonaceous matter in the Murchison meteorite as investigated by Raman imaging. Spectrochim. Acta A 2005, 61, 2049–2056.

26. Zinner, E.; Amari, S.; Anders, E.; Lewis, R. Large amounts of extinct 26Al in interstellar grains from the Murchison meteorite. Nature 1991, 349, 51–54.

27. Carlson, R.W.; Boyet, M. Short-lived radionuclides as monitors of early crust-mantle differentiation on the terrestrial planets. Earth Planet Sci. Lett. 2009, 279, 147–156.

28. Bunch, T.E.; Chang, S. Carbonaceous chondrites: II. Carbonaceous chondrite phyllosilicates and light element geochemistry as indicators of parent body processes and surface conditions. Geochim. Cosmochim. Acta 1980, 44, 1543–1577.

29. Schulte, M.; Shock, E. Coupled organic synthesis and mineral alteration on meteorite parent bodies. Meteorit. Planet. Sci. 2004, 39, 1577–1590.

30. Botta, O.; Bada, J.L. Extraterrestrial organic compounds in meteorites. Surv. Geophys. 2002, 23, 411–467.

31. Gilmour, I.; Structural and isotopic analysis of organic matter in carbonaceaous chondrites. In Meteorites, Comets and Planets. Treatise on Geochemistry; Davis, A.M.; Elsevier: Oxford, U.K., 2005; Volume 1, pp. 269–290.

32. Martins, Z.; Botta, O.; Fogel, M.L.; Sephton, M.A.; Glavin, D.P.; Watson, J.S.; Dworkin, J.P.; Schwartz, A.W.; Ehrenfreund, P. Extraterrestrial nucleobases in the Murchison meteorite. Earth Planet. Sci. Lett. 2008, 270, 130–136.

33. Cronin, J. R.; Pizzarello, S. Enantiomeric excesses in meteoritic amino acids. Science 1997, 275, 951–955.

34. Engel, M.H.; Macko, S.A. The stereochemistry of amino acids in the Murchison meteorite. Precambrian Res. 2001, 106, 35–45.

35. Glavin, D.P.; Dworkin, J.P. Enrichment of the amino acid L-isovaline by aqueous alteration on CI and CM meteorite parent bodies. Proc. Natl. Acad. Sci. USA 2009, 106, 5487–5492.

36. Kobayashi, K.; Kaneko, T; Saito, T.; Ohima, T. Amino acids formation in gas mixtures by high energy particle irradiation. Orig. Life Evol. Biosph. 1998, 28, 155–165.

37. Kerridge, J.F. Formation and processing of organics in the early solar system. Space Science Rev. 1999, 90, 275–288.

38. Takano, Y.; Ohashi, A.; Kaneko T.; Kobayashi, K. Abiotic synthesis of high-molecular-weight organics from an inorganic gas mixture of carbon monoxide, ammonia and water by 3 MeV proton irradiation. Appl. Phys. Lett. 2004, 84, 1410–1412.

39. Takano, Y.; Kaneko, T.; Kobayashi, K.; Hiroishi, D.; Ikeda, H.; Marumo, K. Experimental verification of photostability for free- and bound amino acids exposed to γ-rays and UV irradiation. Earth Planets Space 2004, 56, 669-674.

40. Bernstein, M. Prebiotic material from on and off the early Earth; Phil. Trans. R. Soc. B 2006, 361, 1689–1702.

41. Ehrenfreund, P.; Sephton, M.A. Carbon molecules in space: from astrochemistry to astrobiology. Faraday Discuss. 2006, 133, 277–288.

42. Elsila, J.E.; Dworkin, J.P.; Bernstein, M.P.; Martin, M.P.; Sandford, S.A. Mechanisms of amino acid formation in interstellar ice analogs. Astrophys. J. 2007, 660, 911–918.

43. Takano, Y.; Takahashi, J.; Kaneko, T.; Marumo, K.; Kobayashi, K. Asymmetric synthesis of amino acid precursors in interstellar complex organics by circularly polarized light. Earth Planet. Sci. Lett. 2007, 254, 106–114.

44. Exploring Organic Environments in the Solar System. The National Academies Press: Washington, D.C., USA, 2007.

45. Cleaves, H.J. The prebiotic geochemistry of formaldehyde. Precambrian Res. 2008, 164, 111–118.

46. Nuevo, M.; Auger, G.; Blanot, D.; d'Hendecourt, L. A detailed analysis of the amino acids produced after the vacuum UV irradiation of ice analogs. Orig. Life Evol. Biosph. 2008, 38, 3756.

47. Brown, R.D.; Godfrey, P.D.; Storey, J.W; Bassez, M.-P.; Robinson, B.J.; Batchelor, R.A.; Mc. Culloch, M.G.; Rydbeck, O.E.; Hjalmarson, A.J. A search for interstellar glycine. Mon. Not. R astr. Soc. 1979, 186, 5–8.

48. Bassez, M.P. La structure de l'eau supercritique et l'origine de la vie. In Science et Technologie: Regards Croisés. L'Harmattan: Paris, 1999; pp. 583–591.

49. Takai, K.; Nakamura, K.; Suzuki, K.; Inagaki, F.; Nealson, K.H.; Kumagai, H. UltramaficsHydrothermalism-Hydrogenesis-HyperSLIME (UltraH3 microbial ecosystem in the Archean deep-sea hydrothermal systems.) Paleontolog. Res. 2006, 10, 269–282.
50. Bassez, M.P. Synthèse prébiotique dans les conditions hydrothermales. Comptes Rendus Chimie, 2009, 12, 801–807.
51. Aubrey, A.D.; Cleaves, H.J.; Bada, J.L. The role of submarine hydrothermal systems in the synthesis of amino acids.Orig. Life Evol. Biosph. 2009, 39, 91–108.
52. Fujioka, K.; Futamura, Y.; Shiohara, T.; Hoshino A.; Kanaya, F.; Manome, Y.; Yamamoto, K. Amino acid synthesis in a supercritical carbon dioxide-water mixture, Int. J. Mol. Sci., 2009, 10, 2722–2732.
53. Delacour, A., Institut de Physique du Globe de Paris; Mineral composition of the Logatchev and Ashadze serpentine samples. Private communication 2008.
54. Takano, Y.; Kobayashi, K.; Yamanaka, T.; Marumo, K.; Urabe, T. Amino acids in the 308 °C deep-sea hydrothermal system of the Suiyo Seamount, Izu-Bonin Arc, Pacific Ocean. Earth Planet Sci. Lett. 2004, 219, 147–153.
55. Amelung, W.; Zhang, X. Determination of amino acid enantiomers in soils. Soil Biol. Biochem. 2001, 33, 553–562.
56. Takano, Y.; Kobayashi, K.; Ishikawa, Y.; Marumo, K.; (2006) Emergence of the inflection point on racemization rate constants for D- and L-amino acids in the early stages of terrestrial diagenesis. Org. Geochem. 2006, 37, 334–341.
57. Glynn, S.; Mills, R.A.; Palmer, M.R.; Pancost, R.D.; Severmann, S.; Boyce, A.J. The role of prokaryotes in supergene alteration of submarine hydrothermal sulfides. Earth Planet. Sci. Lett. 2006, 244, 170–185.
58. Proskurowski, G.; Lilley, M.D.; Seewald, J.S.; Früh-Green, G.L.; Olson, E.J.; Lupton, J.E.; Sylva, S.P.; Kelley, D.S. Abiogenis hydrocarbon production at Lost City hydrothermal field. Science 2008, 319, 604–607.
59. Delacour, A.; Früh-Green, G.L.; Bernasconi, S.M.; Schaeffer P.; Kelley D.S. Carbon geochemistry of serpentinites in the Lost City hydrothermal system (30°N, MAR). Geochim. Cosmochim. Acta 2008, 72, 3681–3702.

The Study of Influence of the Teslar Technology on Aqueous Solution of Some Biomolecules

E. Andreev, G. Dovbeshko and V. Krasnoholovets

ABSTRACT

The possibility of recording physical changes in aqueos solutions caused by a unique field generated by the Teslar chip (TC) inside a quartz wristwatch has been studied using holographic interferometry. We show that the refraction index of degassed pure distilled water and aqueous solutions of L-tyrosine and b-alanine affected by the TC does not change during the first 10 minutes of influence. In contrast, a 1% aqueous solution of plasma extracted from the blood of a patient with heart vascular disease changes the refractive index when affected by the TC. The characteristic time of reaction is about 10^2 seconds. Based on our prior research experience, we state that the response of the system studied to the TC's field is similar to that stipulated by the action *of a constant magnetic field with the intensity of 1.1 × 10^{-3} T. Nevertheless, our team have unambiguously proved that the TC generates the inerton field, which is associated with a substructure of the matter waves (and, therefore, it does not relate to the electromagnetic nature). We could unambiguously prove that the TC generates the inerton field.*

Introduction

Numerous experiments fix the influence of electromagnetic field of certain frequency-amplitude ranges on living organisms. For instance, the magnetic field with frequencies in the range 0.3 to 30 Hz and with the intensity that is comparable with the Earth magnetic field can effectively influence the living organism function. It is supposed that the mechanism of influence should be connected with the parametric or Schumann resonance. The first four harmonics of the Schumann resonance are known: 7.8 Hz ± 1.5 Hz, 14.5, 20, 26 Hz (±0.3 Hz) [1–3]. Two main mechanisms of the resonance reaction of the organism to a weak electromagnetic fields are well known. The first one is the alfa rhythm concerned with the thought process; the second one, the parametric resonance of organs, or organ systems, could be responsible for primary human reception [4–6]. A number of physiological processes, such as the reductive-oxidative process in living cells, responsible for the oxygen input, oxygen transport, and so forth, could be taken into account in this case. The parametric resonance of biological tissue and surrounding medium could be also responsible for the medical action of the TC.

The aim of the present study is the following: (1) the influence of the TC on a biological model system and (2) registration of this influence in those cases when it is possible. The inventors [7] of this device state that the chip produces a longitudinal scalar wave/field (the notation was introduced by Nicola Tesla [8]). In this case, a part of the energy is radiated in the form of a scalar longitudinal wave (also known as a Tesla free-standing wave). In more detail, this wave has been studied in [9, 10].

A model object must be sensitive; it has to have a large gain factor and the method must be reproducible and stable, simultaneously. Since our goal is to account for biophysical aspects of the influence of the TC on living organisms, the model system should include components available in hypodermic tissues of the wrist. These conditions allow us to choose, as the model of primary reception, the following:

(i) saturated aqueous solution of amino acids (tyrosine, tryptophane, and alanine);

(ii) diluted aqueous solution of human blood plasma. (Although a solution of human serum albumin is dominating in blood protein, the preparation of its solution by conventional method will mean that we obtain an equilibrium system, and it will be very difficult to move such system from its deep potential minimum. In the case of biomolecules of plasma of blood, which we study in this work, we deal with a nonequilibrium system and even very small stimuli applied to it could be effective.) The parameter under study has become the refraction index n of an aqueous solution.

Materials and Method

Holographic experiments have been carried out with the use of the holographic interferometer IGD-3, developed and produced in the Institute of Physics of Semiconductors of National Academy of Sciences of Ukraine [11–13], whose optical scheme is given and described in Figure 1. The He- Ne laser (1) radiation (power output equals 1mW at λ = 632.8 nm) is divided by the beam splitter cube (2) into two beams: the object beam and the reference beam. In the object beam shoulder there is the mirror (3) and the collimator (4) consisting of negative and positive lenses, which forms a parallel beam, 5 cm in diameter. The beam passes through the object under study (6) and then arrives at the finely dispersed diffuse scatterer (9). According to Lambert's law, its every point is scattering the light in all directions. The light from the whole surface of the scatterer arrives at every point of the light sensitive thermoplastic (10) in which plane the object can be selected. At one inclination of plate (5), the increase of n has to result in the increase of the interference period, that is, in the decrease of the number of bands.

TC (7) has been put onto the top of a quartz cuvette (1× 1 × 4.3 cm3) filled with the solution studied. If the dielectric characteristics of the object studied are the same before and after the TC influence, the fringe pattern remains unaltered and interference bands inside and outside the object's profile continue each other. On the contrary, if an external factor caused changes of n, the fringe pattern within the limits of the object's profile will change.

The amino acids used in our experiments were produced by Sigma, Inc. Aqueous solutions were prepared on the basis of pure bidistilled water. Prepared solutions, before the experiment, were maintained for 24 hours under 25°C. The plasma blood solution was extracted from the blood of a heart vascular disorder patient just after the blood was drawn at hospital by conventional methods. We dilute the solution by distilled water as 1 : 50 and 1 : 100. The time between the blood extraction and the holographic measurement was 4 hours.

The procedure of dynamic measurement consisted of a sequence of records of interference patterns on a special thermoplastic plate, which then was fixed by a digital video camera. Afterward, the images were input into the computer and evaluated. For determination of the interference band center, the 10 points along the horizontal line of cuvette have been chosen.

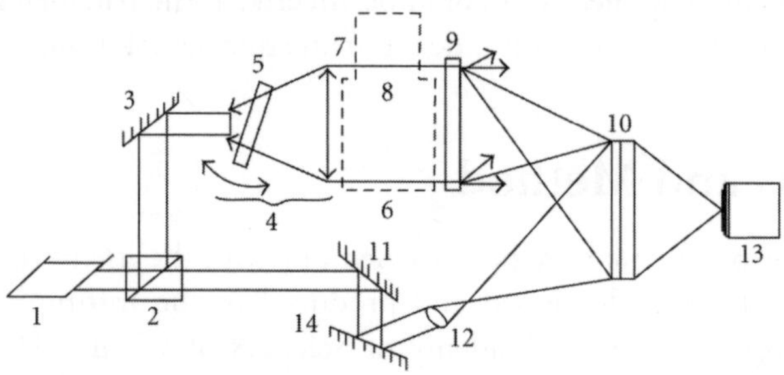

Figure 1: Experimental holographic set: (from 1 to 13) (1) He-Ne laser; (2) beam splitter cube; (3) mirror; (4) collimator; (5) plane parallel plate; (6) quartz flask (cuvette) with the solution; (7) Teslar bracelet; (8) filter that divides two flasks; (9) scattering layer; (10) thermoplastic recording plate; (11) reference beam mirror; (12) reference beam lens; (13) TV camera.

The studies were conducted at temperature 20 ± 1.5°C controlled by the thermocouple accurate to 0.2°C. We suppose that absolute meaning of the temperature does not influence the process under study due to the fact that we have been recording a dynamics of redistribution of the optical density. The most important point in the experiment was to protect the cuvette from the temperature gradient and the airflow. The last two disturbed factors have determined an inner nonstability of the system.

In Figure 2(a), typical interference patterns of the aerial ambient space (Air) and the aqueous solution (Solution) are presented. Vertical black lines show the image of the cuvette corner (its size is 1 × 1 × 4 cm3). Thus, in our experiments we have been able to observe an alteration of the reflective index in the surface zone of the cuvette equal to 1 × 2 cm2 that is determined by the cuvette size and the aperture of laser beam.

In Figure 2(a), typical interference patterns of the aerial ambient space (Air) and the aqueous solution (Solution) are presented. Vertical black lines show the image of the cuvette corner (its size is 1 × 1 × 4 cm3). We have been able to observe an alteration of n in the surface zone 1 × 2 cm2 of the cuvette (the cuvette size) and the aperture of laser beam. Deformations of the interference pattern in different points of the solution have been caused by changes in n in these points. The resolution is defined by a location of the optical wedge, namely, by a sum of horizontal interference lines. The space resolution is about 2 mm.

The method described gives a possibility to follow the response of the solution with the time factor of minimal discontinuous ability equal to 10 seconds. A sequence of pictures of the fringe pattern characterizes the space dynamics of the system studied in any place of the cuvette. As an example, Figure 2(b) shows the fringe pattern formed in about 4 minutes starting from the moment of action of the TC that has been spaced at 2mm from the cuvette. Changes of interference bands occurred during this time are associated with an internal stimulus.

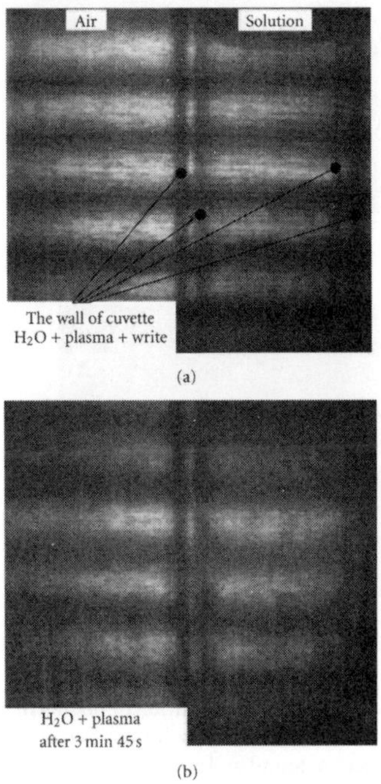

Figure 2: Dynamics of the fringe pattern of the aqueous solution of plasma blood of human without the influence of the TC. The value of the effect is estimated by difference between the shift of the interference band in the cuvette with the solution and the position of same band in the air; we evaluated the shift of interference bands before and after the TC application. It is seen that during 4 minutes the bands in the cuvette have not been deformed and they have essentially not moved relative to those in the air.

Results

A primary series of the experiments was conducted with distilled water, the saturated aqueous solution of L-tyrosine and β-alanine at 25°C. The experiments were conducted both in the morning and afternoon. The results showed typical slight changes of the fringe pattern in 400 seconds or larger time interval. These changes should be associated with the inner drift of liquid parameters. The curve of long-time dynamics does not show any influence on the side of the TC approximate to the cuvette.

The other behavior and picture have been observed in the case of the blood plasma solution. Without the TC action, this solution has shown stable and reproducible characteristics during more than 4 hours.

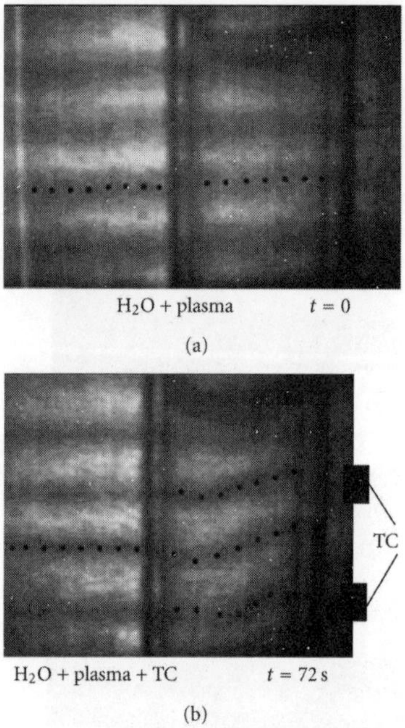

Figure 3: Dynamics of the fringe pattern of the aqueous solution of plasma of human blood after the insertion of 2 TC. The strong disturbance of the optical density of the solution is emerged already in 72 s, right figure. (The back covers of two sections of the bracelet are found at 4mm from the right wall of the cuvette).

During more than one hour, the system of recording and the objects of study (the solution of plasma and water) were stable and reproducible.

In Figure 3, we present the image of the cuvette with plasma blood solution affected by the TC. Black dots indicate the center position of one of the interference bands on the image plane. The value of the shift relating to zero line characterizes the degree of influence of the TC. Black rectangles (Figure 3, right) show the positions of two Teslar chips relating to cuvette; the fringe pattern of the solution relating to the chip is deformed in different ways in different zones (short, mid, and far-distance). A physical mechanism of the change is associated with the increase of n in the short-distance zone; n remains unchanged in the middistance zone and decreases in the far-distance zone.

Moreover, it seems that slow laminar flows have been induced by the TC near the front wall and directed to it.

If the refraction index of the solution changes in one place under the influence of an external factor, the length of optical path will also change. With the purpose of the registration of the changes, the device is designed in such a way that the "starting interferogram" constitutes a family of horizontal bands, bands of equal thickness. Depending on the character of changes of the optical density in the cuvette volume, the bands can be distorted (local changes of n), gaps between bands can expand without umetric decreases of n). Thus, arbitrary deformations of the fringe pattern are caused by a combination of local and global changes of the optical density.

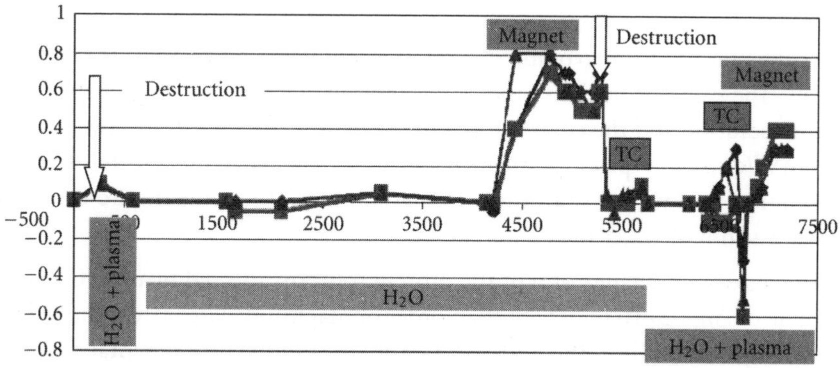

Figure 4: Dynamics of changes of the fringe pattern of the cuvette's volume at different external factors. The horizontal axis shows current time in seconds.

Changes of n are produced by changes in the structure of the network of hydrogen bonds of water, which being under the influence of oxygen, biomolecules, and the inerton field generated by the TC, forms long-lived structures. In the mentioned network, those new structures try to minimize the total energy relative to the volume occupied by the water system. Such kinds of changes (structuring of the aqueous solution) occur sufficiently slowly and therefore allow the recording by optical methods.

The strongest changes in effects associated with the TC have been detected during the first 5 to 15 minutes starting from the moment of influence. The changes in the fringe pattern have been irregular in time. The saturation effect reaches at 10 minutes. Without the TC, the number of interference bands remains the same in the field of the object and around it and is equal to five. Changes in the refractive index n of the sample affected by the TC are estimated from the equation

$$L\Delta n = \lambda\Delta k, \tag{1}$$

where L is the thickness of the sample (the aqueous solution studied), Δn is the change in refractive index, λ is the wavelength of the source of light (laser), Δk is the change in the number of interference bands as a result of an external effect.

Influence of the TC on water leads only to minor changes of the fringe pattern (Δn = 2 × 10–5). The behavior of proteins is mainly determined by the influence of the TC. Effects associated with the TC and heating effects have shown the opposite trend/tendency. The temperature rise in the flask detected by the thermosensor with 1 mW/cm2 power density ranged between 0.2 to 0.5°C. Therefore, heating caused by the laser radiation allows an evaluation of the role of temperature. The estimation of the temperature effect by using the thermal conductivity equation and the thermal balance equations show the following. The maximum heating of the aqueous solution without account of the thermal exchange, that is, under the condition unfavorable for the thermal effect estimation, may amount to 1°C. The calculation shows that even without the heat exchange between the aqueous solution and the environment the radiation effect with the power density of 10mW/cm2 may produce an increase of 1°C of the temperature of the 1.5 cm3 volume of aqueous solution within 6 minutes. The temperature coefficient of changes in n of water makes up Δn = 6 × 10–5.

The maximum change of n of the protein solution affected by the TC reached the value of Δn = 2×10–4, which is an order of magnitude larger than the temperature changes of n. Thus, the numerical estimates and the experimental data show that changes of n caused by the influence of the TC have been conditioned by nonthermal changes of the solution dielectric constant, which may be described as the total contribution of electronic, vibrational and orientational components.

In Figure 4, experimental dots show changes of n of the solution (the vertical axis) at the cuvette's back wall against the upper TC (rectangles) and the lower TC (triangles); recall the two TC are located near the front wall (see Figure 3). Current time, in seconds, is plotted along the horizontal axis. Legends "H2O" (the blue background) and "H2O + plasma" (the orange background) indicate different solutions in the cuvette. Time intervals are singled out for: (1) the magnet ("Magnet" on the green background) is applied to the front wall of the cuvette and (2) two links of the bracelet with two TCs are set into the cuvette (the violet background). Moments of intermix of the solution (Destruction) are shown by means of arrows; the intermix was made by using a medical syringe with 0.2mm needle; the solution was absorbed from the cuvette by the syringe and then poured back.

Conclusion

The TC does not affect distilled water. However, biomolecules of plasma of blood or an ensemble of such biomolecules in a micromolar concentration in water lead to the changes in reological characteristics, which allow the observation by optical methods, in particular, by the holographic interferometer. In the aqueous solution of blood plasma, biomolecules play a role of primary receptors of the TC radiation.

Changes in the aqueous solution affected by the TC cover all the macroscopic volume of the sample studied. This behavior can be associated with both inner convective flows (like in the case of Benar cells) and structural changes of water. The lattermay bring about changes in the reflective index of the solution and the fringe pattern.

Comparative responses of the aqueous solution to the mechanical, magnetic, and TC influences point to a very specific action of the latter. A microscopic physical consideration of the phenomenon of the TC has already been performed in some detail [9, 10, 14]. We could prove [9, 10] that the Teslar's phenomenon belongs to the inerton field effects and hence it does not relate to the electromagnetic nature. The inerton field (the field of inertia) appears as a basic field in the submicroscopic mechanics of canonical particles developed in the real physical space and accounts for the availability of the wave ψ-function in conventional quantum mechanics (see, e.g., [15, 16]). This field transfers local deformations of space, which appear in physical terms as mass. Therefore, the inerton field transfersmass changing potential properties of the environment. Consequently, the defect of mass Δm becomes an inherent property not only of atomic nuclei but also of any physical and physical chemical systems [14] (including biophysical ones). A more extensive study and repeated examinations should be completed to shedmore light upon the mechanism of action of the Teslar chip and the inerton field in general upon living organisms.

References

1. W. R. Adey and A. F. Lawrence, Eds., Nonlinear Electrodynamics in Biological Systems, Plenum Press, New York, NY, USA, 1984.

2. V. V. Novikov and A. V. Karnaukhov, "Mechanism of action of weak electromagnetic field on ionic currents in aqueous solutions of amino acids," Bioelectromagnetics, vol. 18, no. 1, pp. 25–27, 1997.

3. P. P. Belyaev, S. V. Polyakov, E. N. Ermakova, and S. V. Isaev, "Experimental investigations of the ionospheric Alfvén resonator from electromagnetic noise background over solar cycle of 1985–1995," Izvestia VUZov (Radiofizika), vol. 40, pp. 1305–1319, 1997 (Russian).
4. N. A. Temuryanz, B. M. Vladimirsky, and O. G. Tishkin, Extremely Low Frequency Electromagnetic Signals in the Biological World, Naukova Dumka, Kiev, Ukraine, 1992.
5. A. R. Liboff, Interaction between Electromagnetic Fields and Cells, NATO ASI Series A 97, Plenum Press, New York, NY, USA, 1985.
6. G. A. Mikhailova, "A possible biophysical mechanism of the solar activity effect on the central nervous system in man," Biofizika, vol. 46, no. 5, pp. 922–926, 2001 (Russian).
7. "Extremely Low Frequency Laboratories," http://www.teslar.com/.
8. N. Tesla, "The magnifying transmitter," in My Inventions: The Autobiography of Nikola Tesla, B. Johnston, Ed., chapter 6, Hart Brothers, Williston, Vt, USA, 1981.
9. V. Krasnoholovets, S. Skliarenko, and O. Strokach, "On the behavior of physical parameters of aqueous solutions affected by the inerton field of Teslar technology," International Journal of Modern Physics B, vol. 20, no. 1, pp. 111–124, 2006.
10. V. Krasnoholovets, S. Sklyarenko, and O. Strokach, "The study of the influence of a scalar physical field on aqueous solutions in a critical range," Journal of Molecular Liquids, vol. 127, no. 1–3, pp. 50–52, 2006.
11. G. S. Litvinov, N. Y. Gridina, G. Dovbeshko, L. I. Berzhinsky, and M. P. Lisitsa, "Millimeter wave effect on blood plasma solution," Electro- and Magnetobiology, vol. 13, no. 2, pp. 167– 174, 1994.
12. G. Dovbeshko and L. Berezhinsky, "The physical evidence of the weak electromagnetic field action upon biological systems," in Proceedings of the 4th European Symposium on Electromagnetic Compatibility, vol. 2, pp. 41–46, Brugge, Belgium, 2000.
13. L. I. Berezhinskii, N. I. Gridina, G. Dovbeshko, M. P. Lisitsa, and G. S. Litvinov, "Visualization of the effects of millimeter radiation on blood plasma," Biofizika, vol. 38, no. 2, pp. 378– 384, 1993 (Russian).
14. V. Krasnoholovets and J.-L. Tane, "An extended interpretation of the thermodynamic theory including an additional energy associated with a decrease in mass," International Journal of Simulation and Process Modelling, vol. 2, no. 1-2, pp. 67–79, 2006.

15. V. Krasnoholovets, "Submicroscopic deterministic quantum mechanics," International Journal of Computing Anticipatory Systems, vol. 11, pp. 164–179, 2002.

16. V. Krasnoholovets, "On the origin of conceptual difficulties of quantum mechanics," in Developments in Quantum Physics, F. Columbus and V. Krasnoholovets, Eds., pp. 85–109, Nova Science, New York, NY, USA, 2004.

Kinetics and Mechanism of Paracetamol Oxidation by Chromium(VI) in Absence and Presence of Manganese(II) and Sodium dodecyl Sulphate

Mohammed Ilyas, Maqsood Ahmad Malik, Syed Misbah Zahoor Andrabi and Zaheer Khan

ABSTRACT

The kinetics of paracetamol oxidation are first order each in [paracetamol] and [$HClO_4$]. The kinetic study shows that the oxidation proceeds in two steps. The effects of anionic micelles of sodiumdodecyl sulphate (SDS) and complexing agents (ethylenediammine tetraacetic acid (EDTA) and 2,2 -bi-pyridyl (bpy)) were also studied. Fast kinetic spectrophotometric method has been described for the determination of paracetamol. The method is based *on the catalytic effect of manganese(II) on the oxidation of paracetamol by chromium(VI) in the presence of* $HClO_4$ *(= 0.23 mol* dm_{-3}*). Optimum reaction time is 4 to 6 minutes at a temperature of 30° C. The addition of manganese(II) ions largely decreased the absorbance of chromium(VI) at 350 nm. This reaction can be utilized for the determination of paracetamol in drugs.*

Introduction

Spectrophotometric determination of paracetamol in drug formulations has been the subject of several investigators [1–8]. Generally, the same principle, that is, oxidation of paracetamol by metal ion oxidants, has been used for the estimation of paracetamol. The official pharmacopeia [9] and Sultan [10] methods require a 60-minute reflux period and 15 minutes heating of the reaction mixture, respectively. The main disadvantages of the Sultan method are that high concentration (6mol dm^{-3}) of sulphuric acid and high temperature (80°C) are required for the oxidation of paracetamol by chromium(VI).

The kinetic methods of analysis are highly sensitive, selective, simple, accurate, and less expensive. In recent years, several kinetic catalytic techniques have been reported for the detection of biomolecules [11–13]. In search for an alternative to those methods in which high sulphuric acid concentrations are required for paracetamol oxidation by chromium(VI) and to avoid the need for longer heating at higher temperature, complexing agents (manganese(II), EDTA, and bpy) and anionic and cationic surfactants (SDS and CTAB) were added to enhance the decay of chromium(VI) absorbance at 350 nm.

Experimental

Reagents and Solutions

All the reagents were of analytical reagent grade and all the solutions were prepared in doubly distilled (first time from alkaline $KMnO_4$) and CO_2 free deionized water. A solution of paracetamol (99%, Acros organics, NJ, USA) 1.0 °— 10^{-2} mol dm^{-3} was prepared by dissolving 0.151 g of paracetamol in water, and the solution was diluted to the mark in 100 cm3 volumetric flask. Stock Solutions of potassium dichromate (1.0 °— 10^{-4} mol dm^{-3}) and manganese(II) chloride (1.0 °—10^{-2} mol dm^{-3}), disodium salt of ethylenediamine tetraacetic acid (EDTA) (1.0 °— 10^{-2} mol dm^{-3}), and sodiumdodecyl sulphate (SDS) (1.0 °—10^{-2} mol dm^{-3}) were prepared in a similar manner. The solution of EDTA was stored in a polythene bottle as its solution gradually leaches metal ions from glass containers, resulting in a change in the effective [EDTA], and the solution of $K_2Cr_2O_7$ was stored in a dark glass bottle. To maintain hydrogen ion concentration constant, $HClO_4$ (Fisher, 70% reagent grade) was used.

Table 1: Values of pseudo-first-order rate constants for the oxidation of [PCM] (= 1.0 °— 10^{-3} mol dm^{-3}) by [Cr(VI)] (= 1.0 °—10^{-4} mol dm^{-3}) as a function of [complexing agents] at 30°C.

10^4[Mn(II)] (mol dm^{-3})	$10^3 k_{obs1}$ (s^{-1})	10^4[bpy] (mol dm^{-3})	$10^3 k_{obs2}$ (s^{-1})	10^4[EDTA] (mol dm^{-3})	$10^3 k_{obs3}$ (s^{-1})
7.0	1.6	10.0	2.0	10.0	3.1
10.0	2.0	12.0	2.0	14.0	3.9
12.0	2.7	14.0	1.9	20.0	4.1
15.0	3.9	20.0	1.9	26.0	4.1
17.0	4.7	26.0	1.9	30.0	4.2
20.0	7.0	30.0	1.9	34.0	3.1
—	—	—	—	36.0	3.2
—	—	—	—	40.0	3.6
—	—	—	—	40.0	3.0
—	—	—	—	50.0	3.0

Table 2: Effect of temperature on the pseudo-first-order rate constants and activation parameters for the oxidation of paracetamol (=1.0 °— 10^{-3} mol dm^{-3}) by [Cr(VI)] (= 1.0 °— 10^{-4} mol dm^{-3}). [SDS]= 10 °— 10^{-3} mol dm^{-3}.

Temperature °(C)	$10^3 k_{obs}$ (s^{-1}) Aqueous	$10^3 k_{obs}$ (s^{-1}) SDS
30	1.6	1.9
35	2.2	2.3
40	2.7	3.0
45	2.9	3.4
50	2.8	4.0
55	3.2	4.7
Activation parameters:[a]		
E_a (kJmol^{-1})	31	28
$\Delta H^{\#}$ (kJmol^{-1})	29	26
$\Delta S^{\#}$ (JK^{-1}mol^{-1})	−298	−297

[a] With an average linear regression coefficient, $r \geq 0.996$, for all activation parameters.

Kinetic Measurements

An aliquot of the components, potassium dichromate and $HClO_4$, was premixed in a three-necked reaction vessel, thermostated in a water bath at 30°C for 10 minutes, and the required volume of paracetamol (thermally equilibrated) was directly added to the dichromate solution. The course of the reaction was followed by measuring the absorbance of the unreacted chromium (VI) ion from time to time at 350nm against water, using a spectronic 20-D spectrophotometer. The pseudo-first-order rate constants (*k*obs, s−1) were determined from the linear part of the plots of log (absorbance) versus time with a fixed-time method. The same procedure was used to calculate the rate constants in presence of Mn (II), EDTA, bpy, and SDS.

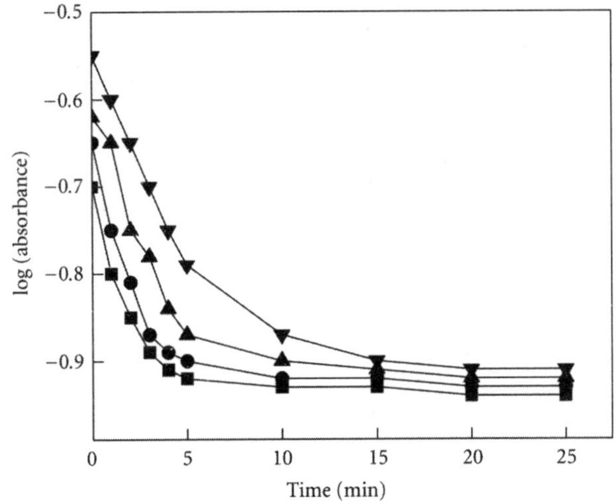

Figure 1: Plots of log (absorbance) versus time for the chromium(VI) + paracetamol reaction. Conditions: [Cr(VI)] = 1.0 °— 10^{-4} mol dm^{-3}; [$HClO_4$] = 0.23 mol dm^{-3}; temperature = 30°C; [PCM] = 1.0 (▼), 2.0 (▲), 3.0 (|), and 4.0 °— 10^{-3} mol dm^{-3}(|). In case of changing [$HClO_4$] and keeping the [PCM] = 1.0 °—10^{-3} mol dm^{-3}, similar behavior was observed.

Results and Discussion

It is well known that paracetamol undergoes redox reaction with dichromate in presence of higher H_2SO_4 amount (6.0 mol dm^{-3}) to form chromium (III) as the reaction product. This reaction is slow, but is sharply increased by the addition of trace amounts of Mn (II) and EDTA. Therefore, in order to take full advantage of the role of Mn (II) and EDTA, the reaction conditions ($HClO_4$, concentration, and temperature) and reagent concentrations (dichromate, paracetamol, Mn (II), and EDTA) must be optimized. Oxidation of paracetamol by dichromate has been studied kinetically as a function of [PCM], [Cr(VI)], [Mn(II)], [EDTA], [bpy], [$HClO_4$], and [SDS]. The results are compiled in Tables 1-2 and Figures 1–5.

Figure 1 represents the changes in the log (absorbance) of dichromate with definite time intervals as paracetamol concentrations changes. As the perchloric acid is added, it results in a sudden decrease in the absorbance of dichromate. In order to see the role of [Mn(II)], a series of kinetic runs were performed under different experimental conditions (Figure 2). The effect of [PCM] on the reaction rate was studied in the absence and presence of SDS anionic micelles. The results show that the kobs increase with increasing [PCM] in both media (Figure 3). The effects of HClO4 and temperature on the sensitivity were also studied. Figure 4 shows that the reaction rate increases with [H+] in absence and presence of SDS micelles. The reaction follows the first, fractional, and first-order kinetics with respect to [Cr(VI)], [PCM], and [Mn(II)], respectively. [EDTA] and [bpy] have zero-order dependence on reaction rate (Table 1). The effect of temperature on the sensitivity was studied in the range 30– 50°C. The results show that as the temperature increases, the reaction rate increases. The value of activation energy (Ea) was calculated from the slope of Arrhenius plots (Table 2). The observation is consistent with the accepted view that a slow reaction would require a higher energy of activation.

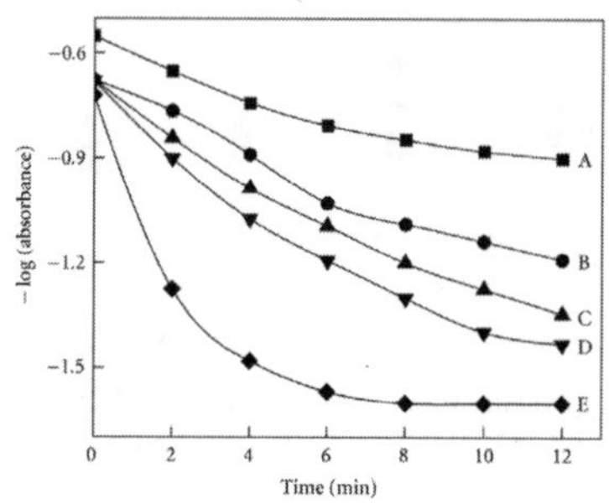

Figure 2: Effect of [Mn(II)] on the plot of log (absorbance) versus time. Conditions: [Cr(VI)] = 1.0 °— 10^{-4} mol dm^{-3}; [$HClO_4$] = 0.23 mol dm^{-3}; [PCM] = 1.0 °— 10^{-3} mol dm^{-3}; temperature = 30°C; [Mn(II)] = (A) 0.0, (B) 1.0, (C) 1.2, (D) 1.4, and (E) 2.0 °—10^{-3} mol dm^{-3}.

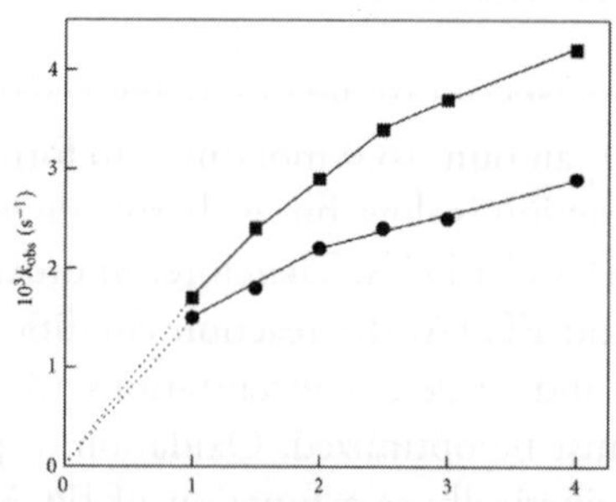

Figure 3: Effect of [PCM] on *k*obs in absence (**|**) and presence (**|**) of SDS. Conditions: [Cr(VI)], [$HClO_4$], and temperature were the same as in Figure 1. [SDS] = 10.0 °— 10^{-3} mol dm^{-3}.

On the basis of above results, Scheme 1 has been proposed for the oxidation of paracetamol by chromium(VI).

In Scheme 1, the reactive species of Cr(VI) and paracetamol readily form chromate ester as the first step in the reduction of Cr(VI) [14]. Chromate ester undergoes oxidative decomposition in the next step (rate determining), leading to the formation of an intermediate and Cr(IV) [15]. The proposed mechanism is further supported by analysis of the products. Ammonia has been detected as ammonium ions in aqueous solution. Benzoquinone and acetic acid were also detected by the spot tests [16].

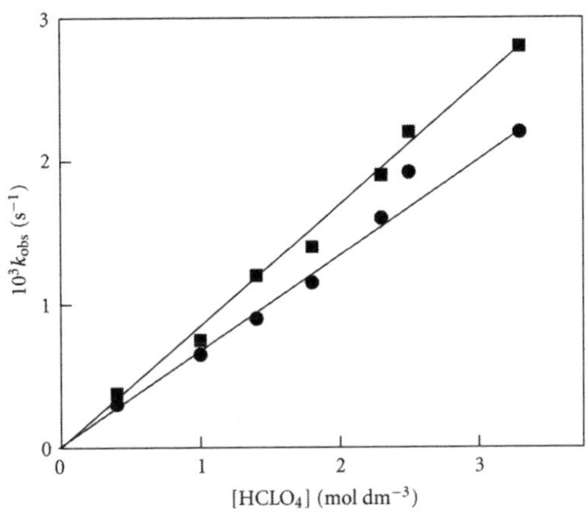

Figure 4: Effect of [$HClO_4$] on *k*obs in absence (|) and presence (|) of SDS. Conditions: [Cr(VI)] = 1.0 °—10^{-4} mol dm^{-3}; [PCM] = 1.0 °—10^{-3} mol dm^{-3}; [SDS] = 10.0 °—10^{-3} mol dm^{-3}; temperature =30°C.

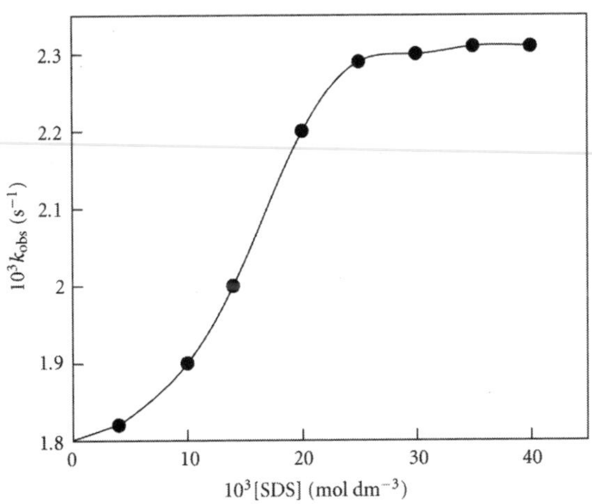

Figure 5: Effect of [SDS] on k_{obs}. Other reaction conditions were the same as in Figures 1 and 2.

Similar products using the same oxidant have been also suggested by Sultan [10]. The positive catalytic effect of Mn (II) (Table 1) is due to a one-step three-electron oxidation of paracetamol directly to chromium(III). One of the electrons transferred is donated by manganese (II) atom and the other two by paracetamol. The observed catalytic effect rules out the possibility of chromium (IV) formation in the rate-determining step [17, 18]. In presence of Mn (II), Scheme 1 mechanism can be modified to Scheme 2.

Scheme 1

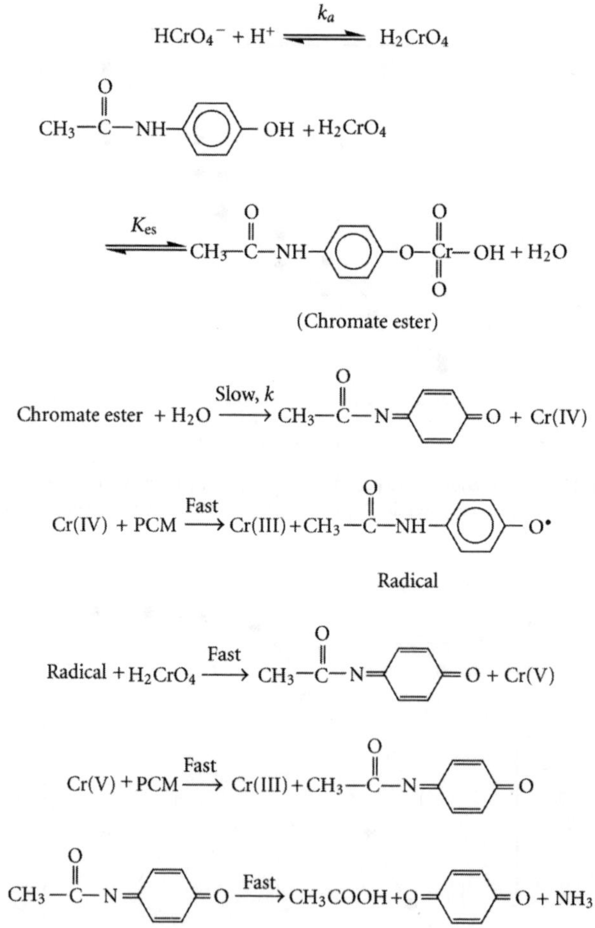

$$k_{obs} = \frac{kK_{es}k_a[H^+][\text{paracetamol}]}{(1 + K_{es}[\text{paracetamol}])}$$

SCHEME 1

Scheme 2

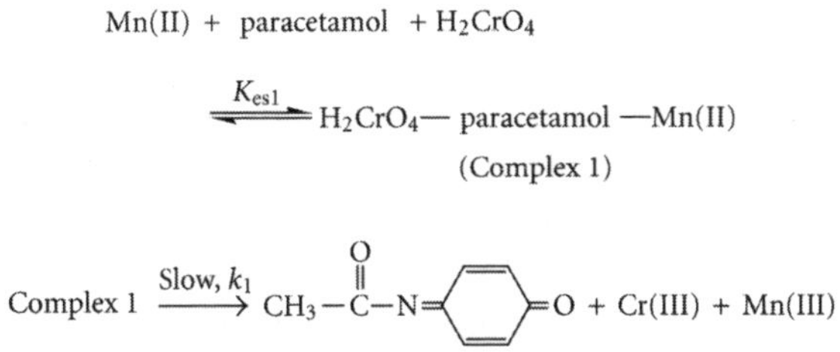

$$k_{obs} = \frac{k_1 K_{es1} k_a [H^+][\text{paracetamol}][\text{Mn(II)}]}{(1 + K_{es}[\text{paracetamol}])}$$

SCHEME 2

Scheme 3

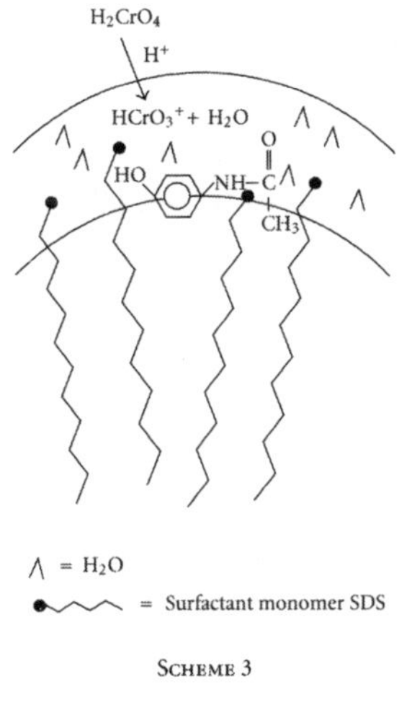

SCHEME 3

In presence of Mn (II), the reaction proceeds through the formation of a termolecular complex between Cr(VI), paracetomal, and Mn (II) (Scheme 2) [19] because the direct oxidation of Mn (II) by chromium (VI) is thermodynamically unfavorable [20]. The positive catalytic effect of Mn (II) is due to a one-step three-electron reduction of chromium (VI), which is in conformity the reduction of Cr(VI) → Cr (III) without passing through formation of Cr(IV) as an intermediate. Table 1 shows the effect of EDTA and bpy on the reaction rate. It was found that whereas the reduction of paracetamol by chromium (VI) is slow, reduction in presence of EDTA/bpy at a similar concentration is fairly fast. It should be emphasized here that the complexing agents (EDTA and bpy) themselves are resistant to oxidation under the exact conditions employed. Addition of even small quantity of these complexing agents gives a pronounced rate enhancement. EDTA gives a higher rate than bpy for the same concentrations.

Micellar catalysis has received considerable attention in view of the analogies drawn between micellar and enzyme catalyses [21, 22]. Micelles increase rates of bimolecular reactions by concentrating both the reactants at their surfaces. Electrostatic-, approximation-, and medium-effects are responsible for the incorporation of reactants into or onto a micelle. In order to verify the role of micelles on the paracetamol oxidation by chromium(VI), cationic and anionic micelles were chosen. Preliminary observations showed that a reaction mixture containing chromium(VI) (=1.0°—10–4 mol dm–3), paracetamol (= 1.0°—10–3 mol dm–3), HClO4 (= 0.23 mol dm–3), and cationic micelles of CTAB became turbid. Therefore, the investigation was confined to verify the effect of anionic SDS micelles. HClO4 is a strong acid which completely dissociates in H+ and ClO4-. In presence of cationic surfactant (CTAB), there are electrostatic interactions between the positive head group of cationic micelles and perchlorate ions, which form water insoluble species.

Figure 5 shows the effect of SDS anionic micelles on the sensitivity for the range 5.0 °— 10–3 mol dm–3 to 40.0 °— 10–3 mol dm–3. The reaction rate increases with increasing [SDS] up to ≥ 30.0 °— 10–3 mol dm–3 and remains constant at higher [SDS]. This may be due to the dilution effect. Therefore, a final [SDS] of 30.0 °— 10–3 mol dm–3 was chosen as the optimum concentration. The role of SDS micelles in catalysis can be explained by incorporation/solubilizition of chromium (VI)/paracetamol in the Stern layer of SDS micelles through electrostatic and hydrophobic interactions (Scheme 3). These results are in good agreement with our previous observations [23].

Conclusion

Although a number of spectrophotometric methods are available for the determination of paracetamol, these are generally associated with some or the other demerits. The use of chromium (VI) for the determination of paracetamol has been suggested but the reaction requires a high concentration of H_2SO_4 and very high temperature for the complete consumption of chromium(VI). The results from this study show that oxidation of paracetamol by Cr (VI) is enhanced in presence of complexing agents (Mn(II), EDTA, and bpy) and surfactant. Of these, Mn (II) is the most effective as only 5–10 minutes are required for the completion of the reaction. This is very significant for any industrial use to avoid or minimize the use of higher acid concentrations. The present method is simple, accurate, rapid, economical, and precise.

References

1. J. E. Wallace, "Determination of ephedrine and certain related compounds by ultraviolet spectrophotometry," *Analytical Chemistry*, vol. 39, no. 4, pp. 531–533, 1967.
2. F. M. Plakogiannis and A. M. Saad, "Spectrophotometric determination of acetaminophen and dichloralantipyrine in capsules," *Journal of Pharmaceutical Sciences*, vol. 64, no. 9, pp. 1547–1549, 1975.
3. K. K. Verma, A. K. Gulati, S. Palod, and P. Tyagi, "Spectrophotometric determination of paracetamol in drug formulations with 2-iodylbenzoate," *The Analyst*, vol. 109, no. 6, pp. 735–737, 1984.
4. S. M. Sultan, I. Z. Alzamil, A. M. Aziz Alrahman, S. A. Altamrah, and Y. Asha, "Use of cerium(IV) sulphate in the spectrophotometric determination of paracetamol in pharmaceutical preparations," *The Analyst*, vol. 111, no. 8, pp. 919–921, 1986.
5. F. A. Mohamed, M. A. AbdAllah, and S. M. Shammat, "Selective spectrophotometric determination of *p*-aminophenol and acetaminophen," *Talanta*, vol. 44, no. 1, pp. 61–68, 1997.
6. J. F. van Staden and M. Tsanwani, "Determination of paracetamol in pharmaceutical formulations using a sequential injection system," *Talanta*, vol. 58, no. 6, pp. 1095–1101, 2002.
7. M. Oliva, R. A. Olsina, and A. N. Masi, "Selective spectrofluorimetric method for paracetamol determination through coumarinic compound formation," *Talanta*, vol. 66, no. 1, pp. 229–235, 2005.
8. M. K. Srivastava, S. Ahmad, D. Singh, and I. C. Shukla, "Titrimetric determination of dipyrone and paracetamol with potassium hexacyanoferrate(III) in an acidic medium," *The Analyst*, vol. 110, no. 6, pp. 735–737, 1985.
9. *British Pharmacopoeia*, Her Majesty's Stationery Office, London, UK, 1980.
10. S. M. Sultan, "Spectrophotometric determination of paracetamol in drug formulations by oxidation with potassium dichromate," *Talanta*, vol. 34, no. 7, pp. 605–608, 1987.
11. S. T. Nandibewoor and V. A. Morab, "Chromium(III)- catalysed oxidation of antimony(III) by alkaline hexacyanoferrate(III) and analysis of chromium(III) in microamounts by a kinetic method," *Journal of the Chemical Society, Dalton Transactions*, no. 3, pp. 483–488, 1995.
12. T. F. Imdadullah and T. Kumamaru, "Catalytic effect of rhodium(III) on the chemiluminescence of luminol in reverse micelles and its analytical application," *Analytica Chimica Acta*, vol. 292, no. 1-2, pp. 151–157, 1994.

13. A. A. Ensafi and M. Keyvanfard, "Kinetic spectrophotometric method for the determination of rhodium by its catalytic effect on the oxidation of o-toluidine blue by periodate in micellar media," *Journal of Analytical Chemistry*, vol. 58, no. 11, pp. 1060–1064, 2003.
14. F. H. Westheimer, "The mechanisms of chromic acid oxidations," *Chemical Reviews*, vol. 45, no. 3, pp. 419–451, 1949.
15. Z. Khan, P. Kumar, and Kabir-ud-Din, "Kinetics and mechanism of the reduction of colloidal manganese dioxide by Dfructose," *Colloids and Surfaces*, vol. 248, no. 1–3, pp. 25–31, 2004.
16. F. Feigl, *Spot Tests in Organic Analysis*, Elsevier, New York, NY, USA.
17. C. F. Huber and G. P. Haight Jr., "The oxidation of manganese(II) by chromium(VI) in the presence of oxalate ion," *Journal of the American Chemical Society*, vol. 98, no. 14, pp. 4128–4131, 1976.
18. Z. Khan, M. Y. Dar, P. S. S. Babu, and Kabir-ud-Din, "A kinetic study of the reduction of chromium(VI) by thiourea in the absence and presence of manganese(II), cerium(IV) and ethylenediaminetetra acetic acid (EDTA)," *Indian Journal of Chemistry*, vol. 42A, no. 5, pp. 1060–1065, 2004.
19. Kabir-ud-Din, K. Hartani, and Z. Khan, "Co-oxidation of malic acid and manganese(II) by chromium(VI) in the presence and absence of ionic surfactants," *Indian Journal of Chemistry*, vol. 41B, no. 12, pp. 2614–2624, 2002.
20. J. F. Perez-Benito and C. Arias, "A kinetic study on the reactivity of chromium(IV)," *Canadian Journal of Chemistry*, vol. 71, no. 5, pp. 649–655, 1993.
21. F. M. Menger and C. E. Portnoy, "Chemistry of reactions proceeding inside molecular aggregates," *Journal of the American Chemical Society*, vol. 89, no. 18, pp. 4698–4703, 1967.
22. C. A. Bunton and G. Savelli, "Organic reactivity in aqueous micelles and similar assemblies," *Advances in Physical Organic Chemistry*, vol. 22, pp. 213–309, 1986.
23. Kabir-ud-Din, A. M. A. Morshed, and Z. Khan, "Influence of sodium dodecyl sulfate/TritonX-100 micelles on the oxidation of D-fructose by chromic acid in presence of HClO4," *Carbohydrate Research*, vol. 337, no. 17, pp. 1573–1583, 2002.

Theoretical Study of Sequence Selectivity and Preferred Binding Mode of Psoralen with DNA

Patricia Saenz-Méndez, Rita C. Guedes, Daniel J. V. A. dos Santos and Leif A. Eriksson

ABSTRACT

Psoralen interaction with two models of DNA was investigated using molecular mechanics and molecular dynamics methods. Calculated energies of minor groove binding and intercalation were compared in order to define a preferred binding mode for the ligand. We found that both binding modes are possible, explaining the low efficiency for monoadduct formation from intercalated ligands. A comparison between the interaction energy for intercalation between different base pairs suggests that the observed sequence selectivity is due to favorable intercalation in 5 -TpA in $(AT)_n$ sequences.

Introduction

The study of the interaction of low molecular weight agents with DNA is essential for a deeper understanding of the biochemistry of the cell and also in the rational design of compounds with desired pharmacological properties. Among the vast amount of molecules that are known or proposed to interact with DNA we have in the current study focused our attention to furocoumarins.

Furocoumarins are tricyclic heterocycles found in a large number of plants [1]. Psoralens (Figure 1) are the best known of these chemical agents and have been employed in the treatment of several skin diseases such as psoriasis, vitiligo, and mycosis fungoides, as well for disorders related to the immune system such as cutaneous T-cell lymphoma [2]. Even though extracts of Ammi majus (Bishop's weed) were used by the Egyptians (1550 BC) in the treatment of vitiligo, the first concise report of the combined use of psoralen derivatives and UV light dates back to 1974 [3], in which Parrish et al. coined the term photochemotherapy to describe the interaction of UVA light and drugs.

The photochemical reactions of psoralens with DNA have been described in detail [1, 4, 5], and is proposed to proceed via a stepwise mechanism. First, the molecule intercalates between the base pairs in double stranded DNA interacting with the π-stack of the nucleobases. After irradiation with UVA light, the photo-excited furocoumarins react with DNA and form covalent adducts through a [2 + 2] pericyclic reaction. The photoaddition takes place mostly between the psoralen 4 , 5 double bond (furan ring) and the C5, C6 double bond of a pyrimidine (usually thymine). The furan-side adduct can absorb an additional UVA photon, forming an interstrand cross-linked derivative through photoaddition at the 3, 4 double bond in the pyrone ring [6–8].

Furocoumarin photoaddition takes place predominantly with thymine. The sequence specificity has been investigated using DNA sequencing methodology. The results revealed that thymines in a GC environment have weak reactivity, while adjacent thymines are better targets. In addition, there is a strong preference for 5 -TpA sites compared with 5 -ApT, and (AT)n sequences are the most reactive towards photoaddition

[9, 10].

There is however no information regarding the factors controlling the sequence selectivity. One possibility is that the preference is due to favorable intercalation into this site. Another explanation is that the molecules are docked between all different base pairs, but photoaddition occurs preferentially in a special environment. A third possibility could be that psoralens prefer both intercalation and photoreaction with a 5 -TpA site in an AT environment.

In this context, computational techniques are an exceptional tool to explore these different factors. Molecular docking is a computational method for predicting ligand-receptor binding when both the binding site and binding mode are unknown [11]. Even tough DNA has an exceptionally wellcharacterized structure, there is a very small number of DNA-ligand docking studies [12–16], and to the best of our knowledge there is only scarce information on docking involving this kind of ligand systems [17].

The aim of this study is to obtain complexes of DNAmodels and psoralen in which the ligand in principle could act either as intercalator or minor groove binder. We used two different DNA fragments, d(CCTTGCTACCTT)2 and d(TATATATATATA)2, as targets for intercalation and minor groove complexes, in order to evaluate the preferential interaction between different base pairs and explore the proposed mechanism. Obtaining detailed information on the possible interaction sites will assist in the development of new and more efficiently intercalating photoactive compounds.

Methods

Duplex B-DNA was constructed with the Molecular Operating Environment (MOE) software [18] and was minimized to 0.00001 kcal mol^{-1} $Å^{-1}$ using the CHARMM27 force field [19]. After minimization, the model was subjected to molecular dynamics (MD) simulations [20] employing an NVT ensemble [21] and the CHARMM27 potential. After the MD simulation, a final energy minimization was performed. The resulting DNA-model was used in the docking studies.

Because of the ring-fused planar structure of furocoumarin, no ligand conformational search is needed. A single energy minimization calculation was performed until the root-mean-square (RMS) gradient was 0.00001 kcal mol–1Å –1 using the same force field.

Manual docking of psoralen to the DNA was performed, in which different orientations for the ligand were considered. The DNA model with the ligand was energy minimized, subjected to MD simulation using the same settings as above except for a longer (250 picoseconds) heating interval, and finally energy minimized, using in all the stages the CHARMM27 force field. All trajectories were rapidly equilibrated during the first 200 picoseconds of the production runs. Hence, a total 1 nanosecond production simulation was considered sufficient to allow for structural relaxation. The potential energy of the final complex was calculated (Ecomplex). The ligand was then moved away from the DNA stack, the noninteracting system energy minimized, and the potential energy of the separated moieties calculated (EDNA + Eligand). Subtracting this from Ecomplex gives the interaction energy (IE) reported herein.

Throughout the calculations, the surrounding was modeled through the distance-dependent dielectricmodel of bulk water [18, 22].

Results and Discussion

Psoralen Binding Modes

El-Gogary and El-Gendy [17] employed spectrophotometric DNA titration to calculate the intercalation affinity. The spectrophotometric data for 8-methoxypsoralen bound to calf-thymus DNA showed that even though the molecule could intercalate, a large fraction (approximately 30%) was also bound to the surface of DNA. No theoretical studies regarding different binding modes of psoralens are available. To this end, the ability of psoralen to intercalate between base pairs in $d(CCTTGCTACCTT)_2$ and $d(TATATATATATA)_2$ sequences, as well to bind to the minor groove, was therefore examined in the current study. The results are shown in Tables 1 and 2.

Psoralen does not present different geometric conformations due to its rigid nature. However, the molecule is asymmetric and may therefore interact with different energy in different poses. Four different orientations for docking into the DNA stack were considered in all systems, numbered as indicated in Figure 1.

First, we examine the ability to intercalate between different base pairs, GC, CT, and TA following the procedure described in Section 2. Each pose was inserted between the considered base pair. In some cases the ligand moved away from the intercalation site to interact with the minor groove, the major groove, or interacting only with one strand of DNA ("single-strand insertion," Table 1). In addition, we calculated the interaction energy when psoralen was placed initially in the minor groove. In this case, all poses remained bound to this region after the MD simulation (Table 2).

In all cases, the interaction between the ligand and the macromolecule lowers the total energy considerably, leading to stable complexes. Our results show that intercalation and minor groove binding are competing processes, in full agreement with the experimental information.When psoralen interacted with the d(CCTTGCTACCTT)2 fragment, the most favorable docking energies for intercalation were obtained for pose 2 intercalated between CT (–36.8 kcal mol–1) and pose 3 between TA (–36.7 kcal mol–1). Minor groove binding into that DNA model was found to be slightly more favorable for poses 1 and 3, yielding interaction energies of –39.1 and –38.5 kcal mol–1, respectively. When considering the poly-TA fragment, intercalation of pose 2 in the 5 -TpA site led to the most stable complex with an interaction energy of -43.7 kcal mol-1. Minor groove binding energies were a few kcalmol-1 less favorable compared with the nonpoly-TA fragment (best docking energy -34.4 kcal mol-1 for pose 1). Those results show that even thoughminor groove binding is possible in both DNA strands, it is comparatively less favored in poly-TA fragments.

The current docking studies also explain other experimental facts. Tessman et al. [6] studied the photochemical reaction of psoralen with calf-thymus DNA. They observed a relatively low efficiency for the conversion of noncovalent complexes to monoadducts (approximately 11%). They suggested that the probability of having the appropriate excited state with the correct geometry alignment for photoaddition is quite low, which could be due to the fact that intercalated psoralen is quite low, which could be due to the fact that intercalated psoralen is in dynamic equilibrium with nonbound ligand outside the helix. From our results, we instead propose an alternative explanation, namely, that the low efficiency is a consequence of the two competing binding modes, that is, intercalation versus minor groove binding.

Table 1: Calculated docking energies IE = $E_{complex}-(E_{DNA}+E_{ligand})$ (kcal mol^{-1}) obtained for different orientations of psoralen interacting with $d(CCTTGCTACCTT)_2$ and $d(TATATATATATA)_2$ fragments.

	$d(CCTTGCTACCTT)_2$ fragment		
Orientation	Docking mode	Docking site	IE
1	Single strand insertion	GC	−38.0
2	Major groove binding	GC	−32.4
3	Single strand insertion	GC	−51.1
4	Major groove binding	GC	−35.5
1	Intercalation	CT	−32.3
2	Intercalation	CT	−36.8
3	Intercalation	CT	−35.5
4	Intercalation	CT	−36.5
1	Intercalation	TA	−36.5
2	Intercalation	TA	−32.7
3	Intercalation	TA	−36.7
4	Intercalation	TA	−30.2
	$d(TATATATATATA)_2$ fragment		
Orientation	Docking mode	Docking site	IE
1	Intercalation	5′-TpA	−24.1
2	Intercalation	5′-TpA	−43.7
3	Intercalation	5′-TpA	−18.2
4	Intercalation	5′-TpA	−21.4
1	Major groove binding	5′-ApT	−28.7
2	Major groove binding	5′-ApT	−29.5
3	Major groove binding	5′-ApT	−26.4
4	Single strand insertion	5′-ApT	−33.7

Table 2: Calculated docking energies IE = $E_{complex}$ − (E_{DNA} + E_{ligand}) (kcal mol^{-1}) obtained for different orientations of psoralen for minor groove binding in $d(CCTTGCTACCTT)_2$ and $d(TATATATATATA)_2$ fragments.

$d(CCTTGCTACCTT)_2$ fragment	
Orientation	IE
1	−39.1
2	−37.7
3	−38.5
4	−35.5
$d(TATATATATATA)_2$ fragment	
Orientation	IE
1	−34.4
2	−31.7
3	−32.7
4	−32.7

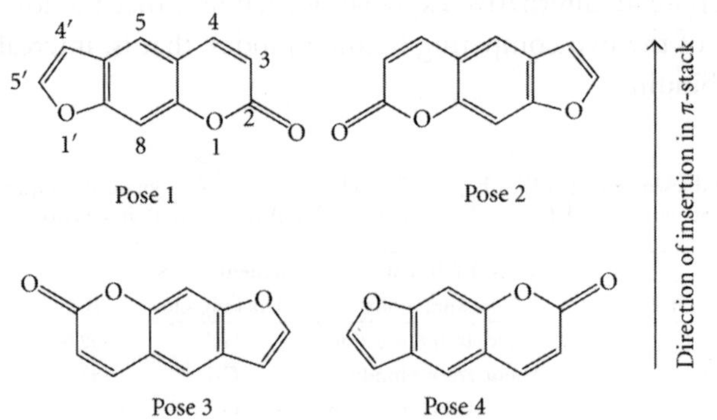

Figure 1: Molecular structure of psoralen and the four different docking poses investigated for intercalation.

Psoralen Sequence Selectivity

The use of two different sequences of bases ($d(CCTTGCTACCTT)_2$ and $d(TATATATATATA)_2$) in our docking studies enables us to evaluate the sequence selectivity. The two strands in principle have the same potential to react with psoralen and form photoadducts. However, the experimental studies reveal a different behavior between the two [9, 10]. Psoralen preferentially photoreact in $(AT)_n$ sequences and especially in 5 -TpA sites. However, it is difficult from those results to conclude whether the preference is a consequence of preferential intercalation, photoreaction, or both.

Thus, we decided to study the ability of psoralen to intercalate between different base pairs in two DNA models (Table 1). In all cases the same four different orientations were considered (Figure 1). Each pose was inserted between the base pair, minimized and subjected to MD simulation. In some cases, the ligand did not remain intercalated, but moved to either the minor or major groove, or became inserted between two bases interacting with one strand of DNA only.

When the ligand interacted with the d(CCTTGCTACCTT) 2 fragment, all three environments (GC, CT, TA) behaved differently.We were unable to get psoralen intercalated between GC, while intercalation between CT turned out to be almost as favorable as between TA. When the ligand is inserted (GC base pair) it interacts with only one strand of DNA, acting as hydrogen bond acceptor via its carbonyl oxygen atom with the cytosine amino group (Figure 2). Such interaction stabilizes the insertion of the ligand to the GC tract. Figure 2 also shows the van derWaals contact surface for pose 3 inserted between GC.

The results of the docking studies with the poly-TA model were completely different. Psoralen intercalates to give stable complexes in 5 -TpA sites that furthermore yields the most favorable interaction energy (–43.7 kcal mol–1). However, when considering 5 -ApT sites, the ligand became bound to the minor or major groove or single-strand inserted, instead of remaining intercalated after MD simulation.

The most stable complex has the ligand perfectly intercalated between T and A, interacting only with those base pairs. During theMDsimulation the psoralen was however rotated 90 degrees from the initial pose to the final one, as is shown in Figure 3. In addition, the van der Waals surface is closed around the ligand (Figure 3).

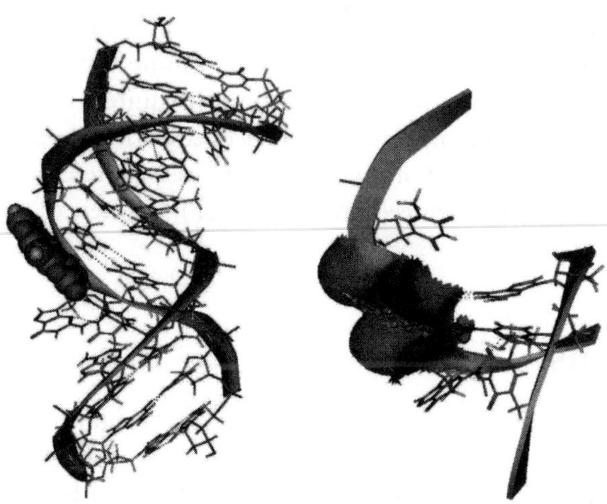

(a) (b)

Figure 2: (a) Space-filling model of psoralen interacting with DNA bases for orientation 3 inserted between GC. (b) van derWaals contact surface between ligand and DNA.

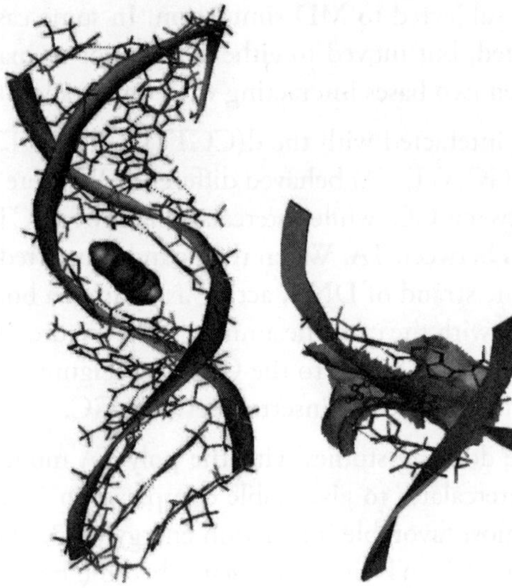

(a) (b)

Figure 3: (a) Space-filling model of psoralen interacting with DNA bases for orientation 2 intercalated in the 5 -TpA site. (b) van der Waals contact surface between ligand and DNA.

The topology of the two sites (5 -TpA and 5 -ApT) is very different in our model. The base stacking is not the same in the two sequences; in the former the base pairs are perfectly parallel while in the latter they are not. These results are in agreement with experimental information. McClellan et al. [23] have found a nonstandard B-helix for alternating (AT)n sequences, as a consequence of poor stacking between the base pairs. The same authors mentioned that (AT)n sequences are sites of great instability, present in promoters where DNA replication is initiated, a process in which unstacking is required. The unstacking and instability compared with nonpoly-TA sequences could lead to a better accessibility of the C5–C6 double bond of thymine due to easier intercalation. Our results suggest that intercalation is indeed favored in (AT)n sequences and especially in 5 -TpA sites.

Conclusions

A computational docking study for the photochemical agent psoralen was performed for the first time, employing two different DNA models ($d(CCTTGCTACCTT)_2$, and $d(TATATATATATA)_2$). We have analyzed different binding modes and suggest that the low efficiency observed for the conversion of the noncovalent complexes to monoadducts are a consequence of two competing binding modes, intercalation and minor groove binding.

It is well known that strong sequence selectivity for furocoumarin photoaddition exists. From our results we can conclude that the preference is due to favorable intercalation in 5 -TpA sites in AT enviroments (poly-TA DNA). Albeit this will control the distribution in different environments, it does however not exclude that photoaddition may take place preferentially in those sites also. We are currently combining additional docking studies with quantum chemical calculations to take these factors into account.

Acknowledgements

The Swedish Science Research Council (VR) and Sparbanksstiftelsen Nya are gratefully acknowledged for financial support.

References

1. T. F. Anderson and J. J. Voorhees, "Psoralen photochemotherapy of cutaneous disorders," Annual Review of Pharmacology and Toxicology, vol. 20, pp. 235–257, 1980.
2. P. S. Song and K. J. Tapley, "Photochemistry and photobiology of psoralens," Photochemistry and Photobiology, vol. 29, no. 6, pp. 1177–1197, 1979.
3. J. A. Parrish, T. B. Fitzpatrick, L. Tanenbaum, and M. A. Pathak, "Photochemotherapy of psoriasis with oral methoxsalen and longwave ultraviolet light," The New England Journal of Medicine, vol. 291, no. 23, pp. 1207–1211, 1974.
4. G. D. Cimino, H. B. Gamper, S. T. Isaacs, and J. E. Hearst, "Psoralens as photoactive probes of nucleic acid structure and function: organic chemistry, photochemistry, and biochemistry," Annual Review of Biochemistry, vol. 54, pp. 1151–1193, 1985.
5. N. Kitamura, S. Kohtani, and R. Nakagaki, "Molecular aspects of furocoumarin reactions: photophysics, photochemistry, photobiology, and structural analysis," Journal of Photochemistry and Photobiology C: Photochemistry Reviews, vol. 6, no. 2-3, pp. 168–185, 2005.
6. J. W. Tessman, S. T. Isaacs, and J. E. Hearst, "Photochemistry of the furan-side 8-methoxypsoralen-thymidine monoadduct inside the DNA helix. Conversion to diadduct and to pyroneside monoadduct," Biochemistry, vol. 24, no. 7, pp. 1669–1676, 1985.
7. M. A. Pathak and T. B. Fitzpatrick, "The evolution of photochemotherapy with psoralens and UVA (PUVA): 2000 BC to 1992 AD," Journal of Photochemistry and Photobiology B: Biology, vol. 14, no. 1-2, pp. 3–22, 1992.

8. I. M. Schmitt, S. Chimenti, and F. P. Gasparro, "Psoralenprotein photochemistry—a forgotten field," Journal of Photochemistry and Photobiology B: Biology, vol. 27, no. 2, pp. 101–107, 1995.
9. E. Sage and E. Moustacchi, "Sequence context effects on 8- methoxypsoralen photobinding to defined DNA fragments," Biochemistry, vol. 26, no. 12, pp. 3307–3314, 1987.
10. V. Boyer, E. Moustacchi, and E. Sage, "Sequence specificity in photoreaction of various psoralen derivatives with DNA: role in biological activity," Biochemistry, vol. 27, no. 8, pp. 3011– 3018, 1988.
11. I. Halperin, B. Ma, H. Wolfson, and R. Nussinov, "Principles of docking: an overview of search algorithms and a guide to scoring functions," Proteins, vol. 47, no. 4, pp. 409–443, 2002.
12. B. G. Feuerstein, N. Pattabiraman, and L. J. Marton, "Spermine-DNA interactions: a theoretical study," Proceedings of theNational Academy of Sciences of theUnited States of America, vol. 83, no. 16, pp. 5948–5952, 1986.
13. K. Gao, B. Tao Fan, N. El Fassi, et al., "Comparative study of activities between verbascoside and rutin by docking method," QSAR & Combinatorial Science, vol. 22, no. 1, pp. 18–28, 2003.
14. T. Tuttle, E. Kraka, and D. Cremer, "Docking, triggering, and biological activity of dynemicin A in DNA: a computational study," Journal of the American Chemical Society, vol. 127, no. 26, pp. 9469–9484, 2005.
15. K. V. Miroshnychenko and A. V. Shestopalova, "Flexible docking of DNA fragments and actinocin derivatives," Molecular Simulation, vol. 31, no. 8, pp. 567–574, 2005.
16. R. Rohs, I. Bloch, H. Sklenar, and Z. Shakked, "Molecular flexibility in ab initio drug docking to DNA: binding-site and binding-mode transitions in all-atom Monte Carlo simulations," Nucleic Acids Research, vol. 33, no. 22, pp. 7048–7057, 2005.
17. T. M. El-Gogary and E. M. El-Gendy, "Noncovalent attachment of psoralen derivatives with DNA: Hartree-Fock and density functional studies on the probes," Spectrochimica Acta, Part A, vol. 59, no. 11, pp. 2635–2644, 2003.
18. "Molecular Operating Environment (MOE)," 2005.06; Chemical Computing Group Inc.; Montreal, Canada, 2005.
19. A. D. Mackerell Jr., M. Feig, and C. L. Brooks III, "Extending the treatment of backbone energetics in protein force fields: limitations of gas-phase quantum mechanics in reproducing protein conformational distributions in molecular

dynamics simulation," Journal of Computational Chemistry, vol. 25, no. 11, pp. 1400–1415, 2004.

20. MD conditions: duration: 1.0 ns, temperature: 300 K, heat:100 ps, initial temperature: 150 K, time step: 0.002 ps, sample period: 0.5 ps, temperature response: 0.1 ps, pressure response: 0.1 ps.
21. S. D. Bond, B. J. Leimkuhler, and B. B. Laird, "The Nosé- Poincaré method for constant temperature molecular dynamics," Journal of Computational Physics, vol. 151, no. 1, pp. 114– 134, 1999.
22. J. Guenot and P. A. Kollman, "Molecular dynamics studies of a DNA-binding protein—2: an evaluation of implicit and explicit solventmodels for themolecular dynamics simulation of the Escherichia coli trp repressor," Protein Science, vol. 1, no. 9, pp. 1185–1205, 1992.
23. J. A. McClellan, E. Palecek, and D. M. Lilley, "(A-T)n tracts embedded in random sequence DNA—formation of a structure which is chemically reactive and torsionally deformable," Nucleic Acids Research, vol. 14, no. 23, pp. 9291–9309, 1986.

Fast Drug Release Using Rotational Motion of Magnetic Gel Beads

Tetsu Mitsumata, Yusuke Kakiuchi and Jun-Ichi Takimoto

ABSTRACT

Accelerated drug release has been achieved by means of the fast rotation of magnetic gel beads. The magnetic gel bead consists of sodium alginate cross-linked by calcium chlorides, which contains barium ferrite of ferrimagnetic particles, and ketoprofen as a drug. The bead underwent rotational motion in response to rotational magnetic fields. In the case of bead without rotation, the amount of drug release into a phosphate buffer solution obeyed non-Fickian diffusion. The spontaneous drug release reached a saturation value of 0.90mg at 25 minutes, which corresponds to 92% of the perfect release. The drug release was accelerated with increasing the rotation speed. The shortest time achieving the perfect release was approximately 3 minutes, which corresponds to 1/8 of the case without rotation. Simultaneous with the fast release, the bead collapsed probably due to the strong water flow surrounding the bead. The beads with high elasticity were hard to collapse and the fast release was not observed. Hence, the fast release of ketoprofen is triggered by the collapse of beads. Photographs of the collapse of beads, time profiles of the drug release, and a pulsatile release modulated by magnetic fields were presented.

Introduction

Many systems of drug release have been fabricated using polymer gel. Polymer gel is a material of open system, and is able to transport drug molecules through crosslinked network of the gel. According to this, spontaneous slow (sustained) release can be achieved. The sustained release is caused by the osmotic pressure difference $\pi_{in} - \pi_{out}$ at the inside and outside of the gel [1]. Furthermore, a novel drug delivery system (DDS) was developed by means of various stimuli-responsive polymers. Polymer gels synthesized with the stimuli-responsive polymers undergo a volume phase transition in response to stimuli such as temperature, solvent, or pH. When the gel achieved the volume phase transition, additional pressures are generated on the gel. Owing to the additional pressure contributed from the mixing π_{mix} and elastic π_{el} energies [1], the drug release is accelerated. Thus, the accelerated release of drugs is attributed to the volume contraction triggered by the additional pressures.

Various stimuli-responsive gels responding to temperature, pH, electric field, chemicals, and UV light have been employed to fabricate the novel drug delivery system. Poly(N-isopropylacrylamide) (PNIPA) gel is well known for a temperature-sensitive gel which exhibits a volume phase transition originating from the lower critical solution temperature. PNIPA gel copolymerized with alkyl methacrylates demonstrates complete on-off release of indometacin in response to stepwise temperature changes between 20 and 30°C [2, 3]. The regulated drug release is explained by the squeezing mechanism [4]. A pulsatile release of ketoprofen occurs when the gel was cycled in buffer solutions between pH 3.0 and pH 6.5 [5]. It has been shown that protein lysozyme is released from the carboxy methyl dextran hydrogel membranes in the pH range of 6.5–9.0 [6]. Electrically stimulated drug release of pilocarpine hydrochloride, glucose, and insulin has been fabricated using chemomechanical shrinking and swelling of polyelectrolyte gels under electric fields [7]. A sharp pulsatile release of glucose from a microcapsule of plant lectin has also been demonstrated based on the concept of competitive binding [8]. An on-off modulation for protein permeation by UV irradiation has been constructed by an azoaromatic polymer membrane [9]. In the present study, we attempted to construct the new system of drug delivery using magnetic gels responding to magnetic fields.

It has been reported that the magnetic gel undergoes a variety of motion under magnetic fields. For example, the magnetic gel consisting of magnetic fluids shows elongation by means of the gradient of magnetic field [10, 11], and it has been applied for a fluid valve [12]. Carrageenan magnetic gel containing magnetic particles presents an isotropic deformation under uniform magnetic fields [13]. Recently, we found that the magnetic gel consisting of sodium alginate and barium ferrite particles demonstrates a rotational motion in response to rotational magnetic fields [14, 15]. Using the rotational motion of the magnetic gel, we have succeeded in the fast and controllable release of ketoprofen from the magnetic bead. The release behavior is briefly reported and the mechanism of the fast release is discussed.

Experimental Procedures

Synthesis of Magnetic Beads

The bead ofmagnetic gels consists of Sodium alginate (Wako Chemicals, Osaka, Japan) as gel matrices, barium ferrite ($BaFe_{12}O_{19}$) (Sigma-Aldrich Co.,Mo, USA) as magnetic particles, and (2RS)-2-(3-Benzoylphenyl)propanoic acid (ketoprofen) (Saitama Daiichi Pharmaceutical Co., LTD., Saitama, Japan) as drugs. The chemical structure of ketoprofen is shown in the inset of Figure 2. A pre-gel solution consisting of sodium alginate (0.3wt.%), barium ferrite (3.0wt.%), and ketoprofen (9.0 wt.%) was prepared. The ketoprofen was dissolved in methanol before mixing these chemicals. The magnetic gel beads were obtained by dropping the pre-gel solution in a methanol/water mixed solvent containing $CaCl_2$ (3.0wt.%). The concentration of methanol in the mixed solvent was 30 vol.%. The obtained beads were stocked in a beaker filled with pure water. The mean diameter of a magnetic particle was determined as 15 μmby using a Particle size analyzer (Mastersizer 2000, Malvern Instruments, Malvern, UK). The bead was irradiated by a 1 T magnetic field for 1 minute in order to give the bead a remanent magnetization.

UV Measurements

The absorbance at 280nm was measured using a UV spectrometer (Shimazu UVmini-1240). The measurement was carried out using a flow cell at room temperature. The total volume of the sample space of the flow cell including a flow tube was 34.9mL. The flow rate was constant to be 46 mL/min. We evaluated the amount of drug released from the gel by the absorbance at 280 nm. The absorbance (Abs) showed linear relationship with the concentration of ketoprofen

c_{KP}; Abs = 284.06 × c_{KP}. The data presented in this paper is the drug release in a phosphate buffer solution. The buffer solution was prepared by dissolving potassium hydrogen phosphate (0.8wt.%, Wako Chemicals) and sodium dihydrogenphosphate (1.5wt.%,Wako Chemicals) in pure water.

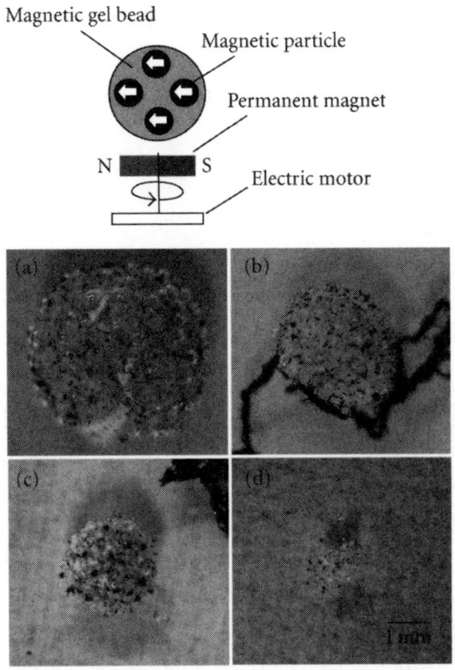

Figure 1: Schematic illustration of the geometry of magnetic gel bead and permanent magnet. Photographs representing the collapse of magnetic gel beads by rotational motion with 5000 rpm: (a) 0 minute, (b) 0.5 minute, (c) 1.0 minute, (d) 1.5 minutes.

Microscope Observations

Microscope observations were carried out using a microscope (Digital microscope VHX-900, VH-Z00R, Keyence Co., Osaka, Japan).

Results and Discussion

Figure 1 shows the time course of the shape of magnetic gel beads showing rotational motion. As seen in Figure 1(a), the bead was turbid and it had a shape of sphere with rough surface. Sodium alginate was soluble in water; however ketoprofen was insoluble in water, but soluble in methanol. Immediately after mixing the sodium alginate aqueous solution and ketoprofen/methanol one, the mixed solution was clouded. The solution was stable in emulsion and it did not show any precipitation. It is considered that the emulsion directly gelled by an addition of calcium chloride. The diameter of the bead decreased with an elapse of the rotation time (Figures 1(b)–1(d)). The bead was collapsed and disappeared 3 minutes after starting rotation. This strongly suggests that the bead is easy to collapse by the strong water flow induced by the rotation. Only magnetic particles tied each other and remained in the buffer solution. Magnetic particles embedded in the bead did not disperse in the solution and made a string because each particle has magnetic poles.

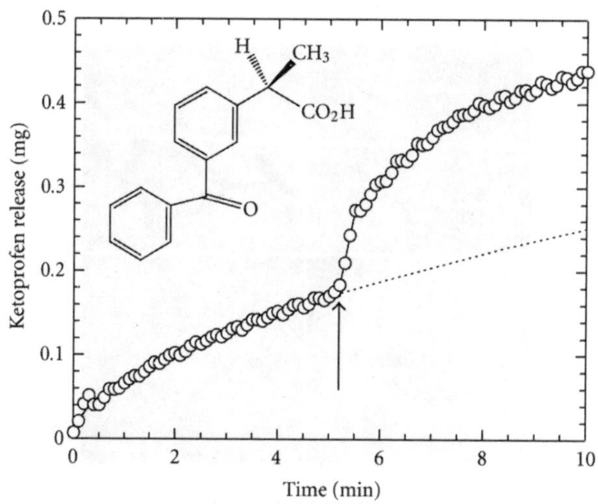

Figure 2: Time profile of the ketoprofen release for a magnetic gel bead (sodium alginate: 0.3wt%, ketoprofen: 5wt%, barium ferrite: 3wt%). The bead rotated (5000 rpm) at the time indicated by the arrow.

Figure 2 shows the effect of rotation on the time profile of the ketoprofen release from the beads. The bead was not rotated till 5 minutes and was rotated at the time indicated by the arrow. Ketoprofen was soluble in not only methanol but also phosphate buffer solution (pH~6). Below the time of 5minutes, the gradual increase in the ketoprofen release was due to spontaneous release from the bead. The bead maintained its spherical shape and the size was not changed. At 5 minutes, it was observed that the ketoprofen release dramatically increased simultaneous with the rotation. Furthermore, the bead was observed to be collapsed remarkably by rotation as shown in the photographs in Figure 1. Therefore, the increase in the ketoprofen release is not attributed to the squeezing effect of drug molecules by volume changes, which has been employed in the drug delivery of polymer gels. Probably the bead was collapsed by strong water flow at the interface between the bead and water generated by the rotation; as a result, the ketoprofen was rapidly released from the bead. Beads were rich in elastic modulus when the concentration of sodium alginate increased. The beads with high elasticity were hard to collapse, and the amount of ketoprofen release has not reached high value seen in Figure 2. The rapid release of ketoprofen was not seen in a magnetized bead without rotation. This means the rapid release was not caused by the collapse due to the magnetic attractive force between magnetic particles. For example, when the concentration of sodium alginate was 2wt.%, the amount of ketoprofen released from the gel was 0.16 mg at 5 minutes, which equals entirely the value without rotation. These results strongly suggest that the fast release of ketoprofen is triggered by the collapse of beads.

Figure 3 shows the time profiles of the amount of ketoprofen released from the beads with various rotation rates. Generally, drug release fromswellable polymer Mt can be described by the following empirical equation [16]:

$$\frac{M_t}{M_\infty} = kt^n$$

where M is the drug release at equilibrium state. Also, k, t, and n are the rate constant, time, and the order of drug release, respectively.

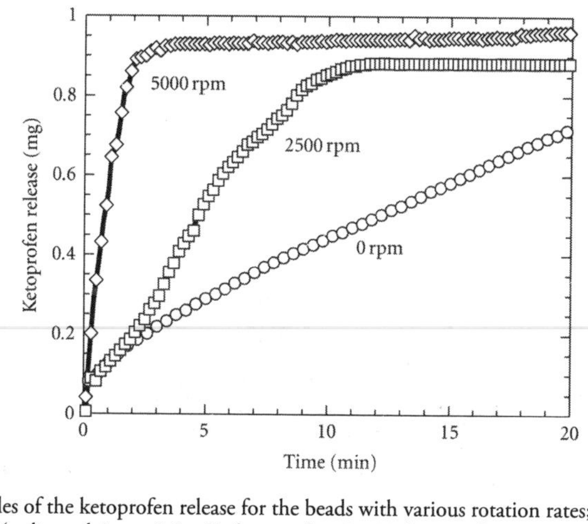

Figure 3: Time profiles of the ketoprofen release for the beads with various rotation rates; (°): 0 rpm, (|): 2500 rpm, (◇): 5000 rpm (sodium alginate: 0.3wt%, ketoprofen: 9wt%, barium ferrite: 3wt%).

In the case of Fickian diffusion, the diffusion obeys Fick's second law and the order n in (1) is equal to 0.5. In the case of bead without rotation, the drug release was explained by the above empirical equation. The order n was estimated to be

0.6, indicating anomalous (non-Fickian) diffusion. The saturation value of the ketoprofen released from the bead was approximately 0.90mg, which was seen at 25 minutes after starting rotation.When the rotation speed was 2500 rpm, the drug release increased and reached the saturation value at 12 minutes. At 5000 rpm of the maximum rotation speed, the drug release was further accelerated and achieved the saturation at only 3 minuttes. All beads collapsed before the ketoprofen release reaches to the saturation value. A bead was ground by a mortar and dissolved in the buffer solution in order to estimate the maximumvalue (perfect release) of ketoprofen in the bead; the estimated value was 0.97 ± 0.03 mg. This value was in good agreement with the saturation values seen in Figure 3. This means that the most of ketoprofen embedded in the bead (>90%) was released into the buffer solution by the fast rotation.

Figure 4 shows the pulsatile release of ketoprofen from the beads upon changes of on/off rotation. Increase in the ketoprofen release during the first 2 minutes is clearly caused by the spontaneous release. It is considered that the release rate due to the spontaneous release corresponds to 4.0 × 10–2 g/g/min. After 2 minutes, a pulsatile drug release synchronized with the rotation of beads was observed. The drug delivery system undergoing pulsatile release has been mainly constructed by using temperature responsive gels. In the temperature responsive drug delivery, the temperature of the environment surrounding the gel has to be changed by heating. Accordingly, it must take a considerable time to modulate the drug release. The drug release controlling by electric stimulus can be modulated within several minutes, however the system needs electrical wires to supply electric power to the gel.

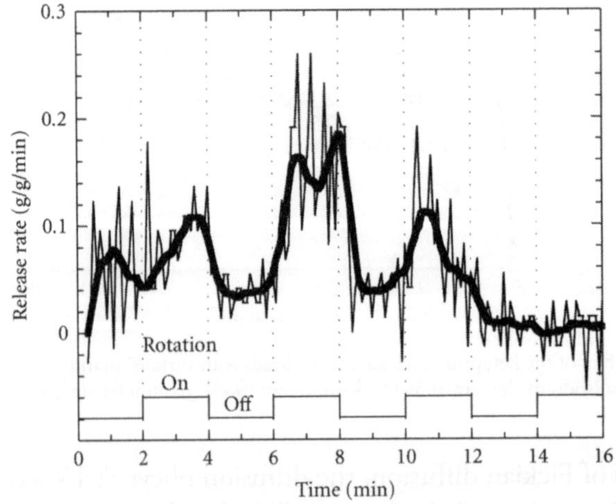

Figure 4: Release of ketoprofen from the beads in response to stepwise rotation speeds between 0 and 5000 rpm (sodium alginate: 0.3wt%, ketoprofen: 9wt%, barium ferrite: 3wt%).

The drug delivery system presented here showed drug release without lag time because the magnetic force directly acts on the gel. Although it is needless to say that the collapse of beads by water flow also plays an important role of the first release. Thus the pulsatile release by using magnetic field was demonstrated by the present study.

Conclusion

Drug delivery system controlled by magnetic fields has been constructed by means of the fast rotation of magnetic gel beads. The magnetic gel bead consists of sodium alginate crosslinked by calcium chlorides, which contains barium ferrite of ferrimagnetic particles, and ketoprofen as a drug.The bead underwent rotational motion in response to rotational magnetic fields. In the case of bead without rotation, the amount of drug release showed non-Fickian diffusion. When the bead was rotated, the drug release rapidly increased and reached the perfect release shortly. The drug release was accelerated with increasing the rotation speed. The shortest time achieving perfect release was approximately 3 minutes, which corresponds to 1/8 of the time without rotation. Simultaneous with the fast release, the bead was collapsed probably due to the strong water flow surrounding the bead. The beads with high elasticity were hard to collapse and the fast release was not observed. Hence, the fast release of ketoprofen is triggered by the collapse of beads, not by the squeezing effect of drug that has been employed in the drug delivery of polymer gels in the past. We also succeeded in pulse release of the drug controlled by the magnetic field. The drug delivery system presented here has an advantage of the drug release without ragtime because the magnetic force directly acts on the gel. In addition, no electric wires for modulating the drug release are needed for this system. If the magnetic bead can be delivered to the target such as cancer tissues, intensive drug release only around the target will be possible. Magnetic beads presented here would have a great potential as a new capsule of drugs without harmful side effects.

Acknowledgements

The authors are grateful to Dr. M. Goto of Saitama Daiichi Pharmaceutical Co. for the useful discussion and his kind offer of ketoprofen. This research is supported by a Grant-in- Aid for Encouragement of Young Scientist from Japan Society for the Promotion of Science (Proposal no. 18750184).

References

1. P. J. Flory, Principles of Polymer Chemistry, Cornell University Press, Ithaca, NY, USA, 1953.
2. R. Yoshida, K. Sakai, T. Okano, Y. Sakurai, Y. H. Bae, and S. W. Kim, "Surface-modulated skin layers of thermal responsive hydrogels as on-off switches—I: drug release," Journal of Biomaterials Science, Polymer Edition, vol. 3, no. 2, pp. 155–162, 1991.
3. R. Yoshida, K. Sakai, T. Okano, and Y. Sakurai, "Surfacemodulated skin layers of thermal responsive hydrogels as on-off switches—II: drug permeation," Journal of Biomaterials Science, Polymer Edition, vol. 3, no. 3, pp. 243–252, 1992.
4. X. S. Wu, A. S. Hoffman, and P. Yager, "Synthesis and characterization of thermally reversible macroporous poly(Nisopropylacrylamide) hydrogels," Journal of Polymer Science Part A, vol. 30, no. 10, pp. 2121–2129, 1992.
5. M. Negishi, A. Hiroki, M. Miyajima, M. Yoshida, M. Asano, and R. Katakai, "In vitro release control of ketoprofen from pH-sensitive gels consisting of poly(acryloyl-L-proline methyl ester) and saturated fatty acid sodium salts," Radiation Physics and Chemistry, vol. 55, no. 2, pp. 167–172, 1999.
6. R. Zhang, M. Tang, A. Bowyer, R. Eisenthal, and J. Hubble, "A novel pH- and ionic-strength-sensitive carboxy methyl dextran hydrogel," Biomaterials, vol. 26, no. 22, pp. 4677–4683, 2005.
7. K. Sawahata, M. Hara, H. Yasunaga, and Y. Osada, "Electrically controlled drug delivery system using polyelectrolyte gels," Journal of Controlled Release, vol. 14, no. 3, pp. 253–262, 1990.
8. K. Makino, E. J. Mack, T. Okano, and S. W. Kim, "A microcapsule self-regulating delivery system for insulin," Journal of Controlled Release, vol. 12, no. 3, pp. 235–239, 1990.
9. K. Ishihara and I. Shinohara, "Photoinduced permeation control of proteins using amphiphilic azoaromatic polymer membrane," Journal of Polymer Science Part C, vol. 22, no. 10, pp. 515–518, 1984.
10. M. Zr´ınyi, L. Barsi, and A. B¨uki, "Ferrogel: a new magnetocontrolled elastic medium," Polymer Gels and Networks, vol. 5, no. 5, pp. 415–427, 1997.
11. D. Szabó, G. Szeghy, and M. Zrínyi, "Shape transition of magnetic field sensitive polymer gels," Macromolecules, vol. 31, no. 19, pp. 6541–6548, 1998.
12. Y. Horikoshi, T. Mitsumata, and J.-I. Takimoto, "Developments for a fluid valve using magnetic gels," Transactions of the Materials Research Society of Japan, vol. 32, no. 3, pp. 843–844, 2007.

13. T. Mitsumata, K. Sakai, and J.-I. Takimoto, "Giant reduction in dynamic modulus of κ-carrageenan magnetic gels," Journal of Physical Chemistry B, vol. 110, no. 41, pp. 20217–20223, 2006.

14. T. Mitsumata, Y. Horikoshi, and J.-I. Takimoto, "Developments for a fluid pump using magnetic gels," Transactions of the Materials Research Society of Japan, vol. 31, no. 3, pp. 803– 805, 2006.

15. T. Mitsumata, Y. Horikoshi, and J.-I. Takimoto, "Flexible fluid pump using magnetic composite gels," e-Polymers, no. 147, pp. 1–10, 2007.

16. Y. Kaneko, K. Sakai, and T. Okano, "Temperature-responsive hydrogels as intelligent materials," in Biorelated Polymers and Gels, T. Okano, Ed., pp. 29–69, chapter 2, Academic Press, Boston, Mass, USA, 1998.

Oxidation of Quercetin by Myeloperoxidase

Tatjana Momić, Jasmina Savić and Vesna Vasić

ABSTRACT

Study of effect of myeloperoxidase on quercetin at pH 6.0 indicated quercetin oxidation via the formation of the oxidation product. The stability of quercetin and oxidation product was investigated as a function of time by using spectrophotometric and HPLC techniques. The apparent pseudo first-order rate constants were calculated and discussed.

Introduction

Flavonol quercetin (Q) (3,5,7,3′,4′-pentahydroxyflavone) (Scheme 1) due to its phenolic structure is a strong antioxidant and free radical scavenger [1]. Oxidation of quercetin during its antioxidative functions is usually accompanied by the production of the quercetin radical anion, superoxide, and hydrogen peroxide [2]. Quercetin undergoes autoxidation—the nonenzymatic reaction with atmospheric oxygen. Also, it is known that one-electron oxidation of quercetin is catalyzed by different peroxidases like lactoperoxidase (LPO) [3] and horseradish peroxidase (HRP) [4–7] in the presence of H_2O_2.

Scheme 1: Oxidation of quercetin by MPO/H_2O_2 system (modified from [15]).

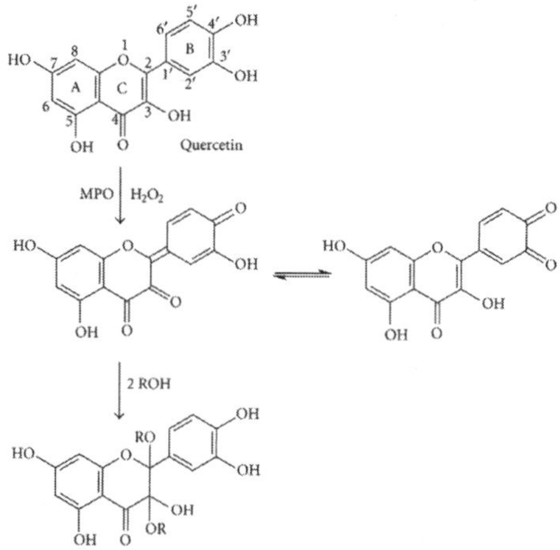

The heme enzyme myeloperoxidase (MPO) is oxidant enzyme in the process of inflammation and atherogenesis [8]. Myeloperoxidase is relatively nonspecific with respect to its reducing substrates. This enzyme is able to oxidize different substrates among which anilines and phenols [9–11]. This letter deals with quercetin oxidation by myeloperoxidase-H2O2 system.

Experimental

Chemicals

Quercetin dihydrate (Sigma-Aldrich) of the highest quality available (98%) was used without purification. 1×10^{-3} M stock solution of quercetin was prepared in methanol immediately before the experiments. For all experiments, freshly prepared solutions of quercetin were made by dilution of the appropriate amount of the stock solution with phosphate buffer at pH 6.0. Myeloperoxidase was purified from human neutrophils to a purity index (A_{430}/A_{280}) greater than 0.70 as described previously [12]. Its concentration was calculated using ε_{430}=91000 $M^{-1}cm^{-1}$ per heme [13]. Hydrogen peroxide solutions were prepared daily by diluting a stock solution, and the concentration was determined using ε_{240}=43.6 $M^{-1}cm^{-1}$ [14]. Redistilled water was used in all experiments.

Oxidation of Quercetin by Myeloperoxidase

Quercetin (5×10^{-5} M) was incubated in 50 mM phosphate buffer, pH 6.0, with various concentrations of MPO. Reaction was started by the addition of 50 μM H_2O_2, and UV absorbance changes were recorded. For the HPLC analysis, reaction was stopped after 30 minutes by adding catalase (100 μg/mL). The reaction mixture was centrifuged for 2 minutes at 10000 rpm. The clear supernatant was analyzed by HPLC.

HPLC Analysis

HPLC equipment consisted of an HP 1100 Series chromatograph coupled with a DAD. Chromatographic separations were run on a C18 Pinnacle ODS column (Restek, 250 mm × 4.6 mm, 5 μm) using an 80 : 20 mixture of 2 vol% H_3PO_4 (A) and acetonitrile (J.T. Baker) (B) as the eluent for the first 2 minutes. The linear gradient was applied from 20% to 45% B from 2 to 7 minutes. An isocratic 55 : 45 mixture was applied between 7 and 13 minutes. The eluent flow rate was 1.0 mL min^{-1}, and the injection volume was 10 μL. The elutions were monitored with DAD at different wavelengths between 200 and 450 nm.

UV-VIS Spectroscopic Studies

UV spectra were recorded on a Perkin Elmer Lambda 35 UV-Vis spectrophotometer equipped with thermostatted quartz cell. The temperature in the cell was kept at 25±0.05°C with a water-thermostatted bath.

Results and Discussion

We investigated the effect of MPOon 5×10^{-5} M quercetin in phosphate buffer, pH 6.0 in the presence of 50 μM H_2O_2 at 25°C, spectrophotometrically and by HPLC. The following concentrations of enzyme were used: 5, 11, 15, 20, 50, 100, 150, and 200 nM.

At the absorption spectra of samples in which concentration of MPO was ≤11 nM, two quercetin absorption bands (254 nm and 367 nm) were observed (Figure 1(a)). The same results were yielded when MPO or H2O2 was omited from the mixture. The absorption spectra of samples in which concentration of MPO was ≥15 nM showed the decay of quercetin absorption bands at 254 nm and 367 nm. At the same time simultaneous rise of the absorption band at 336–342 nm, depended on MPO concentration, was observed (Figure 1(c)), typical of oxidation product formation [16]. Moreover, two well-defined isosbestic points at 290 and 364 nm indicate that there was no significant accumulation of intermediates in the reaction of oxidation product formation (Figure 1(c)). These spectral changes are analogous to those occurring upon autoxidation of quercetin in water [16] and UV irradiation [17] as well as oxidation with some oxidants in organic solvents [16].

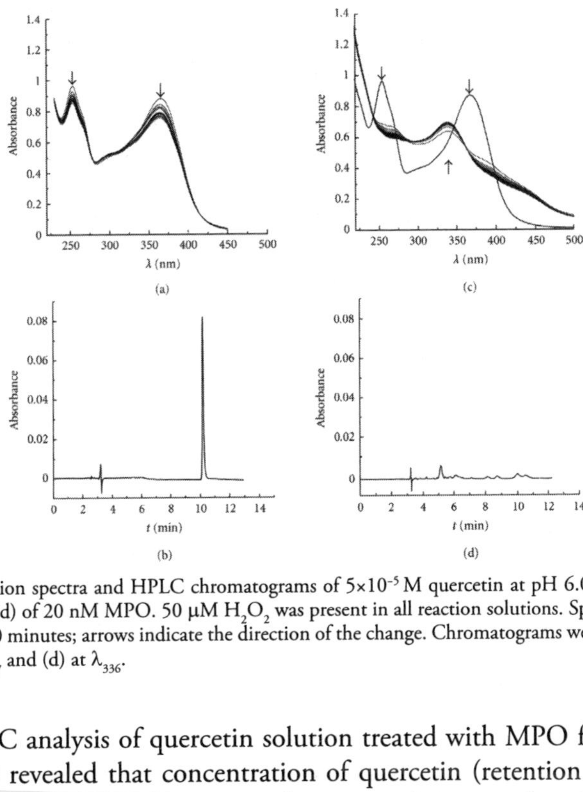

Figure 1: Absorption spectra and HPLC chromatograms of 5×10^{-5} M quercetin at pH 6.0 in absence (a), (b) and presence (c), (d) of 20 nM MPO. 50 μM H_2O_2 was present in all reaction solutions. Spectra were recorded over a period of 30 minutes; arrows indicate the direction of the change. Chromatograms were recorded after 30 minutes (b) at λ_{367} and (d) at λ_{336}.

The HPLC analysis of quercetin solution treated with MPO for 30 minutes (Figure 1(d)) revealed that concentration of quercetin (retention time 10 minutes) was negligible. Also, formation of major oxidation product (retention time 5.26 minutes) which was more polar than quercetin was detected (Figure 1(d)). From literature data and the obtained spectrophotometric and HPLC results in this work, we assume that the oxidation product detectable at 336–342 nm results from H2O addition on the p-quinonemethide formed by H-atom abstraction at 3-OH and 4′-OH of quercetin and subsequent rearrangement of the central ring [16]. The complete sequence of quercetin autoxidation induced by oxidants and catalyzed by metal ions is described earlier [15, 16] (Scheme 1).

According to spectrophotometric and HPLC experimental data, quercetin degradation (1) and oxidation product formation (2) as the function of time followed the relation

$$A_{Q,t} = A_0 e^{-kt}, \tag{1}$$

$$Aoxid,t = A_\infty \left(1 - e^{-kt}\right), \tag{2}$$

where $A_{Q,t}$ and $A_{oxid,t}$ are the absorbances proportional to the quercetin and its oxidation product concentrations, respectively, after an irradiation period of time t and k is the overall pseudo first-order rate constant. A_0 and A_∞ are the absorbance proportional to the initial concentration of quercetin and the concentration of the oxidation product on the plateau of the kinetics curve, respectively.

The kinetic curves that describe the quercetin degradation and the formation of the oxidation product in the presence of different MPO, concentrations are shown in Figure 2. In the presence of MPO concentration that was ≥15 nM oxidation product was formed at the first minute of reaction (Figure 2(a)). The overall apparent first-order rate constants for the oxidation product formation and quercetin transformation to its degradation products were determined from the kinetic curves and are presented in Table 1. For comparison, in Table 1 we also presented values for these constants obtained for quercetin oxidation by UV irradiation [17].

Table 1: Rate constants of quercetin oxidation.

pH	Sample		$k\ (\text{min}^{-1}) \times 10^3$	
		Quercetin degradation	Water adduct formation	Water adduct degradation
6.0	Q	3.61 ± 0.18	—	—
	Q + MPO (20 nM)	346.00 ± 21.15	524.60 ± 32.30	2.67 ± 0.09
10.0*	Non-irradiated Q	8.73 ± 0.34	25.12 ± 0.93	—
	Irradiated Q	110.00 ± 35.25	96.91 ± 0.12	10.13 ± 0.56

*Reference [17].

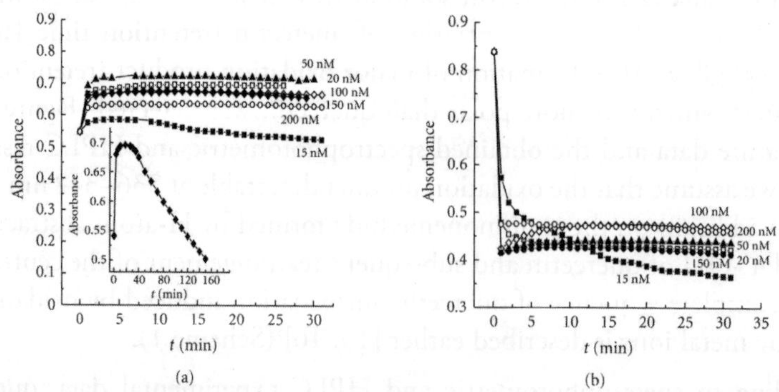

Figure 2: Change of absorbance of quercetin in the presence of various MPO concentrations in the function of time, at pH 6.0 (a) formation of quercetin oxidation product at λ_{336}; Inset: 20 nM MPO and (b) degradation of quercetin oxidation product at λ_{367}.

The rate of quercetin oxidation by MPO at pH 6.0 is 3-fold faster compared with the rate of quercetin oxidation by UV irradiation at pH 10.0. Moreover, product obtained for quercetin oxidation by MPO at pH 6.0 is more stable than product of quercetin oxidation by UV irradiation at pH 10.0.

Acknowledgement

We wish to express our gratitude to the Ministry of Science and Technological Development of the Republic of Serbia supported this work through Project 142051.

References

1. L. Magnani, E. M. Gaydou, and J. C. Hubaud, "Spectrophotometric measurement of antioxidant properties of flavones and flavonols against superoxide anion," Analytica Chimica Acta, vol. 411, no. 1-2, pp. 209–216, 2000.
2. N. Cotelle, J.-L. Bernier, J.-P. Catteau, J. Pommery, J.-C. Wallet, and E. M. Gaydou, "Antioxidant properties of hydroxy-flavones," Free Radical Biology and Medicine, vol. 20, no. 1, pp. 35–43, 1996.
3. D. Metodiewa, A. K. Jaiswal, N. Cenas, E. Dickancaité, and J. Segura-Aguilar, "Quercetin may act as a cytotoxic prooxidant after its metabolic activation to semiquinone and quinoidal product," Free Radical Biology and Medicine, vol. 26, no. 1-2, pp. 107–116, 1999.
4. H. Yamasaki, Y. Sakihama, and N. Ikehara, "Flavonoid-peroxidase reaction as a detoxification mechanism of plant cells against H2O2," Plant Physiology, vol. 115, no. 4, pp. 1405–1412, 1997.
5. G. Galati, T. Chan, B. Wu, and P. J. O'Brien, "Glutathione-dependent generation of reactive oxygen species by the peroxidase-catalyzed redox cycling of flavonoids," Chemical Research in Toxicology, vol. 12, no. 6, pp. 521–525, 1999.
6. D. P. Makris and J. T. Rossiter, "An investigation on structural aspects influencing product formation in enzymic and chemical oxidation of quercetin and related flavonols," Food Chemistry, vol. 77, no. 2, pp. 177–185, 2002.
7. U. Takahama, "Spectrophotometric study on the oxidation of rutin by horseradish peroxidase and characteristics of the oxidized products," Biochimica et Biophysica Acta, vol. 882, no. 3, pp. 445–451, 1986.
8. S. J. Klebanoff, "Myeloperoxidase," Proceedings of the Association of American Physicians, vol. 111, no. 5, pp. 383–389, 1999.

9. J. K. Hurst, "Myeloperoxidase: active site structure and catalytic mechanism," in Peroxidases in Chemistry and Biology, J. Everse, K. E. Everse, and M. B. Grisham, Eds., pp. 37–62, CRC Press, Boca Raton, Fla, USA, 1991.

10. E. Shacter, R. L. Lopez, and S. Pati, "Inhibition of the myeloperoxidase-H2O2-Cl-system of neutrophils by indometacin and other non-steroidal anti-inflammatory drugs," Biochemical Pharmacology, vol. 41, no. 6-7, pp. 975–984, 1991.

11. A. J. Kettle, C. A. Gedye, M. B. Hampton, and C. C. Winterbourn, "Inhibition of myeloperoxidase by benzoic acid hydrazides," Biochemical Journal, vol. 308, no. 2, pp. 559–563, 1995.

12. R. L. Olsen and C. Little, "Purification and some properties of myeloperoxidase and eosinophil peroxidase from human blood," Biochemical Journal, vol. 209, no. 3, pp. 781–787, 1983.

13. T. Odajima and I. Yamazaki, "Myeloperoxidase of the leukocyte of normal blood—I: reaction of myeloperoxidase with hydrogen peroxide," Biochimica et Biophysica Acta, vol. 206, no. 1, pp. 71–77, 1970.

14. R. J. Beers, Jr. and I. W. Sizer, "A spectrophotometric method for measuring the breakdown of hydrogen peroxide by catalase," Journal of Biological Chemistry, vol. 195, no. 1, pp. 133–140, 1952.

15. O. Dangles, C. Dufour, and S. Bret, "Flavonol-serum albumin complexation. Two-electron oxidation of flavonols and their complexes with serum albumin," Journal of the Chemical Society Perkin Transactions, vol. 2, no. 4, pp. 737–744, 1999.

16. H. E. Hajji, E. Nkhili, V. Tomao, and O. Dangles, "Interactions of quercetin with iron and copper ions: complexation and autoxidation," Free Radical Research, vol. 40, no. 3, pp. 303–320, 2006.

17. T. Momić, J. Savić, U. Černigoj, P. Trebše, and V. Vasić, "Protolytic equilibria and photodegradation of quercetin in aqueous solution," Collection of Czechoslovak Chemical Communications, vol. 72, no. 11, pp. 1447–1460, 2007.

Reversible Control of Primary and Secondary Self-Assembly of Poly(4-allyloxystyrene)-Block-Polystyrene

Eri Yoshida and Satoshi Kuwayama

ABSTRACT

The reversible control of primary and secondary self-assemblies was attained using a poly(4-allyloxystyrene)-block-polystyrene diblock copolymer (PASt-b-PSt) through variations in temperature. The copolymer showed no self-assembly in cyclohexane over 35°C and existed as a unimer with a 37.1 nm hydrodynamic diameter. When the temperature was lowered to 30°C, the copolymer formed micelles with 269.9 nm by the primary self-assembly. As the result of further lowering the temperature to 20°C, the secondary self-assembly of the micelles occurred to produce ca. 2975.9 nm aggregates. The aggregates were dissociated into unimers by increasing the temperature up to 40°C. The light scattering studies demonstrated that the thermoresponsivity of the copolymer showed good hysteresis throughout the variation in the temperature in the range between 20 and 40° C, based on the Marquadt analysis of the hydrodynamic diameter distribution. It was found that the primary and secondary self-assemblies of the copolymer were perfectly controlled by the temperature.

Introduction

The secondary self-assembly of molecular aggregates is important for effective control of the activity and function produced by the primary self-assembly of the molecules. The secondary aggregation spontaneously occurs through a number of attractive forces such as van der Waals interaction, solvation, depletion, bridging, π-π stacking, hydrogen bond, and coordination bond [1]. The secondary self-assembly is well known as the formation of the quaternary structure of proteins with a high molecular weight over 100 000 daltons. Examples for the proteins include the hemoglobin [2], microtubules [3], pyruvate dehydrogenase complexes [4], influenza virus [5], tobacco mosaic virus [6–8], tomato bushy stunt virus [9, 10], and aspartate transcarbamylase [11]. Thus, the secondary aggregation has widely been discovered in the natural world, while it has been explored for artificial molecules such as dilauroylphosphatidylcholine with dipalmitoylphosphatidylethanolamine-conjugated biotin [12], polystyrene-block-poly(acrylic acid) [13], polystyrene-block-poly(4-vinylpyridine) onto silica nanoparticles [14], poly(isobutene)-grafted boehmite rods [15], poly(ethylenimine)-graft-poly(ethylene glycol) [16], and a metallo-supramolecular diblock copolymer consisting of polystyrene block connected to a poly(ethylene oxide) by a bis(terpyridine)ruthenium complex [17]. In addition to these experimental results, there are many theoretical results concerning the secondary aggregation [18, 19].

We found that the primary and secondary self-assembly of a poly(4-allyloxystyrene)-block-polystyrene diblock copolymer (PASt-b-PSt) was reversibly controlled by a variation in temperature. The thermoresponsivity of this block copolymer showed good hysteresis in the range between 20 and 40°C. This study describes the reversible control of the primary and secondary self-assemblies of the PASt-b-PSt copolymer in cyclohexane.

Experiment

Instrumentation

The ^{1}H NMR measurements were conducted using a Varian 300 FT NMR spectrometer. Light scattering measurements were performed with a Photal

Otsuka Electronics ELS-8000 electrophoretic light scattering spectrophotometer equipped with a system controller, an ELS controller, and an He-Ne laser operating at λ = 632.8 nm. UV analysis was performed with a Shimadzu UV-160A UV-Vis recording spectrophotometer.

Materials

A poly(4-tert-butoxystyrene)-block-polystyrene diblock copolymer (PBSt-b-PSt) was prepared as reported previously [20]. The molecular weight of the copolymer was Mn(PBSt-b-PSt) = 15400-b-96600 by 1H NMR, while the molecular weight and its distribution were Mn = 58000, and Mw/Mn = 1.36 by gel permeation chromatography based on polystyrene standards. Tetrahydrofuran (THF) and cyclohexane were distilled over sodium. N,N-Dimethylformamide (DMF) was distilled over calcium hydride under reduced pressure. Allyl chloride was also distilled over calcium hydride.

Synthesis of PASt-b-PSt

The PBSt-b-PSt (2.00 g) was dissolved in THF (70 mL). Concentrated hydrochloric acid (7 mL) was added to the copolymer solution. The mixture was heated at 85°C for 4.5 hours. The resulting solution was concentrated to ca. 30 mL by an evaporator, and was poured into water (1 L) to precipitate a polymer. The precipitates were collected by filtration, then freeze-dried with 1,4-dioxane to obtain poly(vinylphenol)-block-polystyrene diblock copolymer (PVPh-b-PSt, 1.593 g). The PVPh-b-PSt (0.70 g) was dissolved in DMF (15 mL). Sodium hydride (0.414 g, 17.3 mmol) was added to the copolymer solution at 0°C under nitrogen atmosphere. The suspension was stirred at 0°C for 5 minutes and was further stirred at room temperature for 1 hour. Allyl chloride (1.41 g, 18.4 mmol) in DMF (5 mL) was added to the suspension at 0°C. The mixture was stirred at 0°C for 5 minutes and was further stirred at room temperature for 20 hours. The resulting solution was poured into methanol (1 L) to precipitate a polymer. The precipitates were collected by filtration, then dried in vacuo for several hours. PASt-b-PSt (0.68 g) was obtained. The molecular weight of the PASt-b-PSt was determined to be Mn(PASt-b-PSt) = 14000-b-96600 by 1H NMR.

Light Scattering Measurements

Cyclohexane (5 mL) was added to PASt-b-PSt (8.2 mg), and the mixture was completely dissolved at 40°C. The solution was injected through a microporous filter into a cell using a syringe. The solution was subjected to light scattering at the angle θ = 90°. The hydrodynamic diameter of the copolymer was estimated by the cumulant analysis, while the scattering intensity distribution of hydrodynamic diameter was obtained by the Marquadt analysis [21].

Results and Discussion

The PASt-b-PSt diblock copolymer showed no self-assembly in cyclohexane at 40°C and existed as a unimer of an isolated copolymer. The light scattering demonstrated that the copolymer formed aggregates by decreasing the temperature. Figure 1 shows a variation in the scattering intensity distribution for the hydrodynamic diameter of the copolymer when the temperature decreased. The scattering intensity distributions were obtained by the Marquadt analysis [21]. The copolymer still existed as a unimer at 35°C, so that only one distribution was observed around 35 nm. When the temperature was lowered to 30°C, another distribution was observed around 270 nm in addition to the unimer distribution. The distribution around 270 nm was attributed to the micelles formed by the primary self-assembly of the copolymer. The unimer distribution completely disappeared at 25°C and only distribution was observed around 700 nm. This distribution was further shifted to the hydrodynamic diameter around 3000 nm at 20°C and did not retain the Gauss distribution any longer. The formation of the aggregates with the hydrodynamic diameter over 1000 nm should be caused by the secondary aggregation of the micelles. In addition, the exact hydrodynamic diameters were estimated by the Marquadt analysis to be 37.1 nm (35°C), 35.1 nm for the unimer and 269.9 nm for the micelles (30°C), 698.0 nm (25°C), and 2975.9 nm (20°C).

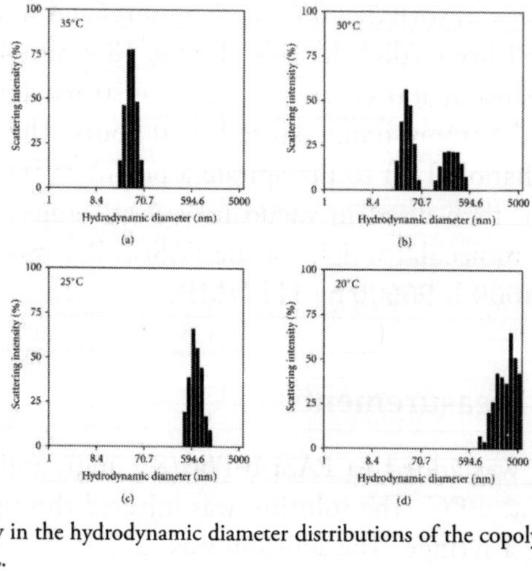

Figure 1: The variability in the hydrodynamic diameter distributions of the copolymer versus the temperature. $[copolymer]_0$ = 1.64 g/L.

The self-assembly of the copolymer was also explored by 1H NMR. Figure 2 shows the 1H NMR spectra obtained at each temperature using cyclohexane-d12. In the spectrum at 35°C, all the protons composing the copolymer were observed, since the copolymer existed as a unimer. The signals at 1.0–2.5, 4.3–4.5, 5.2–5.6, 5.8–6.1, and 6.2–7.5 ppm were assigned to the protons of the main chain, the allyl protons, the α-proton of the vinyl, its β-protons, and the aromatic protons, respectively. The copolymer showed negligible changes at 30°C, although part of the aromatic proton signal at 6.3 ppm was slightly broadened. This signal was more broad at 25°C and was observed as a shoulder. At 20°C, all of the signals were broadened and, in particular, the protons of the allyl groups were barely discerned. The signal broadening was based on the protons shielded by the copolymer self-assembling. The marked effect of the shielding at 20°C indicates that not only the cores of the micelles but also the shells were shielded by the secondary aggregation

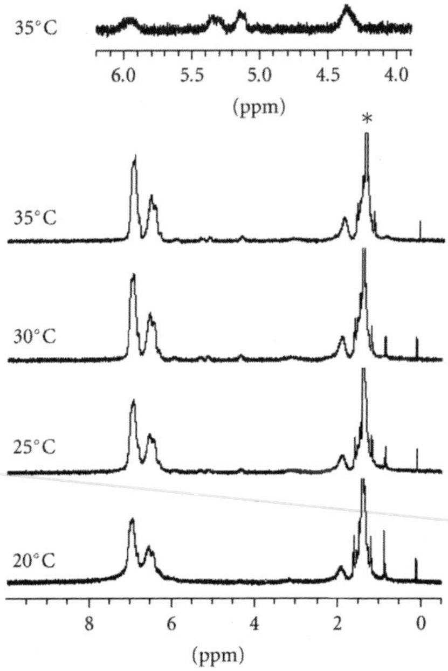

Figure 2: The variation in the ^{1}H NMR spectra of the copolymer versus the temperature. Solvent: cyclohexane-d$_{12}$. [copolymer]$_0$ = 1.6 g/L, * Cyclohexane.

The degree of the shielding was determined by UV analysis. Figure 3 shows the UV spectra of the copolymer at the respective temperatures and the plots of the absorbance at 276 and 326 nm versus the temperature. The absorbance at 326 nm increased as the temperature decreased. Contrarily, the absorbance at 276 nm decreased with the decreased temperature. Both the decrease and increase in the absorbances were accelerated below 30°C. The decrease in the absorbance at 276 nm is based on the shielding of the aromatic groups and the olefins, while the increase at 326 nm is attributed to an increase in the interaction among the aromatic groups and the olefins. This is accounted for by the fact that the delocalization of electrons causes red shift.

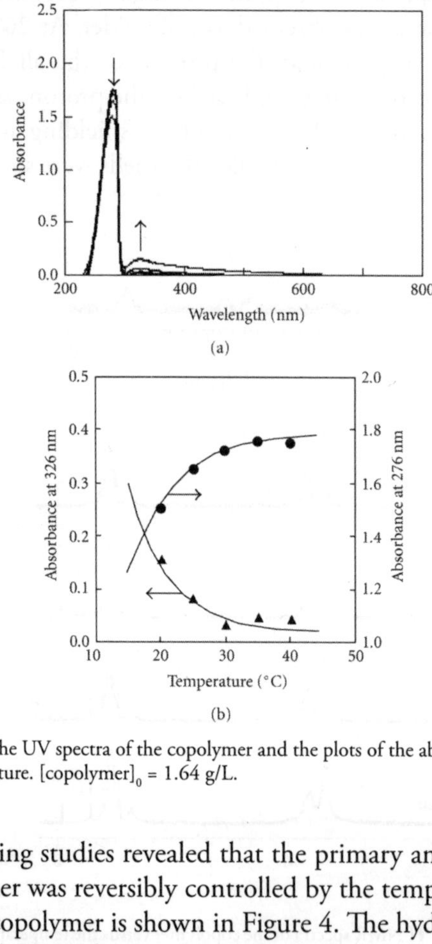

Figure 3: The variation in the UV spectra of the copolymer and the plots of the absorbance at (●) 276 and (▲) 326 nm versus the temperature. [copolymer]$_0$ = 1.64 g/L.

The light scattering studies revealed that the primary and secondary aggregation of the copolymer was reversibly controlled by the temperature. The thermo-responsivity of the copolymer is shown in Figure 4. The hydrodynamic diameters were estimated by the cumulant analysis. The hydrodynamic diameter and scattering intensity increased by lowering the temperature, however, those reverted to the original values as a result of raising the temperature. The aggregates were dissociated into unimers. The course of the dissociation was almost the same as that of the aggregation, indicating that the primary and secondary self-assemblies of the copolymer were perfectly controlled by the temperature.

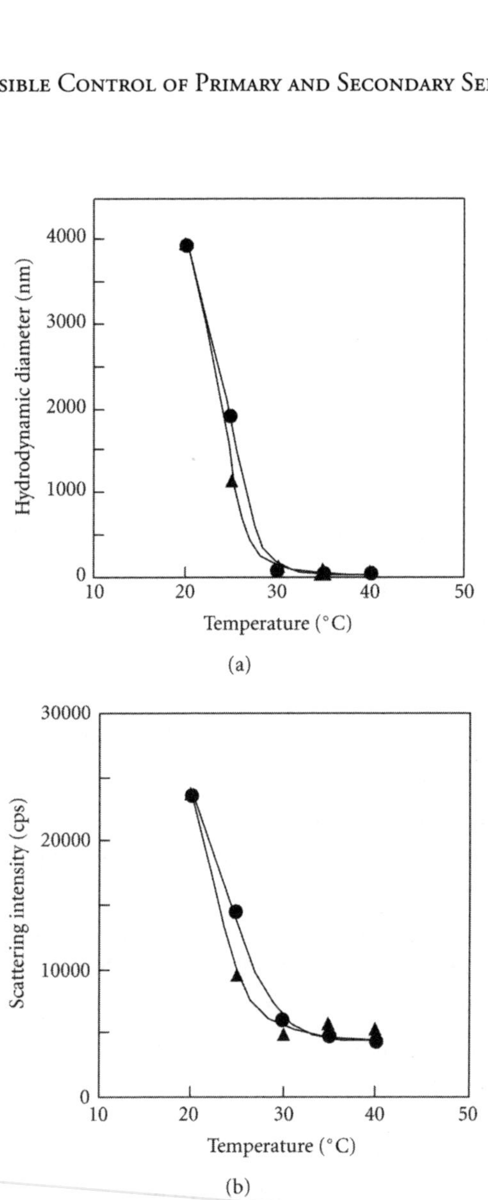

Figure 4: The variation in the hydrodynamic diameter and scattering intensity of the copolymer with (|) the decrease and (▲) increase in the temperature. $[\text{copolymer}]_0$ = 1.64 g/L.

Conclusion

The reversible control of the primary and secondary self-assemblies was attained using the PASt-b-PSt diblock copolymer. The copolymer formed micelles with ca. 270nm hydrodynamic diameter by the primary self-assembly. The micelles were further self-assembled into large aggregates with 3000 nm following the formation of 700 nm aggregates. The aggregates formed by the secondary assembly were dissociated into unimers by increasing the temperature. It was found that the primary and secondary self-assemblies of the copolymer were perfectly controlled by the temperature based on the fact that the thermoresponsivity of the copolymer showed good hysteresis. This is the first study demonstrating that the primary and secondary self-assemblies of the diblock copolymer were reversibly controlled by the variation in the temperature in the narrow range between 20 and 40°C.

References

1. Y. S. Lee, Self-Assembly and Nanotechnology: A Force Balance Approach—Part 1: Self-Assembly, Wiley-Interscience, New York, NY, USA, 2008.
2. M. F. Perutz, "Hemoglobin structure and respiratory transport," Scientific American, New Series, vol. 239, no. 6, pp. 92–125, 1978.
3. P. Dustin, "Microtubules," Scientific American, New Series, vol. 243, no. 2, pp. 59–68, 1980.
4. L. J. Reed, "Multienzyme complexes," Accounts of Chemical Research, vol. 7, no. 2, pp. 40–46, 1974.
5. J. N. Varghese, W. G. Laver, and P. M. Colman, "Structure of the influenza virus glycoprotein antigen neuraminidase at 2.9 Å resolution," Nature, vol. 303, no. 5912, pp. 35–40, 1983.
6. A. C. Bloomer, J. N. Champness, G. Bricogne, R. Staden, and A. Klug, "Protein disk of tobacco mosaic virus at 2.8 Å resolution showing the interactions within and between subunits," Nature, vol. 276, no. 5686, pp. 362–368, 1978.
7. K. Namba and G. Stubbs, "Structure of tobacco mosaic virus at 3.6 Å resolution: implications for assembly," Science, vol. 231, no. 4744, pp. 1401–1406, 1986.
8. P. J. Butler and A. Klug, "The assembly of a virus," Scientific American, New Series, vol. 239, no. 5, pp. 62–69, 1978.
9. S. C. Harrison, A. J. Olson, C. E. Schutt, and F. K. Winkler, "Tomato bushy stunt virus at 2.96 Å resolution," Nature, vol. 276, no. 5686, pp. 368–373, 1978.
10. S. C. Harrison, "Multiple modes of subunit association in the structures of simple spherical viruses," Trends in Biochemical Sciences, vol. 9, no. 8, pp. 345–351, 1984.
11. E. R. Kantrowitz, S. C. Pastra-Landis, and W. N. Lipscomb, "E. coli aspartate transcarbamylase—part II: structure and allosteric interactions," Trends in Biochemical Sciences, vol. 5, no. 6, pp. 150–153, 1980.

12. S. Chiruvolu, S. Walker, J. Israelachvili, F.-J. Schmitt, D. Leckband, and J. A. Zasadzinski, "Higher order self-assembly of vesicles by site-specific binding," Science, vol. 264, no. 5166, pp. 1753–1756, 1994.

13. L. Zhang and A. Eisenberg, "Morphogenic effect of added ions on crew-cut aggregates of polystyrene-b-poly(acrylic acid) block copolymers in solutions," Macromolecules, vol. 29, no. 27, pp. 8805–8815, 1996.

14. M. Motornov, R. Sheparovych, R. Lupitskyy, E. MacWilliams, and S. Minko, "Responsive colloidal systems: reversible aggregation and fabrication of superhydrophobic surfaces," Journal of Colloid and Interface Science, vol. 310, no. 2, pp. 481–488, 2007.

15. J. Buitenhuis, J. K. G. Dhont, and H. N. W. Lekkerkerker, "Static and dynamic light scattering by concentrated colloidal suspensions of polydisperse sterically stabilized boehmite rods," Macromolecules, vol. 27, no. 25, pp. 7267–7277, 1994.

16. H. Wang, X. Chen, and C. Pan, "Linear poly(ethylenimine)-graft-poly(ethylene glycol) copolymers: their micellization and secondary assembly," Journal of Colloid and Interface Science, vol. 320, no. 1, pp. 62–69, 2008.

17. O. Regev, J.-F. Gohy, B. G. G. Lohmeijer, et al., "Dynamic light scattering and cryogenic transmission electron microscopy investigations on metallo-supramolecular aqueous micelles: evidence of secondary aggregation," Colloid and Polymer Science, vol. 282, no. 4, pp. 407–411, 2004.

18. P. van der Schoot, "Remarks on the association of rodlike macromolecules in dilute solution," Journal of Physical Chemistry, vol. 96, no. 14, pp. 6083–6086, 1992.

19. J. Groenewold and W. K. Kegel, "Anomalously large equilibrium clusters of colloids," Journal of Physical Chemistry B, vol. 105, no. 47, pp. 11702–11709, 2001.

20. E. Yoshida and S. Kuwayama, "Micelle formation induced by photolysis of a poly(tert-butoxystyrene)-block-polystyrene diblock copolymer," Colloid and Polymer Science, vol. 285, no. 11, pp. 1287–1291, 2007.

21. D. W. Marquardt, "An algorithm for least-squares estimation of nonlinear parameters," SIAM Journal on Applied Mathematics, vol. 11, no. 2, pp. 431–441, 1963.

Pt/TiO_2 Coupled with Water-Splitting Catalyst for Organic Pollutant Photodegradation: Insight into the Primary Reaction Mechanism

Zizhong Zhang, Xuxu Wang, Jinlin Long, Xianliang Fu, Zhengxin Ding, Zhaohui Li, Ling Wu and Xianzhi Fu

ABSTRACT

A composited system was fabricated by coupling Pt/TiO_2 with water-splitting catalyst for photooxidation of organic pollutants in aqueous solutions. The new composited system exhibits more efficient photocatalytic activity than pure Pt/TiO_2 does under UV light irradiation. The promoting effect is dependent on the photo-produced H_2 over the composited system. The active oxygen species, hydroxyl radical ($\cdot OH$) and hydrogen peroxide (H_2O_2), are *measured by fluorescence spectroscopy and photometric method, respectively. The results reveal that the produced* H_2 *by photocatalytic water splitting over* $NiO/NaTaO_3$*:La transfers to Pt particle of* TiO_2 *surface, then reacts with introducing* O_2 *to generate in situ intermediate* H_2O_2*, and finally translates into* $\cdot OH$ *radical to accelerate the photooxidation of organic pollutants.*

Introduction

Photoinduced charge transfer occurring on semiconductor materials can achieve direct conversion of photo energy to chemical energy, and thus it can be used for elimination of organic pollutants and splitting water into hydrogen. However, the utility of semiconductor-based photocatalytic process is controlled to a large extent by the separation efficiency of the initially formed excited states (h_{vb}^{+} *and* e_{cb}^{-}) [1]. A variety of approaches was made to enhance electron-accepting or electron-donating ability of the material surface to favor the interfacial charge separation and consequently increase the photocatalytic efficiency. One approach involves addition of surface adsorbed redox species capable of scavenging selectively either of the excited states to the photoreaction system [2, 3]. Another promising approach concerns modification of TiO_2 with noble metals, other semiconductors, and coloring matters to improve the separation of the excited states [4–6].

Deposition of platinum on TiO2 has been reported to enhance extremely the photocatalytic efficiency for organic pollutant elimination due to its high electron-trapping effect [7], although an excessive number of platinum particles per grain of TiO2 can be detrimental to the performance of the reaction system [8]. We have recently demonstrated that trace amount of H2 can efficiently improve the activity of benzene photooxidation over Pt/TiO2 [9, 10]. However, the mechanisms have not been fully understood, and a practical approach for the environmental application has not yet to be achieved, due to the difficulties in realizing the integration of H2 gas and photocatalysis into a practical system.

Herein, an alternative system was fabricated by coupling Pt/TiO2 with water-splitting catalyst NiO/NaTaO3:La to supply the in situ H2 to enhance photocatalytic oxidation organic pollutants in an aqueous solution, where the obtained composited system is quite different from the classic coupled semiconductor system. The data show that the high photocatalytic efficiency of the composited system is attributed to the formation of more · OH which is dependent on the generation of in situ H2O2 from the combination between the photo-produced H2 by the NiO/NaTaO3:La and bubbled O2 on Pt/TiO2 surface.

Experimental

Sample Preparation

Titanium dioxide (TiO2) particles were prepared by a sol-gel technique. Titanium isopropoxide (0.1 mol) was first added dropwise to 100 mL of nitric acid aqueous solution. The suspension was stirred to clear and then dialyzed to pH of ca. 4 to obtain the TiO2 sol. The sol was dried at 333 K in an oven for 3 days. The resulting solid powders were ground to fine powders and finally calcined at 623 K for 3 hours.

NaTaO3:La was prepared by the solid state reaction according to the literature [11]. In typical, 0.02 mol Ta2O5, 0.0206 mol Na2CO3, and 0.0004 mol La2O3 were mixed and then calcined in air at 1173 K for 1 hour and 1423 K for 10 hour.

Platinum supported catalyst was prepared by the incipient wetness impregnation method. The calcined TiO2 was impregnated with a 5.22×10–2 M aqueous solution of H2PtCl6. The impregnated sample was dried at 393 K for 6 hours and subsequently reduced with an NaBH4 solution (0.1 M). After reduction, the solid sample was washed with deionized water to remove residual ion, and finally dried in air at 333 K (denoted as Pt/TiO2). The initial ratio of Pt to TiO2 was fixed at 1 wt%.

NiO loaded catalysts were prepared by an impregnation method from a 2.36×10–2 M aqueous solution of Ni(NO3)2 and then dried at 383 K for 2–5 hours. The sample thus obtained was subsequently calcined at 543 K for 1 hour in air using a muffle furnace. The initial ratio of NiO to NaTaO3:La was fixed at 0.2 wt%.

Photocatalytic Reactions and Methods

The photocatalytic reaction was performed at room temperature in a quartz tubal reactor surrounded with 254 nm UV lamps (Philips TUV, 4 W, Holland). The photocatalyst powders were dispersed in the salicylic acid (SA) solution bubbled with oxygen (10 mL min^{-1}). The concentration of SA was analyzed by a high-performance liquid chromatograph (HPLC Waters) equipped with a reverse phase column (Merk, LiChrospher RP-18e, 5 μm) and a UV detector with detection wavelength of 297 nm. The mobile phase consisted of 30 mmol L^{-1} acetate (pH = 4.9) and the flow rate was 1.0 mL min^{-1}. The evolved CO_2 during the reaction was collected with a $Ba(OH)_2$ solution and then determined by a titrate with an oxalic acid ($H_2C_2O_4$) solution (0.02 mol L^{-1}). The evolved H_2 during the reaction was monitored by a hydrogen sensor (Dräger Pac III).

Hydroxyl radical ·OH was captured by terephthalic acid to form fluorescent 2-hydroxyterephthalic acid [12] and then determined with fluorescence spectroscopy (FS/FL920, excitation wavelength: 312 nm, and fluorescence peak: 426 nm). Hydrogen peroxide was analyzed photometrically by the POD (horseradish peroxidase) catalyzed oxidation product of DPD (N,N-diethyl-p-phenylenediamine) at 551 nm [13].

Results and Discussion

Table 1 lists the rate constants of salicylic acid (SA) photodegradation with different catalysts under UV light irradiation in the presence of O_2. The results show that NiO/$NaTaO_3$:La has distinct effect on TiO_2 and Pt/TiO_2 for SA photodegradation. NiO/$NaTaO_3$:La enhances the rate of the SA photodegradation in Pt/TiO_2 reaction system, and yet has no effect on the SA photodegradation over TiO_2. Two controlled experiments are carried out respectively under UV irradiation without catalyst and with NiO/$NaTaO_3$:La. The results show that NiO/$NaTaO_3$:La is photocatalytically inactive for SA degradation despite it was reported to be highly active for photocatalytic splitting water into H_2 even without sacrificial agent. Therefore, it can be deduced that the NiO/$NaTaO_3$:La plays a promoting role for Pt/TiO_2 photocatalytic degradation of SA, and the existence of Pt is indispensable for the promoting effect of NiO/$NaTaO_3$:La.

Table 1: Rate constants for SA photodegradation with different composited catalysts. Catalyst: 0.0500 g, the rate of NiO/$NaTaO_3$:La to Pt/TiO_2 (or TiO_2) is 25 wt%. reactant solution: 120 mL SA (5×10^{-4} mol L-1), with two 254 nm UV lamps irradiation.

Photocatalyst	k (min^{-1})
Pt/TiO_2	0.00308
NiO/$NaTaO_3$:La-Pt/TiO_2	0.00429
TiO_2	0.00289
NiO/$NaTaO_3$:La-TiO_2	0.00286
NiO/Ta_2O_5-Pt/TiO_2	0.00267
NiO/$Sr_2Ta_2O_7$-Pt/TiO_2	0.00415

The conduction band level of the NaTaO3 and TiO2 is −1.03 eV and −0.52 eV, respectively, while the valence band level of the NaTaO3 and TiO2 is 2.97 eV and 2.64 eV, respectively [14, 15]. It is obvious that both the valence and conduction band of TiO2 are sandwiched between the corresponding bands of NaTaO3. Thus in the NiO/NaTaO3:La-Pt/TiO2 composited system, the promoting effect of NiO/NaTaO3:La is unexpected from the viewpoint of coupled semiconductors [5]. It is verified by the fact that NiO/NaTaO3:La has no effect on TiO2 for SA photodegradation (Table 1). Furthermore, simple mechanical addition NiO/NaTaO3:La to Pt/TiO2 suspensions cannot make them intimate contact which was necessary to form coupled semiconductors for an acceleration in photocatalytic reaction rate [16]. Therefore, The results provide a clear conclusion that there are other reasons attributing to NiO/NaTaO3:La promoting effect. NiO/NaTaO3:La was well documented to be a highly efficient photocatalyst for water splitting into H2 under UV light irradiation [11]. In our previous work, the trace amount of H2 was found to significantly increase the activity of Pt/TiO2 for photocatalytic oxidation of volatile organic compounds (VOC's). Therefore, the promoting effect may be attributed to the trace amount of H2 produced from photocatalytic water splitting by NiO/NaTaO3:La to enhance the activity of Pt/TiO2 for SA photodegradation.

In order to check the effect of NiO/NaTaO3:La, the following comparative experiments were carried out under the same conditions. It is experimentally verified that NiO/Ta2O5 is photocatalytically inert for both the SA degradation and water splitting (data not shown here) [17], but has the same band energy level as the NaTaO3 [18]. Adding NiO/Ta2O5 instead of NiO/NaTaO3:La into the Pt/TiO2 suspension, the photodegradation rate of SA shows no change (Table 1). In contrast, replacing NiO/NaTaO3:La with another efficient water-decomposing photocatalyst NiO (0.15 wt.%)/Sr2Ta2O7 [19], SA photodegradation can also be markedly accelerated (Table 1). The above results confirm that the promoting effect is dependent on the water-splitting function of NiO/NaTaO3:La. Furthermore, we examine the photodegradation of other organic contaminations such as phenol with the NiO/NaTaO3:La-Pt/TiO2 suspensions under 125 W high-pressure mercury lamp irradiation for 55 minutes, showing that both the degradation and mineralization of phenol can be enhanced significantly from 63% to 97% and from 54% to 84%, respectively. These results demonstrate that coupling of Pt/TiO2 with a splitting-water photocatalyst is more efficient for the photocatalytic elimination of organic pollutants in aqueous solution than pure what Pt/TiO2 does.

To understand the origin of the promoting effect of NiO/NaTaO3:La, the variety of evolved H2 in the reaction process was monitored. Figure 1 shows the change in H2 yield and SA photodegradation in the composited system with reaction time under O2 bubbling. As H2 evolution reaches a steady state, injecting SA into the system results in a notable decrease of H2 evolution along with quick degradation of SA. However, as the SA is completely decomposed, the production of H2 progressively comes after its former steady state (Figure 1). This indicates that the produced H2 is partly consumed to accelerate the SA photodegradation over Pt/TiO2. This is supported by the result that introducing H2 from an outer bottle instead of NiO/NaTaO3:La into the Pt/TiO2 reaction system, the rate of SA photodegradation was enhanced and comparable. In combination with the results of the SA degradation (Table 1), it is deduced that the promoting effect of NiO/NaTaO3:La to accelerate Pt/TiO2 for SA photodegradation is correlated to the produced H2 consumed by Pt particle on TiO2 in the presence of O2.

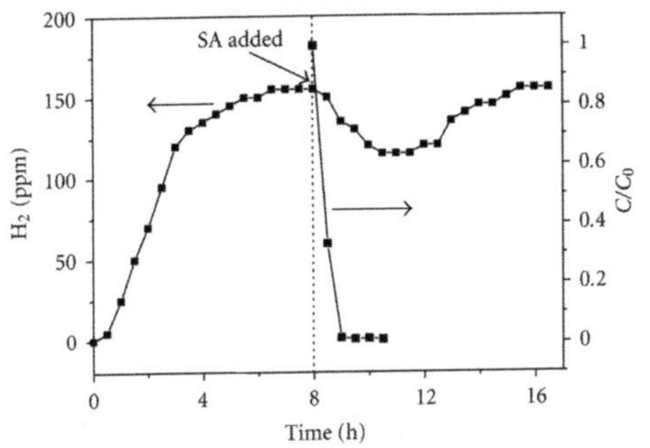

Figure 1: Hydrogen evolution and SA degradation over NiO/NaTaO_3:La-Pt/TiO_2 with oxygen bubbling under UV light illumination.

Photocatalytic degradation of SA and phenol is initial from the attack of · OH radical [20]. Figure 2 shows the plots of increase in fluorescence intensity at 426 nm against illumination time for the reaction system. The linear increase in fluorescence intensity for NiO/NaTaO3:La-Pt/TiO2 system is higher than that for pure Pt/TiO2 system, suggesting that a larger amount of · OH radical was produced in NiO/NaTaO3:La-Pt/TiO2 composited system. Thus we conclude that the consumed H2 is converted to a larger amount of active oxidative species· OH to induce quicker degradation of SA and phenol.

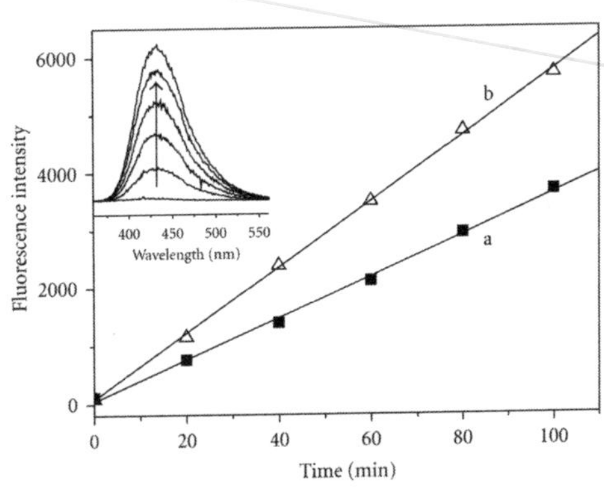

Figure 2: Fluorescence spectra (insert) and induced fluorescence intensity (426 nm) against illumination time for terephthalic acid solution on (a) Pt/TiO_2 and (b) NiO/NaTaO3:La-Pt/TiO2 samples under UV irradiation.

In the presence of H2 and O2, Au supported Ti-based catalysts were reported to selective vapor-phase epoxidation of propylene. The reaction is likely due to the in situ preparation of H2O2 from H2 and O2 at perimeter interface of the catalyst [21]. Thus it is possible that in the NiO/NaTaO3:La-Pt/TiO2 system, the produced H2 by photocatalytic water splitting and bubbled O2 are primarily combined to form H2O2 on Pt/TiO2. Figure 3 shows the absorbance (at 551 nm) of the produced H2O2 against illumination time for the reaction system. It is obvious that a larger amount of H2O2 was produced on NiO/NaTaO3:La-Pt/TiO2 reaction system than that on Pt/TiO2 reaction system. The produced H2 and bubbled O2 can be responsive to the generation of more amount of H2O2 for the composited system. It is confirmed by the result that as the Pt/TiO2 solutions were bubbled both with H2 and O2 in the dark, some amount of H2O2 was detected. The effect of H2O2 on the photocatalytic activity was investigated earlier. Shiraishi and Kawanishi [22] have declared that the photocatalytic activity is closely related to the formation of H2O2. Additional dosage of H2O2 into the TiO2 suspension was often used and found to efficiently enhance the degradation of organic compounds due to the generation of ·OH radical by the direct photolysis or the photoinduced electron reduction of H2O2 [23]. Thereby, in the composited reaction system, the produced H2 and introducing O2 directly combine to form in situ H2O2 on Pt particle of TiO2 firstly, and then the H2O2 traps the photoinduced electron on Pt particle surface or is photocleaved to form ·OH radical. Moreover, the larger amount of H2O2 is produced by the composited reaction system not only simply, but also practicably, and it may be useful for photocatalytic selective oxidation reaction by in situ H2O2 via the NiO/NaTaO3:La and Pt/TiO2 co-deposition on suitable support.

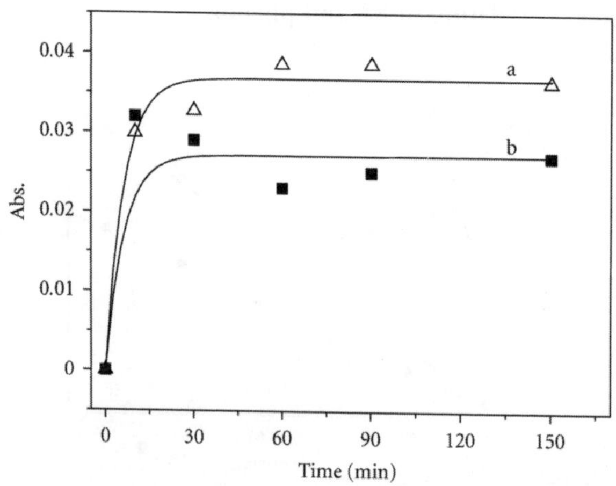

Figure 3: Absorption intensity (551 nm) of DPD/POD reagent after reaction with H_2O_2 against illumination time in the aqueous solution of (a) NiO/$NaTaO_3$:La-Pt/TiO_2 and (b) Pt/TiO_2.

Conclusions

This work opens up a new efficient composited system for improving the efficiency of the photocatalytic process. Coupled with the water-splitting catalyst, NiO/NaTaO$_3$:La can efficiently promote the photocatalytic performance of Pt/TiO$_2$ for organic pollutant elimination in aqueous solutions. It is shown that the in situ H_2O_2 is not only simply, but also practicably formed in the composited system by directly combining H_2 produced by photocatalytic water splitting with introducing O_2 on Pt particle of TiO$_2$, and then the in situ H_2O_2 is photocleaved or reduced by photogenerated electron to produce ·OH radical to accelerate photooxidation reaction. This work is clearly very useful to explore a new efficient and practical route for photocatalytic elimination of organic contaminations.

Acknowledgements

This work was supported financially by NSF of China (Grants no. 20573020, 20373011, and 20537010), the Foundation of Fujian Province Education Department (Grant no. JA05176), and the National Key Basic Research Special Foundation (Grant no. 2004CCA07100).

References

1. T. L. Morkin, N. J. Turro, M. H. Kleinman, C. S. Brindle, W. H. Kramer, and I. R. Gould, "Selective solid state photooxidant," Journal of the American Chemical Society, vol. 125, no. 48, pp. 14917–14924, 2003.
2. D. Bahnemann, A. Henglein, and L. Spanhel, "Detection of the intermediates of colloidal TiO$_2$-catalysed photoreactions," Faraday Discussions of the Chemical Society, vol. 78, pp. 151–163, 1984.
3. P. V. Kamat, "Photoelectrochemistry in particulate systems. 3. Phototransformations in the colloidal titania-thiocyanate system," Langmuir, vol. 1, no. 5, pp. 608–611, 1985.
4. S. W. Lam, K. Chiang, T. M. Lim, R. Amal, and G. K.-C. Low, "The effect of platinum and silver deposits in the photocatalytic oxidation of resorcinol," Applied Catalysis B, vol. 72, no. 3-4, pp. 363–372, 2007.
5. N. Serpone, P. Maruthamuthu, P. Pichat, E. Pelizzeti, and H. Hidaka, "Exploiting the interparticle electron transfer process in the photocatalysed oxidation of phenol, 2-chlorophenol and pentachlorophenol: chemical evidence for electron and hole transfer between coupled semiconductors," Journal of Photochemistry and Photobiology A, vol. 85, no. 3, pp. 247–255, 1995.

6. F. Zhang, J. Zhao, L. Zang, et al., "Photoassisted degradation of dye pollutants in aqueous TiO_2 dispersions under irradiation by visible light," Journal of Molecular Catalysis A, vol. 120, no. 1–3, pp. 173–178, 1997.

7. M. Anpo and M. Takeuchi, "The design and development of highly reactive titanium oxide photocatalysts operating under visible light irradiation," Journal of Catalysis, vol. 216, no. 1-2, pp. 505–516, 2003.

8. P. Pichat, J.-M. Herrmann, J. Disdier, H. Courbon, and M.-N. Mozzanega, "Photocatalytic hydrogen production from aliphatic alcohols over a bifunctional platinum on titanium dioxide catalyst," Nouveau Journal de Chimie, vol. 5, pp. 627–636, 1981.

9. Y. Chen, D. Li, X. Wang, X. Wang, and X. Fu, "H_2–O_2 atmosphere increases the activity of Pt/TiO_2 for benzene photocatalytic oxidation by two orders of magnitude," Chemical Communications, vol. 10, no. 20, pp. 2304–2305, 2004.

10. Y. L. Chen, D. Li, X. Wang, L. Wu, X. Wang, and X. Fu, "Promoting effects of H_2 on photooxidation of volatile organic pollutants over Pt/TiO_2," New Journal of Chemistry, vol. 29, no. 12, pp. 1514–1519, 2005.

11. H. Kato, K. Asakura, and A. Kudo, "Highly efficient water splitting into H_2 and O_2 over lanthanum-doped $NaTaO_3$ photocatalysts with high crystallinity and surface nanostructure," Journal of the American Chemical Society, vol. 125, no. 10, pp. 3082–3089, 2003.

12. K.-I. Ishibashi, A. Fujishima, T. Watanabe, and K. Hashimoto, "Detection of active oxidative species in TiO_2 photocatalysis using the fluorescence technique," Electrochemistry Communications, vol. 2, no. 3, pp. 207–210, 2000.

13. H. Bader, V. Sturzenegger, and J. Hoigné, "Photometric method for the determination of low concentrations of hydrogen peroxide by the peroxidase catalyzed oxidation of N,N-diethyl-p-phenylenediamine (DPD)," Water Research, vol. 22, no. 9, pp. 1109–1115, 1988.

14. H. Kato and A. Kudo, "Water splitting into H_2 and O_2 on alkali tantalate photocatalysts $ATaO_3$ (A = Li, Na, and K)," Journal of Physical Chemistry B, vol. 105, no. 19, pp. 4285–4292, 2001.

15. S. Sakthivel and H. Kisch, "Photocatalytic and photoelectrochemical properties of nitrogen-doped titanium dioxide," ChemPhysChem, vol. 4, no. 5, pp. 487–490, 2003.

16. A. Di Paola, L. Palmisano, M. Derrigo, and V. Augugliaro, "Preparation and characterization of tungsten chalcogenide photocatalysts," Journal of Physical Chemistry B, vol. 101, no. 6, pp. 876–883, 1997.

17. H. Kato and A. Kudo, "New tantalate photocatalysts for water decomposition into H_2 and O_2," Chemical Physics Letters, vol. 295, no. 5-6, pp. 487–492, 1998.

18. M. Metikos-Hukovic and M. Ceraj-Ceric, "Conduction processes in the Ta/ Ta_2O_5-electrolyte system," Thin Solid Films, vol. 145, no. 1, pp. 39–49, 1986.

19. M. Yoshino, M. Kakihana, W. S. Cho, H. Kato, and A. Kudo, "Polymerizable complex synthesis of pure Sr_2NbxTa_2-xO_7 solid solutions with high photocatalytic activities for water decomposition into H_2 and O_2," Chemistry of Materials, vol. 14, no. 8, pp. 3369–3376, 2002.

20. C. Adán, J. M. Coronado, R. Bellod, J. Soria, and H. Yamaoka, "Photochemical and photocatalytic degradation of salicylic acid with hydrogen peroxide over TiO_2/SiO_2 fibres," Applied Catalysis A, vol. 303, no. 2, pp. 199–206, 2006.

21. T. Hayashi, K. Tanaka, and M. Haruta, "Selective vapor-phase epoxidation of propylene over Au/TiO_2 catalysts in the presence of oxygen and hydrogen," Journal of Catalysis, vol. 178, no. 2, pp. 566–575, 1998.

22. F. Shiraishi and C. Kawanishi, "Effect of diffusional film on formation of hydrogen peroxide in photocatalytic reactions," Journal of Physical Chemistry A, vol. 108, no. 47, pp. 10491–10496, 2004.

23. D. F. Ollis, E. Pelizzetti, and N. Serpone, "Photocatalyzed destruction of water contaminants," Environmental Science and Technology, vol. 25, no. 9, pp. 1522–1529, 1991.

Room-Temperature Growth of SiC Thin Films by Dual-Ion-Beam Sputtering Deposition

C. G. Jin, X. M. Wu and L. J. Zhuge

ABSTRACT

Silicon carbide (SiC) films were prepared by single and dual-ion-beam sputtering deposition at room temperature. An assisted Ar^+ ion beam (ion energy Ei = 150 eV) was directed to bombard the substrate surface to be helpful for forming SiC films. The microstructure and optical properties of nonirradicated and assisted ion-beam irradicated films have been characterized by transmission electron microscopy (TEM), scanning electron microscopy (SEM), Fourier transform infrared spectroscopy (FTIR), and Raman spectra. TEM result shows that the films are amorphous. The films exposed to a low-energy assisted ion-beam irradicated during sputtering from a-SiC target have exhibited smoother and compacter surface topography than which deposited with nonirradicated. The ion-beam irradicated improves the adhesion between film and substrate and releases the stress between film and substrate.

With assisted ion-beam irradicated, the density of the Si–C bond in the film has increased. At the same time, the excess C atoms or the size of the sp^2 bonded clusters reduces, and the a-Si phase decreases. These results indicate that the composition of the film is mainly Si–C bond.

Introduction

Amorphous semiconductor alloys are of technological importance for electronic and optoelectronic device application. Amorphous silicon carbide (a-SiC) has been recognized as a semiconductor material with outstanding physical and chemical characteristics. Silicon carbide exhibits a large bandgap, a higher breakdown field, a higher thermal conductivity, and a higher saturatuin velocity, compared to widely used silicon. On the other hand, amorphous SiC films have interest related to their high hardness and optical properties and have potential applications as hard, wear resistant coatings, masking material in Si micromaching technology as well as for the formation of optical windows, filters, and color sensors [1, 2]. Recently, plasma-assisted deposition methods such as plasma enhanced CVD [3, 4], electron cyclotron resonance (ECR) [5, 6], the conventional physical vapour deposition methods (magnetron sputtering [7, 8], pulsed laser deposition [9, 10]), ion implantation [11], and molecular beam epitaxy [12] methods have been used to grow SiC films on Si substrate. However, these methods need high grown temperature, which process defect creation, resulting from high tensile stress generated from a temperature dependent difference in the thermal expansion coefficient between SiC and Si, and involve many pollutions of impurity in the films such as H element.

This letter reports the growth of SiC thin films at room temperature using dual-ion-beam sputtering deposition (DIBSD), a method that has been used to grow many different types of films. Recently, it has also been shown that ion-assisted techniques can greatly improve the adhesion by minimizing the total stress at the interface and compactness of the coatings despite the higher optical absorption near the bandgap of the material [13, 14]. By proper adjustment of dual-ion-beam parameters, the method permits a precise control of film composition, which is almost independent of deposition rate. The microstructural, optical, and electrical properties may also be tailored significantly by ion bombardment during film deposition [15]. The energy given by the ion-beam system is high enough to reduce the temperature needed for the formation of the SiC phase [16, 17]. In the paper, the microstructure and morphology of the films have been measured by Fourier transform infrared spectroscopy (FTIR), scanning electron microscopy (SEM), and Raman spectra. The effect of a low-energy assisted-ion bombardment during deposition on the film properties has been investigated.

Experiment

SiC films were deposited by employing a DIBSD system at room temperature. A DIBSD system consists of a focused Kaufman ion source (main ion source) and a broad-beam Kaufman ion source (assisted-ion source). The Ar gas as pure as 99.999% was used for source gas and for ion-beam generation. Main ion source of 10 cm diameter with the incident angle of 45° served as a sputtering ion source. Normal sintered high-purity SiC target was used as a sputtering target. The sputtering chamber was evacuated to 2×10^{-4} Pa before argon was introduced through a mass flow controller. The working pressure of argon was 4×10^{-2} Pa. Before deposition, the substrates of KBr, Suprasil quatz, and p-type (100) Si wafer (resistivity, 5 Ω-cm) after standard cleaning sequence to remove organic and inorganic surface contaminants were cleaned again with an argon ion beam for 15 minutes using the assisted source (200 eV energy and 10 mA beam intensity). During deposition, the beam energy and beam current of the main ion source which was used to sputter the SiC target were 800 eV and 40 mA, respectively. An assisted-ion source of 10 cm diameter with the incident angle of 60° was used. The beam current of the assisted-ion source which was used to bombardment the substrate was 10 mA, and beam energy was variable parameter 0–150 eV. No intentional heating was applied on substrate during deposition. SiC films were deposited for 2 hours on the substrate. All of the thicknesses of the SiC films were range of 150–260 nm, which were measured with an ET350 surface profilometer (Kosaka Laboratory Ltd.).

The films were studied by scanning electron microscopy (SEM) using Hitachi S-4700 equipped with energy dispersive spectroscopy (EDS), for their morphology and composition, respectively. The bonding configurations and the structure were measured by TEM (Hitachi H600A-II), FTIR (Niolot AVATAR 360), and Raman spectra which were obtained in the backscattering configuration between 200 cm−1 and 1800 cm−1 by JY-HR800 using an argon ion laser at a wavelength of 514 nm. All the measurements were conducted at room temperature.

Results and Discussion

Stucture Analysis

During DIBSD, the assisted ion beam has been utilized to significantly enhance and control the properties of the film. In order to compare the characteristics of films prepared by DIBSD, IBSD, which did not use assisted ion beam to irradiate the substrate, was used to form SiC film. Figure 1 shows our TEM result from which only halos are observed in the selection electron diffraction picture, the samples regarded to be amorphous.

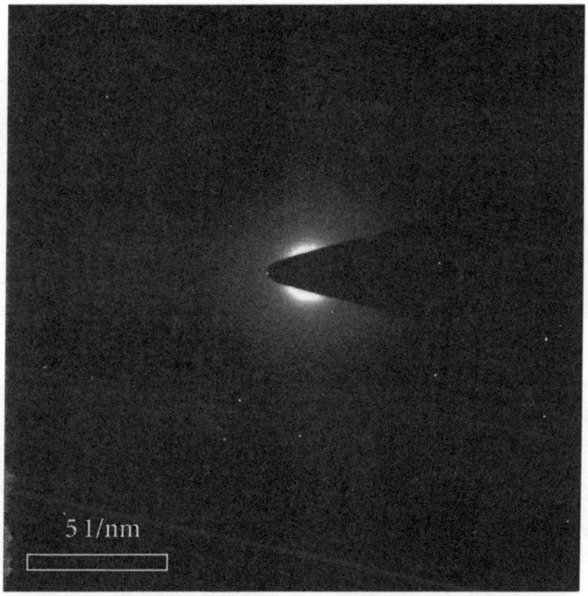

Figure 1: The TEM micrographs of the sample with 150 eV ion-beam irradiated.

At low resolution, the top SEM micrograph of a film prepared on Si substrate nonirradicated (sample a) and irradicated by 150 eV ion beam (sample b) is shown in Figures 2(a) and 2(c), respectively. Compared to Figure 2(a), the average grain size increases, and the surface is found to be denser with ion-beam irradiated in Figure 2(c). The cross-sectional SEM micrographs of sample a and sample b are shown in Figures 2(b) and 2(d), respectively. As shown in Figure 2(b), we see that the film/substrate interface is quite smooth and clear. Moreover, in region A, there are many obvious cracks on both film and substrate, which might originate from the inter-stress between film and substrate. As shown in Figure 2(d), sample b is more compact than sample a. The film/substrate interface is roughness and dark. Nevertheless the quantity of the cracks decreases in region C. There is a buffer layer between SiC film and Si substrate in region B, which releases the stress caused by the large lattice mismatch, around 20%, and indicates very good coating to substrate adhesion. It found that the ion-beam-assisted bombarding enhanced the coating to substrate adhesion. These results indicate that low-energy ion-assisted growth improves the adatom mobility, which is attributed to the distribution of nonequilibrium phonon on surface. At the same time, the ion-beam irradicated is the densification of film, increases the average of the cluster, improves the adhesion between SiC and Si substrate, and releases the stress between SiC and Si substrate.

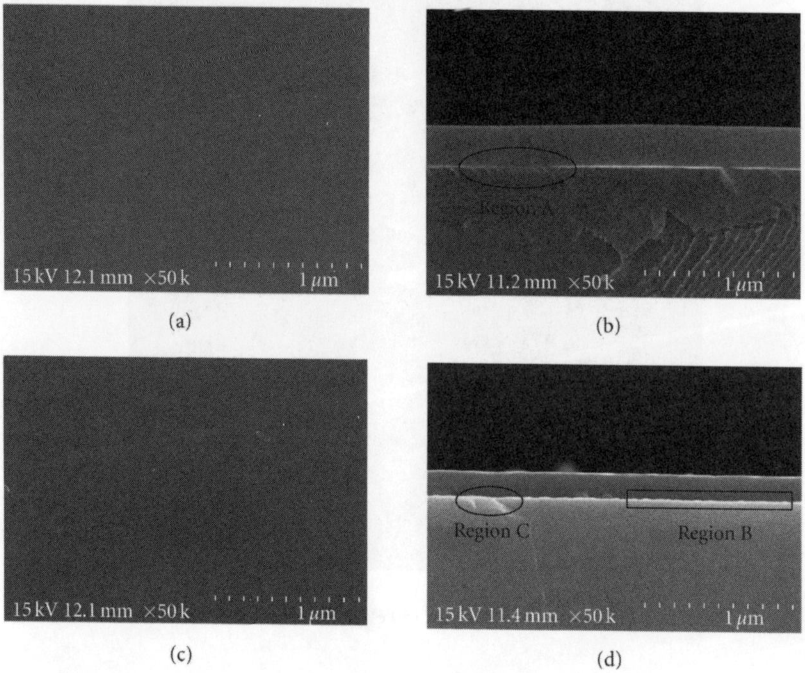

Figure 2: (a) Top-, and (b) cross-sectional SEM micrographs of the sample prepared on Si with nonirradicated, (c) top-, and (d) cross-sectional SEM micrographs of the sample irradiated by 150 eV ion beam.

The EDS was only used to measure the proportion of Si and C in our films grown on KBr substrates. The composition uniformity in the film was studied by analyzing several points (with 1 mm2 analysis area) over the surface of the sample. EDS analyses indeed showed that all the films consist of Si and C. The proportion in sample a is Si:C = 9:11 (at.% ratio), nevertheless it decreases to 2:3 with ion-beam irradicated in sample b. It indicates that all the films are C excess. The main source ion beam sputters the SiC target, then the sputtering particles deposit on the substrate, which include Si, C, and SiC atoms. At the same time, the assisted-source ion beam sputters the deposition, and the secondary ion sputtering rate of Si ion is higher than that of C ion in the films. The increasing of the proportion of C atom might be attributed to resputter on as-deposition film, which maybe due to the sputtering rate of Si ion is higher than that of C ion.

The FTIR is a powerful tool to investigate the bonding structure in a-SiC films. The IR spectra for sample a and the sample b which are deposited on KBr substrates are shown in Figures 3(a) and 3(b), respectively. As can be seen, almost all the measured spectra are characterized by the absorption peak related to the Si–C bond, which is around 800 cm−1, except the weak peaks around 1018 cm−1, 1090 cm−1, 1260cm−1, and 2950 cm−1. The bands corresponding to 1018 cm−1, 1090 cm−1, and 1260 cm−1 are due to oxygen adsorption on the film surface in air or the O atoms at the film/substrate interface [18], while the band corresponding to 2950 cm−1 is attributed to the stretching vibration of C–Hn groups in sp3 configurations [19], which may be due to H2O from the atmosphere adsorbed on the film surface [20]. The 740 cm−1 band as a broad absorption maybe related to these results, just as (a) the high concentration of substitutional C, (b) the appearances of the C–C bonds, and (c) the nonsubstitutional C in the Si substrate [21, 22]. Compared to Figure 3(a), the shape of the 800 cm−1 peak progressively changes and becomes narrow in Figure 3(b). The full width at half maximum (FWHM) of 800 cm−1 peak is smaller than that of other spectra, indicating that the stoichiometric proportion of the SiC films after assisted ion-beam irradicated is improved [23]. It demonstrates a better quality of the SiC film grown with assisted ion-beam irradicated. The full width at half maximum (FWHM) of 800 cm−1 peak is narrower than that of other spectra, indicating that the Si–C band is a majority of the film.

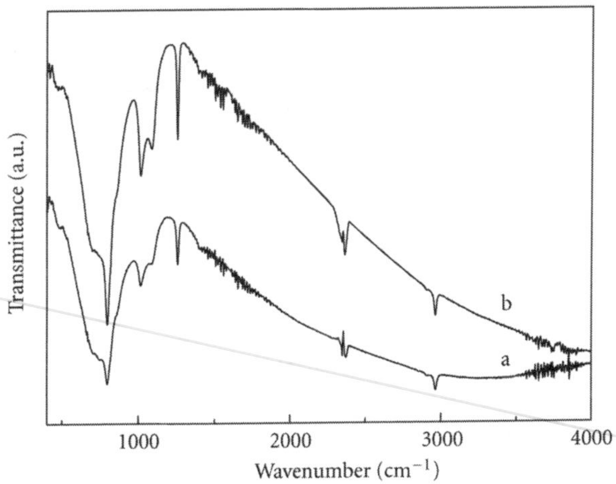

Figure 3: The FTIR spectra of the films on KBr substrates (a) with nonirradicated (b) with 150 eV ion-beam irradiated.

Although FTIR is high efficiency of Si–C band, it is low efficiency of Si–Si and C–C bonds. However, Raman spectra are optimum for researching identic atomic polar bonds, such as Si–Si and C–C bands. So, we used Raman spectra to analyze the changes of Si–Si and C–C bands. The Raman spectra from these samples are characterized by the presence of bonds characteristic of amorphous material, in the 200–600 cm−1 and 1300–1600 cm−1 spectral regions. This can be seen in Figure 4(a), where the spectra measured in these regions from the sample a. These bands are similar to those reported for amorphous Si1-xCx films obtained by different techniques [24–26] and have been interpreted according to a two-mode behavior, related to the different Si–Si and C–C vibrational modes. So, the first band are similar to the Si–Si TO mode one from amorphous Si (at about 480 cm−1). The band appearing in the 1300–1600 spectra region is related to C–C vibrational modes. This C–C signature already observed in Si–C alloys with carbon excess [24, 27–30] corresponds certainly to a specific structure which could be described like a random covalent network of tetrahedral-trigonal bonding carbons with distorted bond angles and bond lengths [31–35]. As a matter of fact, it shows that in mixed sp2–sp3 bonded carbon layer, the overall Raman spectrum is dominated by the G component (E2g Raman mode of perfect graphite crystal), because the cross section of the graphite stretching mode is much higher than that of the 1332 cm−1 diamond mode.

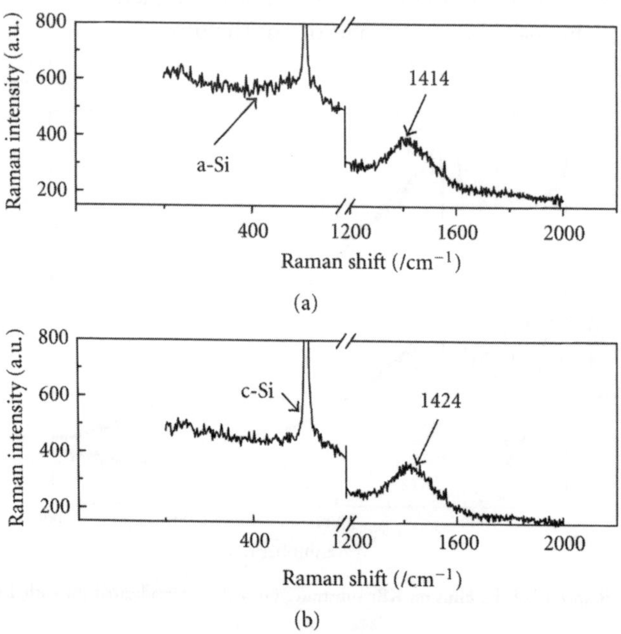

Figure 4: The Raman spectra of the films on Si substrates (a) with nonirradicated (b) with 150 eV ion-beam irradiated.

After assisted ion-beam irradicated, the a-Si contribution in the spectra decreases, the first-order peak from c-Si is restored, and the peak of C–C bond shifts toward to high-frequency while the intensity of the peak decreases as shown in Figure 4(b). It indicates that the sp3/sp2 radio decreases while the excess C atoms or the size of the sp2 bonded clusters reduces.The lack of C–C bands suggests that part of excess carbon atoms are bonded to a-Si instead of forming the graphite-like structure. And that the residual excess C can be present in solid solution.

Conclusion

SiC films were prepared by single and dual-ion-beam sputtering deposition at room temperature. The microstructure and optical properties of nonirradicated and assisted ion-beam irradicated films have been characterized by TEM, EDS, SEM, FTIR, and Raman spectra. The films exposed to a low-energy assisted ion-beam irradicated during sputtering from a SiC target have exhibited smoother and compacter surface topography and different optical behavior than the films deposited with nonirradicated. With assisted ion-beam irradicated, the density of the Si–C bonding in the film has increased. At the same time, the excess C atoms or the size of the sp^2 bonded clusters reduces, and the a-Si phase decreases. These results indicate that the composition of the film is mainly Si–C bond. These date suggest the high stability of a-SiC to be related to the absence of a complete chemical order, which can be attributed to the ion-assisted irradicated at 150 eV during deposition.

Acknowledgements

This work is supported by the National Natural Science Foundation of China (No. 10275047 and No.10575073) and the Natural Science Foundation of Jiangsu Province of China (No.03KJB140116). This work is also partially supported by Hubei Key Laboratory of Plasma Chemistries and Advanced Materials (Wuhan Institute of Technology).

References

1. R. A. Caruso and J. H. Schattka, "Cellulose acetate templates for porous inorganic network fabrication," Advanced Materials, vol. 12, no. 24, pp. 1921–1923, 2000.
2. S. Y. Myong, S. S. Kim, and K. S. Lim, "In situ ultraviolet treatment in an Ar ambient upon p-type hydrogenated amorphous silicon-carbide windows of hydrogenated amorphous silicon based solar cells," Applied Physics Letters, vol. 84, no. 26, pp. 5416–5418, 2004.
3. M. Wang, X. G. Diao, A. P. Huang, P. K. Chu, and Z. Wu, "Influence of substrate bias on the composition of SiC thin films fabricated by PECVD and

underlying mechanism," Surface and Coatings Technology, vol. 201, no. 15, pp. 6777–6780, 2007.

4. C. Ricciardi, A. Primiceli, G. Germani, A. Rusconi, and F. Giorgis, "Microstructure analysis of a-SiC:H thin films grown by high-growth-rate PECVD," Journal of Non-Crystalline Solids, vol. 352, no. 9–20, pp. 1380–1383, 2006.
5. C. Ricciardi, E. Bennici, M. Cocuzza, et al., "Characterization of polycrystalline SiC layers grown by ECR-PECVD for micro-electro-mechanical systems," Thin Solid Films, vol. 427, no. 1-2, pp. 187–190, 2003.
6. C. Ricciardi, G. Fanchini, and P. Mandracci, "Physical properties of ECR-CVD polycrystalline SiC films for micro-electro-mechanical systems," Diamond and Related Materials, vol. 12, no. 3–7, pp. 1236–1240, 2003.
7. S. M. Rajab, I. C. Oliveira, M. Massi, H. S. Maciel, S. G. dos Santos Filho, and R. D. Mansano, "Effect of the thermal annealing on the electrical and physical properties of SiC thin films produced by RF magnetron sputtering," Thin Solid Films, vol. 515, no. 1, pp. 170–175, 2006.
8. Z. D. Sha, X. M. Wu, and L. J. Zhuge, "The structure and optical properties of SiC film on Si (111) substrate with a ZnO buffer layer by RF-magnetron sputtering technique," Physics Letters A, vol. 355, no. 3, pp. 228–232, 2006.
9. M. A. El Khakani, M. Chaker, M. E. O'Hern, and W. C. Oliver, "Linear dependence of both the hardness and the elastic modulus of pulsed laser deposited a-SiC films upon their Si-C bond density," Journal of Applied Physics, vol. 82, no. 9, pp. 4310–4318, 1997.
10. G. Soto, E. C. Samano, R. Machorro, and L. Cota, "Growth of SiC and SiCxNy films by pulsed laser ablation of SiC in Ar and N2 environments," Journal of Vacuum Science and Technology A, vol. 16, no. 3, pp. 1311–1315, 1998.
11. Q. Wang, S. Y. Fu, S. L. Qu, and W. J. Liu, "Enhanced photoluminescence from Si+ and C+ ions co-implanted porous silicon formed by electrochemical anodization," Solid State Communications, vol. 144, no. 7-8, pp. 277–281, 2007.
12. H. Ishihara, M. Murano, T. Watahiki, A. Yamada, M. Konagai, and Y. Nakamura, "Growth of strain relaxed Si1-yCy on Si buffer layer by gas-source MBE," Thin Solid Films, vol. 508, no. 1-2, pp. 99–102, 2006.
13. B. J. Pond, T. Du, J. Sobczak, and C. K. Carniglia, "Comparison of the optical properties of oxide films deposited by reactive-dc-magnetron sputtering with those of ion-beam-sputtered and electron-beam-evaporated films," in Laser-Induced Damage in Optical Materials, vol. 2114 of Proceedings of SPIE, pp. 345–354, Boulder, Colo, USA, October 1993.

14. F. Sarto, M. Alvisi, L. Caneve, L. H. AbuHassan, and S. Scaglione, "Dual ion beam sputtering coating of plastic substrates: improvement of film/substrate adhesion by minimizing the total stress at the interface," in Advances in Optical Interference Coatings, vol. 3738 of Proceedings of SPIE, pp. 66–75, Berlin, Germany, May 1999.

15. J.-E. Sundgren, "Structure and properties of nitride thin films grown by magnetron sputter deposition: effects of ion irradiation during growth," Journal of Vacuum Science and Technology A, vol. 6, no. 3, pp. 1694–1695, 1988.

16. A. Rizzo, M. Alvisi, F. Sarto, and S. Scaglione, "Momentum transfer parameter in argon-assisted carbon coatings," Thin Solid Films, vol. 384, no. 2, pp. 215–222, 2001.

17. Y. C. Zhou, S. G. Long, and Y. W. Liu, "Thermal failure mechanism and failure threshold of SiC particle reinforced metal matrix composites induced by laser beam," Mechanics of Materials, vol. 35, no. 10, pp. 1003–1020, 2003.

18. Y. Sun, T. Miyasato, and J. K. Wigmore, "Characterization of excess carbon in cubic SiC films by infrared absorption," Journal of Applied Physics, vol. 85, no. 6, pp. 3377–3379, 1999.

19. B. P. Swain, "Influence of process pressure on HW-CVD deposited a-SiC:H films," Surface and Coatings Technology, vol. 201, no. 3-4, pp. 1132–1137, 2006.

20. V. Jousseaume, N. Rochat, L. Favennec, O. Renault, and G. Passemard, "Mechanical stress in PECVD a-SiC:H: aging and plasma treatments effects," Materials Science in Semiconductor Processing, vol. 7, no. 4–6, pp. 301–305, 2004.

21. J. Kouvetakis, D. Chandrasekhar, and D. J. Smith, "Growth and characterization of thin Si80C20 films based upon Si4C building blocks," Applied Physics Letters, vol. 72, no. 8, pp. 930–932, 1998.

22. C. Serre, L. Calvo-Barrio, A. Pérez-Rodríguez, et al., "Ion-beam synthesis of amorphous SiC films: structural analysis and recrystallization," Journal of Applied Physics, vol. 79, no. 9, pp. 6907–6913, 1996.

23. Y. Sun and T. Miyasato, "Infrared absorption properties of nanocrystalline cubic SiC films," Japanese Journal of Applied Physics, vol. 37, no. 10, part 1, pp. 5485–5489, 1998.

24. A. Chehaidar, R. Carles, A. Zwick, C. Meunier, B. Cros, and J. Durand, "Chemical bonding analysis of a-SiC:H films by Raman spectroscopy," Journal of Non-Crystalline Solids, vol. 169, no. 1-2, pp. 37–46, 1994.

25. J. Pezoldt, B. Stottko, G. Kupris, and G. Ecke, "Sputtering effects in hexagonal silicon carbide," Materials Science and Engineering B, vol. 29, no. 1–3, pp. 94–98, 1995.

26. A. Pérez-Rodríguez, C. Serre, L. Calvo-Barrio, et al., "Ion-beam-induced amorphization and recrystallization processes in SiC: Raman-scattering analysis," in International Conference on Optical Diagnostics of Materials and Devices for Opto-, Micro-, and Quantum Electronics, S. V. Svechnikov and M. Y. Valakh, Eds., vol. 2648 of Proceedings of SPIE, pp. 481–494, Kiev, Ukraine, May 1995.

27. Y. Inoue, S. Nakashima, A. Mitsuishi, S. Tabata, and S. Tsuboi, "Raman spectra of amorphous SiC," Solid State Communications, vol. 48, no. 12, pp. 1071–1075, 1983.

28. M. Gorman and S. A. Solin, "Direct evidence for homonuclear bonds in amorphous SiC," Solid State Communications, vol. 15, no. 4, pp. 761–765, 1974.

29. N. Laidani, R. Capelleti, M. Elena, et al., "Spectroscopic characterization of thermally treated carbon-rich Si1-xCx films," Thin Solid Films, vol. 223, no. 1, pp. 114–121, 1993.

30. S. Scordo and M. Nadal, "On the nature of microwave deposited hard silicon-carbon films," in Proceeding of the 13th International Conference on Chemical Vapor Deposition, Electrochemical Society Proceedings 96-5, p. 656, Los Angeles, Calif, USA, May 1996.

31. D. Beeman, J. Siverman, R. Lynds, and M. R. Anderson, "Modeling studies of amorphous carbon," Physical Review B, vol. 30, no. 2, pp. 870–875, 1984.

32. N. Savvides, "Optical constants and associated functions of metastable diamondlike amorphous carbon films in the energy range 0.5–7.3 eV," Journal of Applied Physics, vol. 59, no. 12, pp. 4133–4145, 1986.

33. M. A. Tamor and C. H. Wu, "Graphitic network models of "diamondlike" carbon," Journal of Applied Physics, vol. 67, no. 2, pp. 1007–1012, 1990.

34. F. Rossi, B. André, A. van Veen, et al., "Effect of ion beam assistance on the microstructure of nonhydrogenated amorphous carbon," Journal of Applied Physics, vol. 75, no. 6, pp. 3121–3129, 1994.

35. H. Efstathiadis, Z. Akkerman, and F. W. Smith, "Atomic bonding in amorphous carbon alloys: a thermodynamic approach," Journal of Applied Physics, vol. 79, no. 6, pp. 2954–2967, 1996.

Homogenization of Mutually Immiscible Polymers Using Nanoscale Effects: A Theoretical Study

Sarah Montes, Agustín Etxeberria, Javier Rodriguez and Jose A. Pomposo

ABSTRACT

A theoretical study to investigate homogenization of mutually immiscible polymers using nanoscale effects has been performed. Specifically, the miscibility behavior of all-polymer nanocomposites composed of linear-polystyrene (PS) chains and individual cross-linked poly(methyl methacrylate)-nanoparticles (PMMA-NPs) has been predicted. By using a mean field theory accounting for combinatorial interaction energy and nanoparticle-driven effects, phase diagrams were constructed as a function of PMMA-NP size, PS molecular weight, and temperature. Interestingly, complete miscibility (i.e., homogeneity) was predicted from room temperature to 675 K for PMMA-nanoparticles with radius less than ~7 nm blended with PS chains (molecular weight 150 kDa, nanoparticle volume fraction 20%) in spite of the well-known immiscibility between PS and PMMA. Several nanoscale effects affecting miscibility in PMMA-NP/PS nanocomposites involving small PMMA-nanoparticles are discussed.

Introduction

Nanoscale effects are responsible for the growing scientific interest in both nanomaterials and nanocomposites [1]. Often, nanoobjects display interesting physical phenomena such as surface plasmon resonance in metallic nanoparticles (NPs), ballistic transport in carbon nanotubes (CNTs), or fluorescence emission in quantum dots (QDs), among others. Most of them are useful for a huge of potential applications covering from nanoelectronic to improved sensors devices [2]. Also, nanocomposites composed of well-dispersed nanoobjects into a polymer matrix lead to unexpected mechanical [3] and/or rheological properties [4]. A representative example is the large decrease (up to 80%) in melt viscosity upon nanoparticle addition first observed in polystyrene-nanoparticle (PS-NP)/linear-polystyrene (PS) blends by Mackay et al. [5]. The underlying physics of this unexpected behavior is being currently explored [6] and should be rationalized in terms of scaling concepts at the nanoscale [7]. Blend miscibility (homogeneity) is a necessary condition to observe such a viscosity drop in all-polymer nanocomposites (i.e., polymer-nanoparticle/linear-polymer blends).

Recently, we have introduced an entropic model for predicting the miscibility behavior of PS-NP/PS nanocomposites with very good agreement between theory and experiment [8]. Additionally, the theory has been employed for the prediction of the interaction parameter, the miscibility behavior, and the melting point depression of athermal poly(ethylene) (PE)-nanoparticle/linear-PE nanocomposites using chain dimensions data from Monte-Carlo (MC) simulations [9]. Our main findings indicate that dilution of contact, hard sphere-like, nanoparticle-nanoparticle interactions plays a key role in explaining the miscibility behavior of polymer-nanoparticles dispersed in a chemically identical linear-polymer matrix [8, 9].

In a recent work, the athermal model has been extended to calculate the phase diagram of weakly interacting all-polymer nanocomposites by accounting for combinatorial interaction energy and nanoparticle-driven effects [10]. Complete miscibility was predicted for PS-nanoparticles with radius < 6 nm blended with poly(vinyl methyl ether) (PVME) at low concentrations. When compared to linear-PS/PVME blends displaying phase splitting at T>375 K, the miscibility improving effect of sub-10 nm PS-nanoparticles was clearly highlighted [10].

In this letter, we explore theoretically the conditions for homogenization of two mutually immiscible polymers by changing all the linear-polymer chains of one of the components by cross-linked polymer nanoparticles. Specifically, we consider the PS/poly(methyl methacrylate) (PMMA) pair as a model system. Immiscibility between PS and PMMA is well known in the literature as a result of unfavorable interactions between styrene (S) and methyl methacrylate (MMA) repeat units [11–13]. Here, miscibility diagrams for PMMA-NP/PS nanocomposites are reported as a function of PMMA-NP size, PS molecular weight (Mn) and temperature. Finally, several nanoscale effects affecting the miscibility behavior of PMMA-NP/PS nanocomposites are also discussed.

Theoretical Approach

For a binary blend to be thermodynamically stable against phase separation, the following well-known conditions must be fulfilled:

$$\Delta g_{mix} < 0,$$

$$\Delta g_{mix}^{(2)} \equiv \left(\frac{\partial^2 \Delta g_{mix}}{\partial \phi_1^2} \right) > 0.$$

In a mean field, theoretical framework, Δgmix, the free energy of mixing (per unit volume) for an all-polymer nanocomposite composed of spherical polymer-nanoparticles (component 1) of volume fraction ϕ 1, radius Rp, nanoparticle volume υ1, and linear-polymer chains of degree of polymerization N2(≫1) and monomer volume υ2 is given by [10]

$$\Delta g_{mix} = \Delta g_{mix}^{co} + \Delta g_{mix}^{np} + \Delta g_{mix}^{eg}, \quad (1)$$

$$\Delta g_{mix}^{co} = kT \left[\frac{\phi_1}{v_1} \ln \phi_1 + \frac{\phi_2}{N_2 v_2} \ln \phi_2 \right], \quad (1a)$$

$$\Delta g_{mix}^{np} = kT \left[\frac{\phi_1}{v_1} \left(\frac{4\phi_1 - 3\phi_1^2}{(1-\phi_1)} \right) + \frac{3}{2} \times \frac{\phi_1 \phi_2}{N_2 v_2} \left(\frac{R_p}{R_{g2}} \right)^2 \right], \quad (1b)$$

$$\Delta g_{mix}^{ex} = \left(\frac{r_2}{R_p} \right) \left[\frac{kT}{v_2} \left(X_s + \frac{X_H}{T} \right) \phi_1 \phi_2 \right] \quad (1c)$$

where k is the Boltzmann constant, T is the absolute temperature, R_{g2} is the radius of gyration of linear-polymer 2, r_2 is the radius of a repeat unit 2, and χ_S and χ_H are the entropic and enthalpic components of the blend interaction parameter ($\chi=\chi_S+\chi_H/T$), respectively.

Equation (1a) provides the contribution to the total free energy of mixing due to the combinatorial entropy of mixing, whereas (1b) gives the contribution to the free energy of mixing due to nanoparticle-driven effects. Equation (1b) takes into account the dilution of hard sphere-like nanoparticle-nanoparticle interactions upon mixing (first RHS term) [8, 9] and the stretching of the linear-polymer chains due to the presence of the nanoparticles (second RHS term). Referring to this latter term, a Ginzburg-type expansion term is adopted to account for the fact that the polymer-nanoparticles cause stretching of the polymer chains in their vicinity [14].

Equation (1c) accounts for the temperature-dependent and nanoparticle size-dependent interaction effects in the all-polymer nanocomposite. A prefactor (r2/Rp) is introduced in (1c) since the number of surface contacts with monomers 2 for each nanoparticle becomes smaller as one increases the nanoparticle radius [14]. This ratio tends toward unity inasmuch Rp→r1 as it should, r1 being the radius of a repeat unit 1.

For the sake of simplicity, in (1) we have omitted a term (~constant× ϕ 1) arising from the (non-zero) reference free energy of the pure component 1 in the disordered molten state at ϕ_1^{fr} =0.494 (maximum packaging of monodisperse spheres) as dicussed in [9] which has no effect in the resulting spinodal equation (second derivative of the free energy of mixing with respect to ϕ 1).

The condition $\Delta g_{mix}^{(2)} = 0$ just determines the spinodal miscibility boundary in the phase diagram, which is given by

$$\Delta g_{mix}^{(2)} = \Delta g_{mix}^{co(2)} + \Delta g_{mix}^{np(2)} + \Delta g_{mix}^{ex(2)} = 0, \tag{2}$$

$$\Delta g_{mix}^{co(2)} = kT\left[\frac{1}{v_1\phi_1} + \frac{1}{N_2 v_2 \phi_2}\right], \tag{2a}$$

$$\Delta g_{mix}^{np(2)} = kT\left[\frac{2}{v_1}\left(\frac{4-\phi_1}{(1-\phi_1)^4}\right) - \frac{9}{2v_2}\left(\frac{R_p}{r_2 N_2}\right)^2\right], \tag{2b}$$

$$\Delta g_{mix}^{ex(2)} = -2\left(\frac{r_2}{R_p}\right)\left[\frac{kT}{v_2}\left(X_s + \frac{X_H}{T}\right)\right], \tag{2c}$$

where we have employed $R_{g2} \approx 2r_2 N_2^{1/2} / \sqrt{6}$.

Equations (1)–(2) are presumably valid only for ϕ 1<0.494 (the theoretical maximum nanoparticle packaging volume fraction without freezing) and for Rp>~5r2 (it should be noted that r2 is typically around 0.3 nm) [10].

Results and Discussion

It is well known that PS and PMMA lead to heterogeneous (phase separated) blends due to the immiscibility between components at high molecular weights [11–13]. The S/MMA temperature-dependent interaction parameter ($\chi=\chi_S+\chi_H/T$) has been determined by several experimental techniques such as small angle neutron scattering (SANS) (using deuterated block copolymers) [11, 12] and cloud-point (CP) measurements (using oligomer mixtures and taking into account end-group effects) [13] with good agreement between them. As an example, χ=0.028+3.9/T by SANS measurements [11] and χ=0.021+2.8/T by CP experiments [13]. Since the entropic contribution, χ_S, is much greater than the (χ_H/T) term, the temperature dependence of χ was found to be relatively weak.

We have employed the temperature-dependent interaction parameter determined by CP measurements to calculate the spinodal miscibility boundary of PMMA-NP/PS composites as a function of nanoparticle radius at constant blend composition (see Figure 1). Complete miscibility across the 275–675 K temperature range was predicted for PMMA-nanoparticles with radius less than 6.8 nm in spite of the well-known immiscibility between PMMA and PS homopolymers. Conversely, for PMMA-nanoparticles with radius higher than 7.2 nm, complete immiscibility (phase separation) is expected. For PMMA-NP of radius in between 6.8 and 7.2 nm, partial miscibility was predicted as a function of temperature (the blends displaying upper critical solution temperature (UCST)-type behavior). No significant changes was observed when χ values from SANS experiments [11, 12] were employed in the calculations.

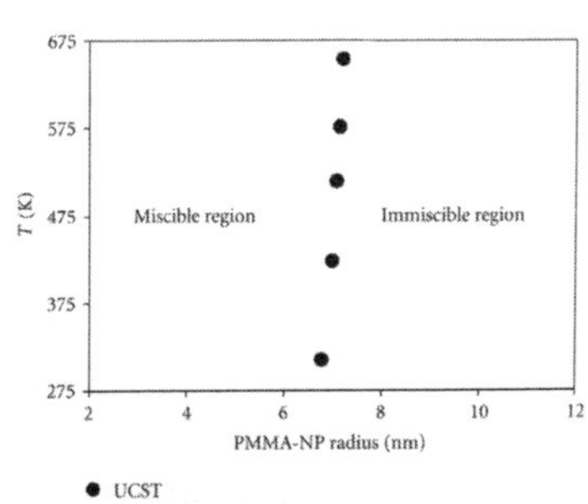

Figure 1: Predicted phase diagram for PMMA-NP/PS (M_n=150 kDa) nanocomposites as a function of nanoparticle size as calculated from (2) by using χ=0.021+2.8/T [13], ϕ 1=0.2, N_2=1443, υ_2=99 cm^3/mol, and r_2=0.32 nm.

In order to rationalize these nanoscale-driven results, we have examined the values of the different factors governing (2) (second derivative of the energy of mixing). $\Delta g_{mix}^{co(2)}$ (arising from combinatorial effects) was found to be positive (favorable to mixing) and increased linearly with temperature. Conversely, $\Delta g_{mix}^{ex(2)}$ (arising from interactions) was negative and increased in absolute value with T. In general, $\left|\Delta g_{mix}^{ex(2)}\right| > \Delta g_{mix}^{co(2)}$ so miscibility was conditioned by favorable values of the $\Delta g_{mix}^{np(2)}$ term. At constant ø 1, the size of the PMMA-nanoparticles was the main factor affecting the $\Delta g_{mix}^{np(2)}$ term (see (2b)). At a given temperature, $\Delta g_{mix}^{np(2)}$ was found to decrease upon increasing the PMMA-NP size leading, respectively, to partial and complete immiscibility at PMMA-NP radius of 7 and 7.5 nm. As a result, miscibility in PMMA-NP/PS nanocomposites can be mainly attributed to two combined effects: (1) reduced (unfavorable) PS chain stretching by smaller PMMA-nanoparticles and (2) favorable dilution of (hard sphere-like) nanoparticle-nanoparticle interactions upon mixing.

The effect of PS molecular weight on the predicted phase diagram for PMMA-NP/PS nanocomposites is illustrated in Figure 2. As expected, a reduction in PS molecular weight shifts the miscibility boundary toward smaller nanoparticle sizes due to the large entropy penalty paid for nanoparticle inclusion into short linear-polymer chains. Hence, the critical PMMA-NP radius for complete immiscibility (R_P^C) changes from 5.4 to 7.2 and 7.8 nm upon changing the PS molecular weight from 50 to 150 and 500 kDa, respectively.

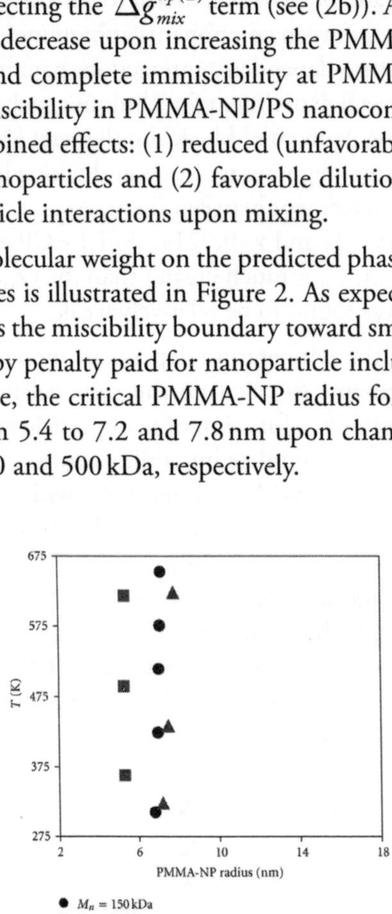

Figure 2: Influence of PS molecular weight on the calculated phase diagram for PMMA-NP/PS nanocomposites (ø 1=0.2): M_n=50 kDa (solid squares), M_n=150 kDa (solid circles), and M_n=500 kDa (solid triangles).

Concerning the effect of blend composition on R_P^C for PMMA-NP/PS (Mn=150 kDa) nanocomposites, a linear increase in R_P^C (up to ~29%) on going from ø 1=0.15 (R_P^C=6.9nm) to ø 1=0.35 (R_P^C=8.9 nm) was observed.

Conclusions

A mean field theoretical model, accounting for combinatorial interaction energy and nanoparticle-driven effects, has been employed to investigate homogenization of mutually immiscible polymers using nanoscale effects. The PS/PMMA pair has been selected as a model system since immiscibility between PS and PMMA is well documented in the literature, and reliable values of the S/MMA interaction parameter are available as a function of T. Specifically, we have investigated the effect on blend miscibility (homogeneity) of the replacement of linear-PMMA chains by cross-linked individual PMMA-nanoparticles.

Hence, phase diagrams have been constructed for PMMA-NP/PS nanocomposites as a function of PMMA-NP size, PS molecular weight, blend composition, and temperature. Interestingly, complete miscibility across the 275–675 K temperature range was predicted for PMMA-nanoparticles with radius less than 6.8 nm blended with PS (Mn=150 kDa, ∅ 1=0.2). Increasing PS molecular weight and nanoparticle content was found to have a small positive effect on PMMA-NP/PS nanocomposite miscibility.

Homogenization of PMMA-NP/PS nanocomposites was mainly attributed to two combined nanoeffects: reduced PS chain stretching by the smaller PMMA-nanoparticles, and favorable dilution of contact (hard sphere-like) nanoparticle-nanoparticle interactions upon mixing.

Acknowledgements

Financial support by MEC (Grant no. CSD2006-53), Basque Government (Grupos Consolidados IT-274-07), and Diputación de Gipuzkoa through C. I. C. Nanogune—Consolider and Nanotron Project is gratefully acknowledged.

References

1. A. C. Balazs, T. Emrick, and T. P. Russell, "Nanoparticle polymer composites: where two small worlds meet," Science, vol. 314, no. 5802, pp. 1107–1110, 2006.
2. E. Ochoteco, N. Murillo, J. Rodriguez, J. A. Pomposo, and H. Grande, "Conducting polymer-based electrochemical sensors," in Encyclopedia of Sensors, vol. 2, pp. 259–278, American Scientific Publishers, Stevenson Ranch, Calif, USA, 2006.

3. X. Wang, J. E. Hall, S. Warren, et al., "Synthesis, characterization, and application of novel polymeric nanoparticles," Macromolecules, vol. 40, no. 3, pp. 499–508, 2007.
4. A. Tuteja, P. M. Duxbury, and M. E. Mackay, "Multifunctional nanocomposites with reduced viscosity," Macromolecules, vol. 40, no. 26, pp. 9427–9434, 2007.
5. M. E. Mackay, T. T. Dao, A. Tuteja, et al., "Nanoscale effects leading to non-Einstein-like decrease in viscosity," Nature Materials, vol. 2, no. 11, pp. 762–766, 2003.
6. A. Tuteja, M. E. Mackay, S. Narayanan, S. Asokan, and M. S. Wong, "Breakdown of the continuum Stokes-Einstein relation for nanoparticle diffusion," Nano Letters, vol. 7, no. 5, pp. 1276–1281, 2007.
7. F. Brochard Wyart and P. G. de Gennes, "Viscosity at small scales in polymer melts," The European Physical Journal E, vol. 1, no. 1, pp. 93–97, 2000.
8. J. A. Pomposo, A. Ruiz de Luzuriaga, A. Etxeberria, and J. Rodríguez, "Key role of entropy in nanoparticle dispersion: polystyrene-nanoparticle/ linear-polystyrene nanocomposites as a model system," Physical Chemistry Chemical Physics, vol. 10, no. 5, pp. 650–651, 2008.
9. A. Ruiz de Luzuriaga, A. Etxeberria, J. Rodríguez, and J. A. Pomposo, "Phase diagram and entropic interaction parameter of athermal all-polymer nanocomposites," Polymers for Advanced Technologies. In press.
10. A. Ruiz de Luzuriaga, H. Grande, and J. A. Pomposo, "A theoretical investigation of polymer nanoparticles as miscibility improvers in all-polymer nanocomposites," Journal of Nano Research, In press.
11. T. P. Russell, R. P. Hjelm, Jr., and P. A. Seeger, "Temperature dependence of the interaction parameter of polystyrene and poly(methyl methacrylate)," Macromolecules, vol. 23, no. 3, pp. 890–893, 1990.
12. T. P. Russell, "Changes in polystyrene and poly(methyl methacrylate) interactions with isotopic substitution," Macromolecules, vol. 26, no. 21, p. 5819, 1993.
13. T. A. Callaghan and D. R. Paul, "Interaction energies for blends of poly(methyl methacrylate), polystyrene, and poly(α-methyl styrene) by the critical molecular weight method," Macromolecules, vol. 26, no. 10, pp. 2439–2450, 1993.
14. V. V. Ginzburg, "Influence of nanoparticles on miscibility of polymer blends. A simple theory," Macromolecules, vol. 38, no. 6, pp. 2362–2367, 2005.

Copyrights

Index

B

D

E

S

T

U

V